Selected Titles in This Series

689 M. A. Dickmann and F. Miraglia, Special groups: Boolean-theoretic methods in the theory of quadratic forms, 2000

688 Piotr Hajłasz and Pekka Koskela, Sobolev met Poincaré, 2000

687 Guy David and Stephen Semmes, Uniform rectifiability and quasiminimizing sets of arbitrary codimension, 2000

686 L. Gaunce Lewis, Jr., Splitting theorems for certain equivariant spectra, 2000

685 Jean-Luc Joly, Guy Metivier, and Jeffrey Rauch, Caustics for dissipative semilinear oscillations, 2000

684 Harvey I. Blau, Bangteng Xu, Z. Arad, E. Fisman, V. Miloslavsky, and M. Muzychuk, Homogeneous integral table algebras of degree three: A trilogy, 2000

683 Serge Bouc, Non-additive exact functors and tensor induction for Mackey functors, 2000

682 Martin Majewski, ational homotopical models and uniqueness, 2000

681 David P. Blecher, Paul S. Muhly, and Vern I. Paulsen, Categories of operator modules (Morita equivalence and projective modules, 2000

680 Joachim Zacharias, Continuous tensor products and Arveson's spectral C^*-algebras, 2000

679 Y. A. Abramovich and A. K. Kitover, Inverses of disjointness preserving operators, 2000

678 Wilhelm Stannat, The theory of generalized Dirichlet forms and its applications in analysis and stochastics, 1999

677 Volodymyr V. Lyubashenko, Squared Hopf algebras, 1999

676 S. Strelitz, Asymptotics for solutions of linear differential equations having turning points with applications, 1999

675 Michael B. Marcus and Jay Rosen, Renormalized self-intersection local times and Wick power chaos processes, 1999

674 R. Lawther and D. M. Testerman, A_1 subgroups of exceptional algebraic groups, 1999

673 John Lott, Diffeomorphisms and noncommutative analytic torsion, 1999

672 Yael Karshon, Periodic Hamiltonian flows on four dimensional manifolds, 1999

671 Andrzej Rosłanowski and Saharon Shelah, Norms on possibilities I: Forcing with trees and creatures, 1999

670 Steve Jackson, A computation of δ^1_5, 1999

669 Seán Keel and James McKernan, Rational curves on quasi-projective surfaces, 1999

668 E. N. Dancer and P. Poláčik, Realization of vector fields and dynamics of spatially homogeneous parabolic equations, 1999

667 Ethan Akin, Simplicial dynamical systems, 1999

666 Mark Hovey and Neil P. Strickland, Morava K-theories and localisation, 1999

665 George Lawrence Ashline, The defect relation of meromorphic maps on parabolic manifolds, 1999

664 Xia Chen, Limit theorems for functionals of ergodic Markov chains with general state space, 1999

663 Ola Bratteli and Palle E. T. Jorgensen, Iterated function systems and permutation representation of the Cuntz algebra, 1999

662 B. H. Bowditch, Treelike structures arising from continua and convergence groups, 1999

661 J. P. C. Greenlees, Rational S^1-equivariant stable homotopy theory, 1999

660 Dale E. Alspach, Tensor products and independent sums of $\mathcal{L}_p$-spaces, $1 < p < \infty$, 1999

659 R. D. Nussbaum and S. M. Verduyn Lunel, Generalizations of the Perron-Frobenius theorem for nonlinear maps, 1999

658 Hasna Riahi, Study of the critical points at infinity arising from the failure of the Palais-Smale condition for n-body type problems, 1999

(*Continued in the back of this publication*)

Special Groups

Boolean-Theoretic Methods
in the Theory of Quadratic Forms

of the
American Mathematical Society

Number 689

Special Groups

Boolean-Theoretic Methods
in the Theory of Quadratic Forms

M. A. Dickmann
F. Miraglia

May 2000 • Volume 145 • Number 689 (second of 4 numbers) • ISSN 0065-9266

American Mathematical Society
Providence, Rhode Island

2000 *Mathematics Subject Classification.*
Primary 11E81, 06E99; Secondary 11E70, 11E04, 11E10, 12D15, 12J15, 12G05, 03C60, 03C65.

Library of Congress Cataloging-in-Publication Data

Dickmann, M. A., 1940–
  Special groups : boolean-theoretic methods in the theory of quadratic forms / M. A. Dickmann, F. Miraglia.
    p. cm. — (Memoirs of the American Mathematical Society, ISSN 0065-9266 ; no. 689)
  "May 2000, volume 145, number 689 (second of 4 numbers)."
  Includes bibliographical references and index.
  ISBN 0-8218-2057-5 (alk. paper)
  1. Forms, Quadratic. 2. Algebra, Boolean. I. Miraglia, Francisco. II. Title. III. Series.
QA3.A57 no. 689
[QA243]
510s—dc21
[512′.71]   00-020862

Memoirs of the American Mathematical Society

  This journal is devoted entirely to research in pure and applied mathematics.

  Subscription information. The 2000 subscription begins with volume 143 and consists of six mailings, each containing one or more numbers. Subscription prices for 2000 are $466 list, $419 institutional member. A late charge of 10% of the subscription price will be imposed on orders received from nonmembers after January 1 of the subscription year. Subscribers outside the United States and India must pay a postage surcharge of $30; subscribers in India must pay a postage surcharge of $43. Expedited delivery to destinations in North America $35; elsewhere $130. Each number may be ordered separately; *please specify number* when ordering an individual number. For prices and titles of recently released numbers, see the New Publications sections of the *Notices of the American Mathematical Society.*

  Back number information. For back issues see the *AMS Catalog of Publications.*

  Subscriptions and orders should be addressed to the American Mathematical Society, P. O. Box 5904, Boston, MA 02206-5904. *All orders must be accompanied by payment.* Other correspondence should be addressed to Box 6248, Providence, RI 02940-6248.

  Copying and reprinting. Individual readers of this publication, and nonprofit libraries acting for them, are permitted to make fair use of the material, such as to copy a chapter for use in teaching or research. Permission is granted to quote brief passages from this publication in reviews, provided the customary acknowledgment of the source is given.

  Republication, systematic copying, or multiple reproduction of any material in this publication is permitted only under license from the American Mathematical Society. Requests for such permission should be addressed to the Assistant to the Publisher, American Mathematical Society, P. O. Box 6248, Providence, Rhode Island 02940-6248. Requests can also be made by e-mail to reprint-permission@ams.org.

  Memoirs of the American Mathematical Society is published bimonthly (each volume consisting usually of more than one number) by the American Mathematical Society at 201 Charles Street, Providence, RI 02904-2294. Periodicals postage paid at Providence, RI. Postmaster: Send address changes to Memoirs, American Mathematical Society, P. O. Box 6248, Providence, RI 02940-6248.

  © 2000 by the American Mathematical Society. All rights reserved.
  This publication is indexed in *Science Citation Index*®, *SciSearch*®, *Research Alert*®, *CompuMath Citation Index*®, *Current Contents*®/*Physical, Chemical & Earth Sciences.*
  Printed in the United States of America.
  ∞ The paper used in this book is acid-free and falls within the guidelines established to ensure permanence and durability.
  Visit the AMS home page at URL: http://www.ams.org/

  10 9 8 7 6 5 4 3 2 1   05 04 03 02 01 00

Contents

Preface	xi
Chapter 1. Special Groups	1
1. Special Groups and their Morphisms. Examples	1
2. Characterizations of Special Groups	14
3. Fields and Special Groups	20
Chapter 2. Pfister Forms, Saturated Subgroups and Quotients	33
1. Pfister Forms	33
2. Saturated Subgroups	35
3. Pfister quotients	39
Chapter 3. The Space of Orders of a Reduced Group. Duality	50
Chapter 4. Boolean Algebras and Reduced Special Groups	59
1. Boolean Algebras as Reduced Special Groups	60
2. The Boolean Hull of a Reduced Special Group	68
Chapter 5. Embeddings	74
1. Complete Embeddings	75
2. SG-morphisms with a retract	87
3. Pure Embeddings	91
4. Isotropy-Reflecting Morphisms	95
Chapter 6. Special Groups of Continuous Functions	100
1. Filtered Powers	100
2. SG-filtered Powers	103
3. Stalks and Germs	111

4.	The Space of Orders of a SG-filtered Power	119
5.	The Boolean Hull of SG-Filtered Powers	125

Chapter 7. Horn-Tarski and Stiefel-Whitney Invariants   135

1. The Horn-Tarski conditions   136
2. Basic Properties   141
3. Some applications   149
4. The invariants of Pfister forms   154
5. The invariants of linear combinations of Pfister forms.   163

Chapter 8. Algebraic K-theory of Fields and Special Groups   173

1. Milnor's Algebraic K-theory   174
2. Isometry and K-Theoretic Stiefel-Whitney Invariants   178

Chapter 9. Marshall's Conjecture for Pythagorean Fields   182

1. Introduction   185
2. The inductive limit of graded rings of exponent two   187
3. The isomorphisms $k(F) \approx B(F)$ and $\mathcal{W}(G) \approx B_G$ 196
4. The equivalence of Marshall's conjecture to the weak Marshall conjecture for $\mathcal{AP}$ groups.   201
5. Marshall's conjecture for Pythagorean fields   203

Chapter 10. The category of special groups   209

1. Monics and Epics   209
2. Free, injective and projective objects   211
3. Products and Coproducts   213

Chapter 11. Some Model Theory of Special Groups   217

1. The first-order theory of fans   217
2. Existentially closed special groups   219
3. The model-completion of the theory of reduced special groups 221
4. The atomless hull of a Boolean algebra   221
5. The atomless hull of a reduced special group   228
6. Quantifier Elimination.   228

Appendix A. The Universal Theory of Reduced Special Groups   231
             by M. Dickmann and A. Petrovich

Appendix B. Table of References for [DM1] and [DM2]   237

Bibliography   239

Index   242

ABSTRACT. This monograph presents a systematic study of **Special Groups**, a first-order universal-existential axiomatization of the theory of quadratic forms, which comprises the usual theory over fields of characteristic different from 2, and is dual to the theory of abstract order spaces.

The heart of our theory begins in Chapter 4 with the result that Boolean algebras have a natural structure of reduced special group. More deeply, every such group is canonically and functorially embedded in a certain Boolean algebra, its **Boolean hull**. This hull contains a wealth of information about the structure of the given special group, and much of the later work consists in unveiling it.

Thus, in Chapter 7 we introduce two series of invariants "living" in the Boolean hull, which characterize the isometry of forms in any reduced special group. While the multiplicative series —expressed in terms of meet and symmetric difference— constitutes a Boolean version of the Stiefel-Whitney invariants, the additive series —expressed in terms of meet and join—, which we call **Horn-Tarski invariants**, does not have a known analog in the field case; however, the latter have a considerably more regular behaviour. We give explicit formulas connecting both series, and compute explicitly the invariants for Pfister forms and their linear combinations.

In Chapter 9 we combine Boolean-theoretic methods with techniques from Galois cohomology and a result of Voevodsky to obtain an affirmative solution to a long standing conjecture of Marshall concerning quadratic forms over formally real Pythagorean fields.

Boolean methods are put to work in Chapter 10 to obtain information about categories of special groups, reduced or not. And again in Chapter 11 to initiate the model-theoretic study of the first-order theory of reduced special groups, where, amongst other things we determine its model-companion.

The first-order approach is also present in the study of some outstanding classes of morphisms carried out in Chapter 5, e.g., the pure embeddings of special groups. Chapter 6 is devoted to the study of special groups of continuous functions.

Preface

The present extended monograph contains most of the results obtained in joint research work carried out during the period 1993-98 in Paris, France and São Paulo, Brazil.

The notion of **special group** occurred to the first author in 1991. The underlying motivation was to produce a system of axioms sufficient to develop an abstract version of the theory of quadratic forms,

i) Comprising as a particular case the usual theory over arbitrary fields of characteristic $\neq 2$ (as expounded, for instance, in Lam's book [L1]);

ii) Capable of being formalized by first-order axioms in a mathematically natural language.

It was natural for this purpose to take as primitives the isometry relation of forms in each dimension. However, the existence of an inductive characterization of the isometry of n-forms in terms of the isometry of 2-forms, expressed by an existential (first-order) formula (cf. Marshall [Ma1]; Thm. 1.13]) showed that it was enough to give an axiom system for the isometry of forms of dimension 2. Elementary properties of the isometry of binary forms, known in the field case, led to the axioms [SG0] – [SG6] given in Definition 1.2 below.

To be sure, ours is not the first attempt to develop the algebraic theory of quadratic forms on an axiomatic basis. Most notable amongst our ancestors are :

i) Marshall's **abstract Witt rings**, whose beautiful theory is developed at length in [Ma1];

ii) The **quaternionic structures**, also presented in [Ma1];

iii) The **quaternionic schemes**, studied by Kula, Szczepanik and Szymiczek in [KSS].

All these axiomatic treatments are actually equivalent to our special groups in the sense that models of any one system can be canonically (and functorially) constructed from the models of the other (for the pair abstract Witt rings – special groups, this construction is done in [D]; for quaternionic structures – special groups, in [LP3]). Furthermore, each of them illuminates the common underlying theory from a different angle, and all are complementary in one way or another.

The most systematically developed of the axiomatic approaches listed above, the abstract Witt rings, has a drawback from our point of view : its axioms are not first-order, a non-negligible inconvenience for those wishing to explore the incidence of model-theoretic techniques in quadratic form theory.

Quaternionic structures, on the other hand, can easily be presented in a first-order clothing. However, they have a drawback of a methodological character : the concept taken as primitive in this axiomatization — essentially that of isometry of Pfister forms of dimension 4— appears only after having traveled a substantial stretch into quadratic form theory.

This is why it was decided to take a fresh start from first principles, the result of which is the notion of special group presented here.

We shall mention at this point that a notion along similar lines, the **higher level form schemes**, was introduced in [MP] —independently and, presumably, with a different motivation. This notion is used to carry out an algebraic study of higher level forms over fields. Unfortunately, [MP] did not consider the powerful implications of their notion within the **classical** theory of quadratic forms itself.

The axioms for special groups were chosen so as to make the development of quadratic form theory as smooth as possible. This development is outlined in Chapter 1, and proceeds very much along the lines of [Ma1; Ch.2] The second section of this chapter contains a number of examples and some characterizations of the crucial 3-transitivity axiom [SG6]. We pay some attention, as well, to the consequences of the (universal) axioms [SG0] – [SG5]. In the last section we indicate how quadratic form theory over fields fits in the axiomatic framework of special groups, and illustrate the use of our tools by proving a general, special group version of the Baer-Krull theorem.

The second chapter deals with the notion of **saturated subgroup** of a special group and the theory of quotients arising thereof. Saturated subgroups are the kernels of special group homomorphisms with values in a **reduced** special group; they also fit together in a very natural way with the Boolean-theoretic set up introduced later on (they are exactly the traces of the ideals of the Boolean hull of a reduced special group). Our study of saturated subgroups yields, principally, a Hahn-Banach type separation theorem (Theorem 2.11) and a version of Pfister's local-global principle (Theorem 2.30 and Proposition 2.31).

The real interest of the special group approach lies in that it unveils hitherto unknown or neglected aspects of the reduced theory of quadratic forms. The first of them is the existence of an algebraic-topological duality between the category of reduced special groups and that of **abstract order spaces**, introduced and systematically studied by Marshall in [Ma2] – [Ma5]; see also [Ma7]. In Chapter 3 we include a proof of this result, which appears in Lira's thesis [LP3] (it was first reported in [LP2]); its main ingredients are the separation theorem 2.11, our version of Pfister's local-global principle (Theorem 2.30) and (a special case of) Pontrjagin's duality. Stone's duality was the source of inspiration for the ideas leading to the duality theorem 3.19; the results of Chapter 4 show that Theorem 3.19 literally generalizes Stone's duality theorem. In [LP1] – [LP3], Lira carries out a thorough examination of the abovementioned duality.

The heart of the theory presented in this monograph begins with the result, proved in Chapter 4, that every Boolean algebra has a natural structure of reduced special group. Furthermore, Craven's well-known result that any Boolean space is homeomorphic to the space of orders of a SAP field (cf. [P; Thms 6.7 - 6.9 and 9.4]) shows, then, that (the special group of) any Boolean algebra is isomorphic to the (reduced) special group of some SAP field (the proof also uses Proposition 7.17). However, the connection between reduced special groups and Boolean algebras is much deeper: any reduced special group is embedded in a canonically determined Boolean algebra, its **Boolean hull**. The Boolean hull B_G of G is the Boolean algebra of clopens of the space of orders (= special group characters) X_G; equivalently, it is the set $\mathcal{C}(X_G, \pm 1)$ of continuous functions of X_G into $\{\pm 1\}$.

The fact that G is embedded in $\mathcal{C}(X_G, \pm 1)$ has been known for a long time. We just took this embedding seriously and —using the useful terminologies of special groups and Boolean algebras— we show :

i) That this construction is canonical;

ii) That it has the functorial properties one may expect being, in fact, an adjunction to the forgetful functor from Boolean algebras to reduced special groups (see Theorem 4.17).

The construction just explained is at the heart of our application of Boolean-theoretic techniques in quadratic form theory. The existence of the Boolean hull and its basic functorial properties are the subject matter of Chapter 4. In Chapter 10 we use these properties to establish some results concerning the category of reduced special groups; our Boolean methods allow the reduction of a number of categorical questions about reduced special groups to problems about the category of Boolean algebras, where the solution is known. Generally speaking, our results show that the category of reduced special groups is much less "symmetric" than that of Boolean algebras.

In Chapter 5 we study several classes of embeddings among special groups : complete, those possessing a retract, pure, and isotropy-reflecting.

We start by showing that the embedding $\varepsilon_G : G \longrightarrow B_G$ of a reduced special group into its Boolean hull has the remarkable property that forms of any dimension over G are isometric **iff** their images under ε_G are isometric in B_G (Corollary 5.4). We call **complete** any embedding of reduced special groups with this property. Completeness turns out to be equivalent to the injectivity of the associated Boolean homomorphism (Theorem 5.2). This powerful result underlies many later applications. For example, we employ it to show, with little effort, that the embeddings associated to many standard constructions are complete. Theorem 5.2 also gives a characterization of complete embeddings in terms of the isometry of Pfister forms, a result which seems to lend additional support to the widespread feeling that "Pfister forms determine everything" (this intuition finds a precise formulation in Milnor's conjecture [Mi; Question 4.3], now proved).

In the remaining sections of Chapter 5 we consider the other classes of embeddings mentioned above and present natural examples, some of which arise in the context of fields. We only scratch the surface of this matter here; some of these classes of morphisms deserve deeper examination, especially in regard with connections to real algebraic geometry.

Chapter 6 contains methods for constructing special groups by considering certain subgroups of continuous functions, defined in a Boolean space and with values in a special group, called **filtered Boolean powers**. Besides characterizing when these structures are special groups (Theorem 6.10), we also determine their spaces of orders (Theorem 6.27) and Boolean hulls (Theorem 6.34).

We shall now discuss the theory of the Horn-Tarski invariants presented in Chapter 7. Its motivation was the question whether the knowledge of the Boolean algebra $B_G = \mathcal{C}(X_G, \pm 1)$ gives information about the isometry of forms over a reduced special group G. We were led to a solution while looking for a way to lift characters (i.e., special group homomorphisms into $\mathbb{Z}_2$) from a reduced special group to its Boolean hull. Although completeness of the embedding ε_G did answer this problem (cf. Corollary 5.4), the question led us to [HT], a paper dealing with the extension of 2-valued (and real-valued) functions partially defined on a Boolean algebra, to finitely and σ-additive measures defined on the whole algebra. We discovered that the conditions required by Horn and Tarski for a function to extend —conditions involving certain symmetric Boolean polynomials— could be transformed into a characterization of isometry in G. Thus, we obtained the series of **additive** Horn-Tarski invariants, defined in terms of the join and meet operations in B_G. We also realized that there is also a **multiplicative** series, written in terms of symmetric difference and meet. Only after the theory presented in Chapter 7 was fully developed did we realize that the multiplicative version of the invariants is closely connected with Milnor's Stiefel-Whitney invariants, [Mi; §3]. In fact, in Chapter 8 it is shown that, for formally real fields, there is a graded ring homomorphism from Milnor's K-theory to the Boolean hull, taking the Stiefel-Whitney invariants to the multiplicative Horn-Tarski invariants (Theorem 8.7). This motivated us to name the multiplicative series as *Boolean-theoretic Stiefel-Whitney invariants*.

Each series is a complete system of invariants for isometry (Theorem 7.1), containing as many terms as the dimension of the form. Paraphrasing Milnor [Mi; Rmk., p. 329], we may say that every <u>reduced</u> quadratic module (over a field of characteristic $\neq 2$) is determined, up to isomorphism, by its dimension and the series of its Horn-Tarski or Boolean-theoretic Stiefel-Whitney invariants; in particular, this holds for arbitrary quadratic modules over a Pythagorean field.

On the other hand, the series of additive Horn-Tarski invariants does not seem to have a meaningful analog in Milnor's algebraic K-theory context. However, these invariants are, as a rule, easier to compute than the multiplicative (i.e., Stiefel-Whitney) invariants; for example, the former are decreasing in the order of B_G (Proposition 7.5, [HT 3]), while the latter are not (essentially because join is a monotonous operation, while symmetric difference is not). Nevertheless, we give explicit formulas, expressing each series of invariants in terms of the other (Theorem 7.6).

Illustrating the extent to which the preceding invariants lend themselves to explicit computations, we prove some very useful addition formulas (Proposition 7.8) and present, in sections 7.4 and 7.5, the explicit computation of both series of invariants for arbitrary Pfister forms (Theorems 7.18 and 7.22), as well as for linear combinations of them (Theorems 7.26 and 7.29). As an application we give a new, purely combinatorial proof of the Arason-Pfister Hauptsatz for the reduced case, using the additive Horn-Tarski invariants (Theorem 7.31).

Another important observation is the close link, revealed by this circle of ideas, between Pfister forms and the Boolean join operation, see Theorem 7.18 and Corollary 7.20; representation by arbitrary forms is also connected with the fundamental Boolean operations of meet and join, see Corollary 7.13.

In Chapter 8 it is shown that, for a formally real field F there is a graded ring homomorphism from Milnor's mod 2 K-theory of F to the Boolean hull of the reduced special group of classes of non-zero elements of F by sums of squares (Theorem 8.7). As an application, we give a new proof of a result of Elman and Lam showing that the K-theoretic Stiefel-Whitney invariants classify isometry iff the third power of the fundamental ideal is torsion free (Theorem 8.8). The main results in this Chapter where first reported in [DM2].

Chapter 9 contains an exposition —included here by suggestion of the referee— of the affirmative answer to Marshall's conjecture for Pythagorean fields, that appeared in [DM1].

This conjecture proposed an arithmetical criterion for deciding whether a quadratic form φ over a (formally real) Pythagorean field F is (Witt-equivalent to) a linear combination of Pfister forms of a given degree n, in terms of this integer and the signature of φ at the various orders of the underlying field F.

The solution is achieved in two steps. Firstly, by use of the Boolean-theoretic techniques introduced in Chapter 4, Marshall's conjecture is shown to be equivalent to one of its particular instances, the Weak Marshall conjecture (Theorem 9.21); this reduction works for a wide class of formally real special groups, including the reduced ones and those arising from formally real fields (Lemma 9.6). In a second step, techniques from Galois cohomology together with (a part of) Voevodsky's recent and celebrated result, [V], on Milnor's conjecture lead to a proof of Marshall's weak conjecture for Pythegorean fields (§5).

On the way to the reduction of Marshall's conjecture to its weak form we establish results which underscore, once more, the central character of the Boolean hull construction. Indeed, in §3 we prove that the inductive limits of certain systems of groups which occur in the theory of Witt rings and in Milnor's mod 2 K-theory of fields are **isomorphic** to Boolean hulls (of the relevant special groups). The Horn-Tarski invariants are of much help in the computations involved in these proofs.

Chapter 11 is devoted to model-theoretic results about the first-order theory of reduced special groups. We show first that the theory of infinite fans admits quantifier-elimination in the natural language L_{SG} for special groups. Next, we prove that the existentially closed reduced special groups are the atomless Boolean algebras; it follows that the first-order theory of (the special groups of) these algebras (in L_{SG}) is the model-companion of that of reduced special groups. The amalgamation property for complete embeddings (Proposition 5.9) implies that it is also the model-completion, provided the language L_{SG} is enlarged with predicates denoting representation by Pfister forms. We also prove that the theory of (special groups of) atomless Boolean algebras admits quantifier-elimination in this enlarged language (Theorem 11.18). This is achieved through the construction of the canonical "atomless hull" of an arbitrary Boolean algebra, a construction presumably known which, due to the lack of a suitable reference, we carry out in detail (sections 11.4, 11.5). In Appendix A, a set of axioms is given, which characterizes the universal part of the first-order theory of reduced special groups.

The crucial role of Boolean algebras in the theory of reduced special groups is, once again, forcefully underlined by the model-theoretic results proved in Chapter 11.

M. A. Dickmann   F. Miraglia   Paris, February, 1999

CHAPTER 1

Special Groups

In this Chapter we introduce the notion of Special Group and discuss their basic properties. Section 1 contains the fundamental concepts, while section 2 gives useful criteria for a pre-special group to be a special group, including a transcription of Witt's celebrated chain equivalence theorem to the context of special groups. The third and last section is devoted to showing how special groups relate to the theory of quadratic forms over fields and to the proof of a generalization of the Baer-Krull theorem.

1. Special Groups and their Morphisms. Examples

If A is a set, $card(A)$ denotes the cardinal of A. A family F of subsets of A is said to be **up (down) directed** if for all $C, D \in F$ there is $E \in F$ such that $C, D \subseteq E$ (resp., $E \subseteq C, D$).

We shall be dealing most of the time with groups G of exponent 2 ($x^2 = 1$, $\forall\, x \in G$). Although such a G is commutative, we shall indicate its operation multiplicatively. Thus, we write 1 for the neutral element of the operation in G.

NOTATION 1.1. a) If G is a group of exponent 2, let $\mathcal{X}(G)$ be the set of all group homomorphisms from G to the multiplicative group $2 = \{\pm 1\} \subseteq \mathbb{Z}$; $\mathcal{X}(G)$ is a group under pointwise multiplication called the **group of characters** of G. As a closed subset of 2^G with the product topology, it inherits a compact, totally disconnected topology, with which it is a topological group. To describe a clopen basis for this topology, define for $a \in G$ and $\delta \in 2$

$$[a = \delta] = \{\sigma \in \mathcal{X}(G) : \sigma(a) = \delta\}.$$

Then a clopen basis for $\mathcal{X}(G)$ is given by

[⋆] $\quad \mathcal{B} = \{\bigcap_{i=1}^{n} [a_i = \delta(i)] : \{a_1, \ldots, a_n\} \subseteq G \text{ and } \delta \in 2^{\{1,\ldots,n\}}\}.$

Received by the editor November 9, 1997; and in revised form February 9, 1999.

This work was partially supported by grants from the Equipe de Logique Mathématique, University of Paris VII and the University of São Paulo.

b) If S, T are subsets of a group G, set $ST = \{ab \in G : a \in S \text{ and } b \in T\}$. When $S = \{a\}$, write aT for ST.

c) We indicate the field with 2 elements by $\mathbb{F}_2$.

d) If G is a group of exponent 2 with a distinguished element -1, write $-a$ for the product of $a \in G$ by -1. It is clear that the subgroup $\{1, -1\}$ of G is isomorphic to 2 iff $1 \neq -1$ in G.

e) Let A be a set and $\equiv$ a binary relation on $A \times A$. We extend $\equiv$ to a binary relation $\equiv_n$ on A^n, by induction on $n \geq 2$, as follows :

   i) $\equiv_2 \ = \ \equiv$

   ii) $\langle a_1, \ldots, a_n \rangle \equiv_n \langle b_1, \ldots, b_n \rangle$ iff there are $x, y, z_3, \ldots, z_n$ in A such that $\langle a_1, x \rangle \equiv \langle b_1, y \rangle$, $\langle a_2, \ldots, a_n \rangle \equiv_{n-1} \langle x, z_3, \ldots, z_n \rangle$ and $\langle b_2, \ldots, b_n \rangle \equiv_{n-1} \langle y, z_3, \ldots, z_n \rangle$.

Whenever clear from the context, we frequently abuse notation and indicate the aforedescribed extension of $\equiv$ by the same symbol. Thus, if $a_1, \ldots, a_n, b_1, \ldots, b_n \in A$, we write $\langle a_1, \ldots, a_n \rangle \equiv \langle b_1, \ldots, b_n \rangle$ to mean that these two sequences are related by $\equiv_n$. ◇

We now identify the class of objects that shall be of interest in what follows.

DEFINITION 1.2. A **pre-special group (psg)** is a group G of exponent 2, together with a distinguished element -1 and a binary relation $\equiv_G$ on G^2 such that for all $a, b, c, d \in G$,

[SG0] : $\equiv_G$ is an equivalence relation.

[SG1] : $\langle a, b \rangle \equiv_G \langle b, a \rangle$.   [SG2] : $\langle a, -a \rangle \equiv_G \langle 1, -1 \rangle$.

[SG3] : $\langle a, b \rangle \equiv_G \langle c, d \rangle$ implies $ab = cd$.

[SG4] : $\langle a, b \rangle \equiv_G \langle c, d \rangle$ implies $\langle a, -c \rangle \equiv_G \langle -b, d \rangle$.

[SG5] : $\langle a, b \rangle \equiv_G \langle c, d \rangle$ implies $\langle xa, xb \rangle \equiv_G \langle xc, xd \rangle$, $\forall\ x \in G$.

$(G, \equiv_G, -1)$ is a **reduced pre-special group (rpsg)** if $-1 \neq 1$ and $\equiv_G$ satisfies

[red] (**reduction**) : $\forall\ a \in G$, $\langle a, a \rangle \equiv_G \langle 1, 1 \rangle$ iff $a = 1$.

A pre-special group is a **special group (sg)** if it satisfies

[SG6] (**3-transitivity**) :   The extension of $\equiv_G$ to G^3, (as in Notation 1.1.(e)), is a transitive relation.

A relation satisfying conditions [SG1] – [SG5] ([SG6]) is called a **pre-special (resp. special) relation** on G, referred to as **isometry** in G.

When context allows, the name of G will be dropped from the notation. A special group is **reduced (rsg)** if its special relation is reduced. ◇

All the classic terminology of quadratic forms adapts to the present context easily. We summarize the pertinent definitions in

DEFINITION 1.3. Let G be a psg. A **form** φ on G is an n-tuple, $\langle a_1, \ldots, a_n \rangle$, of elements of G; n is called the **dimension** of φ, $dim(\varphi)$; we also call φ a **n-form**.

By convention, two forms of dimension 1 are isometric iff they have the same coefficients. If $\varphi = \langle a_1, \ldots, a_n \rangle$ is a form on G, define

(1) The **set of elements represented by** φ as

$$D_G \varphi = \{b \in G : \exists\, z_2, \ldots, z_n \in G \text{ such that } \varphi \equiv_G \langle b, z_2, \ldots, z_n \rangle\}.$$

If $a \in G$, then $D_G(\langle a \rangle) = \{a\}$. When $\varphi = \langle a, b \rangle$, we write $D_G(a, b)$ for $D_G(\langle a, b \rangle)$.

(2) The **discriminant** of φ as $d(\varphi) = \prod_{i \leq n} a_i$.

For forms $\varphi = \langle a_1, \ldots, a_n \rangle$ and $\psi = \langle b_1, \ldots, b_m \rangle$ we define the

(3) **Direct sum** as $\varphi \oplus \psi = \langle a_1, \ldots, a_n, b_1, \ldots, b_m \rangle$

(4) **Tensor product** as $\varphi \otimes \psi = \langle a_1 b_1, \ldots, a_i b_j, \ldots, a_n b_m \rangle$

If $a \in G$, the product $\langle a \rangle \otimes \varphi$ is written $a\varphi$.

A form φ on G is **isotropic** if there is a form ψ over G such that $\varphi \equiv_G \langle 1, -1 \rangle \oplus \psi$; otherwise it is said to be **anisotropic**. We say that φ is **universal** if $D_G(\varphi) = G$.

A form φ is **hyperbolic** if $\varphi \equiv_G \bigoplus_{i=1}^m \langle 1, -1 \rangle$; note that $dim(\varphi)$ must be even. ◇

REMARK 1.4. a) One can show by induction on $n \geq 2$, that [SG1] – [SG6] imply that the extension of $\equiv_G$ to G^n is transitive, for all $n \geq 3$. This is done in Theorem 1.23 below.

b) It is easily verified, using [SG5], that a psg is reduced iff

$$\forall\, x, a \in G, \quad x \in D_G(a, a) \Rightarrow x = a. \quad \diamond$$

c) If φ, ψ are forms over G and $a, b \in G$, it is straightforward to see that

[ten] $\qquad\qquad ab(\varphi \otimes \psi) = a\varphi \otimes b\psi. \qquad\qquad \diamond$

LEMMA 1.5. Let $(G, \equiv_G, -1)$ be a pre-special group. Let a, b, c, d be elements of G and φ, ψ be n-forms on G. Then

a) $\langle a, b \rangle \equiv \langle c, d \rangle$ iff $ab = cd$ and $ac \in D_G(1, cd)$. Further, $c \in D_G(1, a)$ iff $\langle c, ac \rangle \equiv \langle 1, a \rangle$.

b) $\varphi \equiv \psi$ implies $d(\varphi) = d(\psi)$.

PROOF. a) It follows easily from [SG3] and [SG5] that $\langle a, b \rangle \equiv \langle c, d \rangle$ implies $ab = cd$ and $ac \in D_G(1, cd)$. For the converse, if $ac \in D_G(1, cd)$, then there is $u \in G$ such that $\langle ac, u \rangle \equiv \langle 1, cd \rangle$. By [SG5], $\langle a, cu \rangle \equiv \langle c, d \rangle$; [SG3] and $cd = ab$ yield $acu = cd = ab$ and so $cu = b$, as desired. The last statement in (a) follows directly from what was proven and the definition of $D_G(1, a)$.

b) Let $\varphi = (a_1, \ldots, a_n)$ and $\psi = (b_1, \ldots, b_n)$; we proceed by induction on $n \geq 2$, the result being clear for $n = 2$ by [SG3]. Assume that $n \geq 3$ and that $\varphi \equiv_G \psi$. Then there are x, y and $\vec{z} = (z_3, \ldots, z_n)$ in G such that

$$\langle a_1, x \rangle \equiv_G \langle b_1, y \rangle, \langle a_2, \ldots, a_n \rangle \equiv_G \langle x, \vec{z} \rangle \text{ and } \langle b_2, \ldots, b_n \rangle \equiv_G \langle y, \vec{z} \rangle.$$

Let $\alpha = d(\langle a_2, \ldots, a_n \rangle)$, $\beta = d(\langle b_2, \ldots, b_n \rangle)$ and $\gamma = d(\vec{z})$. By the induction hypothesis, we have $a_1 x = b_1 y$, $\alpha = x\gamma$ and $\beta = y\gamma$. Thus, $x\alpha = y\beta$ and we have $d(\varphi) = a_1 \alpha = a_1 xx\alpha = b_1 yy\beta = d(\psi)$. □

The following Proposition describes some of the basic properties of the isometry of forms in special groups. It will be important in all that follows.

PROPOSITION 1.6. *Let $(G, \equiv_G, -1)$ be a pre-special group and φ, ψ and θ be forms on G. Then,*

a) *The direct sum and the tensor product of isometric forms are isometric.*

If G is a special group, then we also have

b) *(Witt cancellation) If $\varphi \oplus \theta \equiv \psi \oplus \theta$, then $\varphi \equiv \psi$.*

c) *For all $a \in G$ and forms $\varphi_1, \ldots, \varphi_n$ on G,*

[rep] $\quad a \in D_G(\bigoplus_{i=1}^n \varphi_i) \quad$ iff $\quad \begin{cases} \exists\, x_i \in D_G(\varphi_i),\ 1 \leq i \leq n, \\ \text{such that } a \in D_G(\langle x_1, \ldots, x_n \rangle). \end{cases}$

d) *$\varphi \oplus \psi$ is isotropic iff there is $x \in G$ such that $x \in D_G(\varphi)$ and $-x \in D_G(\psi)$.*

In particular, if a is an element of G, $a \in D_G(\varphi)$ iff $\langle -a \rangle \oplus \varphi$ is isotropic.

e) *The following are equivalent :*

   (1) G is a reduced special group.

   (2) For any form ψ on G, $D_G(\psi \oplus \psi) = D_G(\psi)$.

   (3) For any form ψ on G, $\psi \oplus \psi$ isotropic $\Rightarrow \psi$ isotropic.

   (4) For all forms ψ, θ on G, $\psi \oplus \psi \equiv \theta \oplus \theta \Rightarrow \psi \equiv \theta$.

   (5) For any form ψ of even dimension on G, $\psi \oplus \psi$ hyperbolic $\Rightarrow$ ψ hyperbolic.

PROOF. a) Suppose that φ_i, ψ_i, $i = 1, 2$ are forms over G such that $\varphi_1 \equiv \varphi_2$ and $\psi_1 \equiv \psi_2$.

We prove

(S) $\varphi_1 \oplus \psi_1 \equiv \varphi_2 \oplus \psi_2$ and (T) $\varphi_1 \otimes \psi_1 \equiv \varphi_2 \otimes \psi_2$,

by induction on $n = dim(\varphi_i)$, $i = 1, 2$. For $n = 1$, $\varphi_1 = \varphi_2 = \langle a \rangle$, for some $a \in G$. Let $\psi_1 = \langle c_1, \ldots, c_m \rangle$ and $\psi_2 = \langle d_1, \ldots, d_m \rangle$. The isometries

$$\langle a, c_1 \rangle \equiv \langle a, c_1 \rangle, \ \psi_1 \equiv \psi_1 \text{ and } \psi_2 \equiv \psi_1,$$

show that $\langle a, c_1, \ldots, c_m \rangle \equiv \langle a, d_1, \ldots, d_m \rangle$, as required.

For (T), observe that the result holds when $dim(\psi_i) = 2$, by [SG5]. We proceed by induction on $m = dim(\psi_i)$. Assume that the result holds true for $(m-1)$ and let $x, y, \vec{z} = (z_3, \ldots, z_m)$ be elements of G that are witnesses to the fact that $\psi_1 \equiv \psi_2$, i.e.

(*) $\langle c_1, x \rangle \equiv \langle d_1, y \rangle$, $\langle c_2, \ldots, c_m \rangle \equiv \langle x, \vec{z} \rangle$ and $\langle d_2, \ldots, d_m \rangle \equiv \langle y, \vec{z} \rangle$.

By induction, all isometries in (*) can be multiplied by a to yield the isometries that correspond to $a\psi_1 \equiv a\psi_2$.

Assume the result true for $dim(\varphi_i) = n$, and suppose $\varphi_1 = \langle a, a_1, \ldots, a_n \rangle$ and $\varphi_2 = \langle b, b_1, \ldots, b_n \rangle$. Let $x, y, \vec{z} = (z_2, \ldots, z_n)$ be witnesses to the isometry between the φ_i's, that is

(**) $\langle a, x \rangle \equiv \langle b, y \rangle$, $\langle a_1, \ldots, a_n \rangle \equiv \langle x, \vec{z} \rangle$ and $\langle b_1, \ldots, b_n \rangle \equiv \langle y, \vec{z} \rangle$.

The isometries in (**) and the induction hypothesis give

$$\langle a_1, \ldots, a_n \rangle \oplus \psi_1 \equiv \langle x, \vec{z} \rangle \oplus \psi_1 \text{ and } \langle b_1, \ldots, b_n \rangle \oplus \psi_2 \equiv \langle y, \vec{z} \rangle \oplus \psi_1.$$

But these isometries together with the first one in (**) yield

$$\langle a, a_1, \ldots, a_n \rangle \oplus \psi_1 \equiv \langle b, b_1, \ldots, b_n \rangle \oplus \psi_2,$$

as needed to prove (S) in full.

Before dealing with the general situation of tensor products, we state

Fact. If $\langle a, b \rangle \equiv \langle c, d \rangle$, then $\langle a, b \rangle \otimes \psi_1 \equiv \langle c, d \rangle \otimes \psi_2$.

PROOF. As a first step, suppose that $\psi_1 = \langle u, v \rangle \equiv \langle w, t \rangle = \psi_2$. Since multiplication by an element of G and sum preserve isometry, we have,

$$\langle a, b \rangle \otimes \langle u, v \rangle = a\langle u, v \rangle \oplus b\langle u, v \rangle \equiv a\langle w, t \rangle \oplus b\langle w, t \rangle = \langle a, b \rangle \otimes \langle w, t \rangle$$
$$\equiv w\langle a, b \rangle \oplus t\langle a, b \rangle \equiv w\langle c, d \rangle \oplus t\langle c, d \rangle \equiv \langle c, d \rangle \otimes \langle w, t \rangle.$$

Now proceed by induction on the dimension m of ψ_i. From (*) come

(A) $\langle a, b \rangle \otimes \langle c_1, x \rangle \equiv \langle c, d \rangle \otimes \langle d_1, y \rangle$;
(B) $\langle a, b \rangle \otimes \langle c_2, \ldots, c_m \rangle \equiv \langle a, b \rangle \otimes \langle x, \vec{z} \rangle$;
(C) $\langle c, d \rangle \otimes \langle d_2, \ldots, d_m \rangle \equiv \langle c, d \rangle \otimes \langle y, \vec{z} \rangle$;
(D) $\langle a, b \rangle \otimes \langle \vec{z} \rangle \equiv \langle c, d \rangle \otimes \langle \vec{z} \rangle$.

These isometries and the preservation of isometry by sums yield

$$\begin{aligned}
\langle a,b \rangle \otimes \psi_1 &= c_1\langle a,b \rangle \oplus [\langle a,b \rangle \otimes \langle c_2, \ldots, c_m \rangle] \\
&\equiv c_1\langle a,b \rangle \oplus [\langle a,b \rangle \otimes \langle x, \vec{z} \rangle] \\
&= c_1\langle a,b \rangle \oplus x\langle a,b \rangle \oplus [\langle a,b \rangle \otimes \langle \vec{z} \rangle] \\
&\equiv d_1\langle c,d \rangle \oplus y\langle c,d \rangle \oplus [\langle c,d \rangle \otimes \langle \vec{z} \rangle] \\
&= d_1\langle c,d \rangle \oplus [\langle c,d \rangle \otimes \langle y, \vec{z} \rangle] \\
&\equiv d_1\langle c,d \rangle \oplus [\langle c,d \rangle \otimes \langle d_2, \ldots, d_m \rangle] = \langle c,d \rangle \otimes \psi_2.
\end{aligned}$$

This ends the proof of the Fact. $\square$

To handle (T), we shall use the preservation of isometry by sums, and the Fact above. Note that we may suppose that $dim(\varphi_i) = n \geq 3$. By induction, let $\varphi_1 = \langle a, a_1, \ldots, a_n \rangle$ and $\varphi_2 = \langle b, b_1, \ldots, b_n \rangle$. Just as above, let x, y, $\vec{z} = (z_2, \ldots, z_n)$ be elements of G satisfying (**). From the isometries in (**) we get, by the induction hypothesis and the Fact,

(A') $\quad \langle a, x \rangle \otimes \psi_1 = a\psi_1 \oplus x\psi_1 \equiv \langle b, y \rangle \otimes \psi_2 = b\psi_2 \oplus y\psi_2;$

(B') $\quad \langle a_1, \ldots, a_n \rangle \otimes \psi_1 \equiv \langle x, \vec{z} \rangle \otimes \psi_1;$

(C') $\quad \langle b_1, \ldots, b_n \rangle \otimes \psi_2 \equiv \langle y, \vec{z} \rangle \otimes \psi_2;$

(D') $\quad \langle \vec{z} \rangle \otimes \psi_1 \equiv \langle \vec{z} \rangle \otimes \psi_2.$

With (A') - (D') at hand and the result on sums, we compute just as in the proof of the Fact to finish the verification of (a).

b) Let $n = dim(\varphi) = dim(\psi)$. First suppose that $\theta = \langle a \rangle$, $a \in G$. Then, the hypothesis in this case reads $\langle a \rangle \oplus \varphi \equiv \langle a \rangle \oplus \psi$. Thus, there are x, y, $\vec{z} = (z_3, \ldots, z_n)$ in G such that

$$\langle a, x \rangle \equiv \langle a, y \rangle, \quad \varphi \equiv \langle x, \vec{z} \rangle \quad \text{and} \quad \psi \equiv \langle y, \vec{z} \rangle.$$

The first isometry tells us that $ax = ay$ and so $x = y$. But then the transitivity of $\equiv$ (Theorem 1.23.(2)) yields $\varphi \equiv \psi$. To finish the proof, use induction on $dim(\theta)$, noting that the isometry $\theta \oplus \varphi \equiv \theta \oplus \psi$ can be written as $\langle a \rangle \oplus \varphi' \equiv \langle a \rangle \oplus \psi'$, for suitable φ', ψ'.

c) It is enough to prove the statement for $n = 2$ and use induction. First, we verify that if $a \in D_G(\varphi)$ and $b \in D_G(\psi)$, then $D_G(a, b) \subseteq D_G(\varphi \oplus \psi)$. Let $n = dim(\varphi)$, $m = dim(\psi)$ and let z be an element of $D_G(a,b)$. Then, there are w, $\vec{t} = (t_2, \ldots, t_n)$ and $\vec{z} = (z_2, \ldots, z_m)$ in G such that

$$\langle a, b \rangle \equiv \langle z, w \rangle, \langle a, \vec{t} \rangle \equiv \varphi \text{ and } \langle b, \vec{z} \rangle \equiv \psi.$$

But then,

$$\varphi \oplus \psi \equiv \langle a, \vec{t} \rangle \oplus \langle b, \vec{z} \rangle \equiv \langle a, b \rangle \oplus \langle \vec{t}, \vec{z} \rangle \equiv \langle z, w \rangle \oplus \langle \vec{t}, \vec{z} \rangle,$$

showing that $z \in D_G(\varphi \oplus \psi)$. For the converse, we use induction on $dim(\varphi) = n$. If $\varphi = \langle b \rangle$ and $a \in \langle b \rangle \oplus \psi$, there is $\vec{t} = (t_1, \ldots, t_m)$ in G such that $\langle a, \vec{t} \rangle \equiv \langle b \rangle \oplus \psi$. This means that we can find x, y and $\vec{z} = (z_1, \ldots, z_m)$ in G such that (among other things), $\langle a, x \rangle \equiv \langle b, y \rangle$ and $\psi \equiv \langle y, \vec{z} \rangle$. Since $y \in D_G(\psi)$, this is exactly what was to be proved.

Now suppose $\varphi = \langle b, \vec{v} \rangle$, where $\vec{v} \in G^n$. If $a \in D_G(\varphi \oplus \psi)$, then there is $\vec{t} = (t_2, \ldots, t_l)$ in G, with $l = n + m + 1$, such that $\langle a, \vec{t} \rangle \equiv \langle b, \vec{v} \rangle \oplus \psi$. Just as before, there are x, y and $\vec{z} = (z_3, \ldots, z_l)$ in G such that

$$\langle a, x \rangle \equiv \langle b, y \rangle \quad \text{and} \quad \langle y, \vec{z} \rangle \equiv \langle \vec{v} \rangle \oplus \psi.$$

By induction, since y is represented by $\langle \vec{v} \rangle \oplus \psi$, there are $u \in D_G(\langle \vec{v} \rangle)$ and $w \in D_G(\psi)$ such that $y \in D_G(u, w)$. Now note that we have $a \in D_G(b, y)$ and $y \in D_G(u, w)$.

By what was proven above, we may conclude that $a \in D_G(\langle b, u, w \rangle) = D_G(\langle b, u \rangle \oplus \langle w \rangle)$. Using the first step in the induction, we get the existence of $t \in D_G(b, u)$ such that $a \in D_G(t, w)$. But again, by the first part of the proof, $D_G(b, u) \subseteq D_G(\langle b \rangle \oplus \vec{v}) = D_G(\varphi)$, and the proof of (c) is complete.

d) If there is $x \in D_G(\varphi)$ such that $-x \in D_G(\psi)$, then there are $\vec{t} = \langle t_1, \ldots, t_n \rangle$ and $\vec{z} = \langle z_1, \ldots, z_m \rangle$ such that $\langle x, \vec{t} \rangle \equiv \varphi$ and $\langle -x, \vec{z} \rangle \equiv \psi$. But then

$$\varphi \oplus \psi \equiv \langle x, -x \rangle \oplus (\langle \vec{t} \rangle \oplus \langle \vec{z} \rangle) \equiv \langle 1, -1 \rangle \oplus (\langle \vec{t} \rangle \oplus \langle \vec{z} \rangle),$$

i.e., $\varphi \oplus \psi$ is isotropic. For the converse, we proceed by induction on $n = dim(\varphi)$. If $\varphi = \langle a \rangle$, then we have

$$\langle a \rangle \oplus \psi \equiv \langle 1, -1 \rangle \oplus \theta \equiv \langle a, -a \rangle \oplus \theta,$$

and so cancelling a on both sides yields $\psi \equiv \langle -a \rangle \oplus \theta$, which shows that $-a \in D_G(\psi)$. By induction, write $\varphi = \langle a \rangle \oplus \gamma$, thus

$$\varphi \oplus \psi = \langle a \rangle \oplus \gamma \oplus \psi \equiv \langle 1, -1 \rangle \oplus \theta \equiv \langle a, -a \rangle \oplus \theta,$$

which yields, by cancellation of a on both sides, $\gamma \oplus \psi \equiv \langle -a \rangle \oplus \theta$. By (c) above, there are $x \in D_G(\gamma)$ and $y \in D_G(\psi)$ such that $-a \in D_G(x, y)$, i.e. for some $z \in G$, $\langle -a, z \rangle \equiv \langle x, y \rangle$. Using [SG1] and [SG4], $\langle z, -y \rangle \equiv \langle a, x \rangle$. By (c), $-y \in D_G(\langle a \rangle \oplus \gamma) = D_G(\varphi)$, completing the proof.

e) (1) $\Rightarrow$ (2). Let $\psi = \langle a_1, \ldots, a_n \rangle$ and $x \in D_G(\psi \oplus \psi)$. Clearly, $\psi \oplus \psi \equiv \bigoplus_{i=1}^{n} \langle a_i, a_i \rangle$. Thus, (c) yields $x_i \in D_G(a_i, a_i)$, such that $x \in D_G(\langle x_1, \ldots, x_n \rangle)$. By Remark 1.4.(b) $x_i = a_i$, $1 \leq i \leq n$, proving that $x \in D_G(\psi)$.

(2) $\Rightarrow$ (3). Let $\psi' = \langle a_2, \ldots, a_n \rangle$; then $\psi \oplus \psi \equiv \langle a_1, a_1 \rangle \oplus (\psi' \oplus \psi')$. Since $\psi \oplus \psi$ is isotropic, by (d), there is $x \in D_G(a_1, a_1)$ such that $-x \in D_G(\psi' \oplus \psi')$. By (2), $x \in D_G(\langle a_1 \rangle)$, whence $x = a_1$ and $-a_1 =$

$-x \in D_G(\psi')$. Invoking (d) again, we conclude that $\psi = \langle a_1 \rangle \oplus \psi'$ is isotropic.

(3) $\Rightarrow$ (4). Let ψ and ψ' be as above and set $\theta = \langle b_1, \ldots, b_n \rangle$, with $\theta' = \langle b_2, \ldots, b_n \rangle$. We proceed by induction on n. Assume $\psi \oplus \psi \equiv \theta \oplus \theta$; then, $\langle a_1, a_1 \rangle \oplus (\psi' \oplus \psi') \equiv \langle b_1, b_1 \rangle \oplus (\theta' \oplus \theta')$, whence

(*) $\quad \langle a_1, a_1 \rangle \oplus \langle -b_1, -b_1 \rangle \oplus (\psi' \oplus \psi') \equiv \langle 1, -1 \rangle \oplus \langle 1, -1 \rangle \oplus (\theta' \oplus \theta'),$

i.e. the form $\langle 1, 1 \rangle \otimes (\langle a_1, -b_1 \rangle \oplus \psi')$ is isotropic. By (3), the same is true of $(\langle a_1, -b_1 \rangle \oplus \psi') \equiv \langle -b_1 \rangle \oplus \psi$. Thus, there is a form θ_0 of dimension $n - 1$ such that

(**) $\qquad \langle -b_1 \rangle \oplus \psi \equiv \langle 1, -1 \rangle \oplus \theta_0 \equiv \langle -b_1, b_1 \rangle \oplus \theta_0,$

where we have used the preservation of isometry by sum and [SG2]. By Witt cancellation (item (b)), $\psi \equiv \langle b_1 \rangle \oplus \theta_0$. From (*) and (**) we also have

$$\langle 1, 1 \rangle \otimes (\langle 1, -1 \rangle \oplus \theta') \equiv \langle 1, 1 \rangle \otimes (\langle -b_1 \rangle \oplus \psi)$$
$$\equiv \langle 1, 1 \rangle \otimes (\langle 1, -1 \rangle \oplus \theta_0);$$

cancelling out $\langle 1, 1 \rangle \otimes \langle 1, -1 \rangle$, gives $\langle 1, 1 \rangle \otimes \theta' \equiv \langle 1, 1 \rangle \otimes \theta_0$. By the induction hypothesis, $\theta' \equiv \theta_0$, which yields

$$\theta \equiv \langle b_1 \rangle \oplus \theta' \equiv \langle b_1 \rangle \oplus \theta_0 \equiv \psi,$$

as desired.

(4) $\Rightarrow$ (5). Assume that $\psi \oplus \psi$ is hyperbolic. Since $dim(\psi)$ is even, (say $2l$), our assumption comes down to

$$\psi \oplus \psi \equiv \bigoplus_{i=1}^{n} \langle 1, -1 \rangle = (2l) \times \langle 1, -1 \rangle \equiv \theta \oplus \theta$$

with $\theta = \bigoplus_{i=1}^{l} \langle 1, -1 \rangle$. By (4), $\psi \equiv \theta$, i.e. ψ is hyperbolic.

(5) $\Rightarrow$ (1). Assume that $\langle a, a \rangle \equiv \langle 1, 1 \rangle$. Then the form $\langle a, -1 \rangle \oplus \langle a, -1 \rangle$ is hyperbolic. By (5), so is $\langle a, -1 \rangle$, that is, $\langle a, -1 \rangle \equiv \langle 1, -1 \rangle$, which implies $a = 1$ by [SG3]. $\square$

Remark : Most of the content of Proposition 1.6 can be found dispersed in Chapter 1 of [Ma1] (sometimes with different proofs), except for the equivalence of (1)–(3) in (e), which appears as Theorem 4.27, pp. 89-90. The equivalence of items (4) and (5) with the other conditions in (e) appears to be new. The proofs are presented here for the convenience of the reader and to convince him that the axioms set down for special groups make it possible to derive the fundamental relations valid in the classic theory of quadratic forms over fields. $\diamond$

We now give some examples of special groups.

EXAMPLE 1.7. We present the notion of **fan**. Let G be a group of exponent 2 with a distinguished element $-1 \neq 1$. For each $a \in G$, $a \neq -1$, define $G_a = \{1, a\}$, setting $G_{-1} = G$. We now define a relation $\equiv_{fan}$ on $G \times G$ by

[fan] $\qquad \langle a, b \rangle \equiv_{fan} \langle c, d \rangle$ iff $ab = cd$ and $ac \in G_{cd}$.

A proof of the following result can be found in [LP3] :

LEMMA 1.8. *Let G be a group of exponent 2 and let $-1 \neq 1$ be a distinguished element in G. Then $(G, \equiv_{fan}, -1)$ is a reduced special group. Moreover, for $a \in G$, $D_G(1, a) = G_a$.* ◇

As an example consider the multiplicative group $2 = \{\pm 1\}$ with -1 as the distinguished element. It is easily verified that, for $a, b, c, d \in 2$,

$$\langle a, b \rangle \equiv_{fan} \langle c, d \rangle \quad \text{iff} \quad a + b = c + d,$$

(computed in $\mathbb{Z}$); with this structure $\{\pm 1\}$ is a reduced special group with $D(1, 1) = \{1\}$ and $D(1, -1) = 2$.

If t is a form over 2 of dimension n (i.e., a sequence of 1's and -1's of length $n \geq 1$), let p_t = number of 1's in t and n_t = number of -1's in t. Clearly, $p_t + n_t = n$. If s, t are forms of dimension n in 2, then the definition of isometry of n-forms and induction, yields

$s \equiv_{fan} t$ iff $\sum_{i \leq n} s(i) = \sum_{i \leq n} t(i)$ (in $\mathbb{Z}$) iff $p_s = p_t$ and $n_s = n_t$.

A moment of thought will convince the reader that this is the only structure of reduced (pre-)special group on $\{\pm 1\}$, with $1 \neq -1$, **to be indicated by $\mathbb{Z}_2$.** ◇

EXAMPLE 1.9. (The trivial special relation)

Let G be a group of exponent 2 and -1 any element of G distinct from 1. For $a, b, c, d \in G$, define : $\langle a, b \rangle \equiv_t \langle c, d \rangle$ iff $ab = cd$. It is easily verified that $(G, \equiv_t, -1)$ satisfies [SG1] – [SG5]. For [SG6] we have

Fact. *For forms $\varphi = \langle a_1, \ldots, a_n \rangle$, $\psi = \langle b_1, \ldots, b_n \rangle$ on G, $\varphi \equiv_t \psi$ iff $d(\varphi) = d(\psi)$.*

PROOF. It follows from Lemma 1.5.(b) that $\varphi \equiv_t \psi$ implies the equality of the discriminants. For the converse, we use induction on $n \geq 2$, observing that for $n = 2$ the equality of discriminants is the definition of $\equiv_t$.

Assume $n \geq 3$ and that $d(\varphi) = d(\psi)$. Set $\alpha = d(\langle a_2, \ldots, a_n \rangle)$ and $\beta = d(\langle b_2, \ldots, b_n \rangle)$. Let $\vec{z} = (z_3, \ldots, z_n) = (1, \ldots, 1)$. Then, using the induction hypothesis,

(i) $a_1 \alpha = b_1 \beta$ yields $(a_1, \alpha) \equiv_t (b_1, \beta)$;

(ii) $\alpha \, d(\langle \vec{z} \rangle) = \alpha$ yields $\langle a_2, \ldots, a_n \rangle \equiv_t \langle \alpha, \vec{z} \rangle$;

(iii) $\beta \, d(\langle \vec{z} \rangle) = \beta$ yields $\langle b_2, \ldots, b_n \rangle \equiv_t \langle \beta, \vec{z} \rangle$.

The three isometries above imply $\varphi \equiv_t \psi$, as desired. $\square$

By the Fact, $\equiv_t$ is transitive. We refer to the relation $\equiv_t$ as the **trivial special group** structure on G, denoting it by G_t.

It is straightforward to verify that G_t is never reduced, all binary forms are universal and all forms of dimension ≥ 3 are isotropic. $\diamond$

EXAMPLE 1.10. We now describe a construction known as **extension** of a special group by a group of exponent two. In [LP3], the reader can find a proof that this construction yields a special group using only the axioms of sg's. In the case that G is a rsg, a proof of this, in the dual language of abstract order spaces, appears in [ABR], Prop. IV.2.13, p. 93.

Let $(G, \equiv, -1)$ be a sg and Δ be a group of exponent 2. Write $G[\Delta]$ for the group $G \times \Delta$ with its usual (coordinate-wise) group structure, with $\mathbb{1} = (1, 1)$ as identity and $-\mathbb{1} = (-1, 1)$ as distinguished element. We write $g \cdot \delta$, instead of (g, δ), for a typical element of $G[\Delta]$. For each $g \cdot \delta$ in $G[\Delta]$ we define a subgroup $E_{g \cdot \delta}$ of $G[\Delta]$ as follows :

[ext] $$E_{g \cdot \delta} = \begin{cases} D_G(1, g) \times \{1\} & \text{if } g \neq -1 \text{ and } \delta = 1 \\ G[\Delta] & \text{if } g = -1 \text{ and } \delta = 1 \\ \{\mathbb{1}, g \cdot \delta\} & \text{if } \delta \neq 1 \end{cases}$$

Define a relation $\equiv_{ext}$ on $G[\Delta] \times G[\Delta]$ by

$$(g_1 \cdot \delta_1, g_2 \cdot \delta_2) \equiv_{ext} (g_3 \cdot \delta_3, g_4 \cdot \delta_4) \text{ iff } \begin{cases} g_1 g_2 = g_3 g_4, \quad \delta_1 \delta_2 = \delta_3 \delta_4 \\ \text{and} \\ g_3 g_1 \cdot \delta_3 \delta_1 \in E_{g_1 g_2 \cdot \delta_1 \delta_2}. \end{cases}$$

Then, $\equiv_{ext}$ is a special relation on $G[\Delta]$ and $(G[\Delta], \equiv_{ext}, -\mathbb{1})$ is a special group. Moreover, $G[\Delta]$ is reduced iff G is reduced, and for each $g \cdot \delta$ in $G[\Delta]$, $D_{G[\Delta]}(\langle 1, g \cdot \delta \rangle) = E_{g \cdot \delta}$. $\diamond$

We now present the notion of morphism of special groups.

DEFINITION 1.11. A map $(G, \equiv_G, -1) \xrightarrow{f} (H, \equiv_H, -1)$ between pre-special groups is said to be:

(i) A **morphism of psg's** iff it is a group homomorphism such that $f(-1) = -1$, and for all $a, b, c, d \in G$ we have

[mor] $\quad \langle a, b \rangle \equiv_G \langle c, d \rangle \quad \text{implies} \quad \langle f(a), f(b) \rangle \equiv_H \langle f(c), f(d) \rangle.$

(ii) An **isomorphism of psg's** iff it is bijective and both f and f^{-1} are morphisms of psg's.

A morphism (resp., isomorphism) of special groups, also called an **SG-morphism** (resp., **SG-isomorphism**) is a morphism of their underlying psg's.

If $\varphi = \langle a_1, \ldots, a_n \rangle$ is a form on G, the image form by f is denoted $f \star \varphi = \langle f(a_1), \ldots, f(a_n) \rangle$. ◇

Clearly, special groups (reduced special groups) and their morphisms are categories, denoted respectively by **SG** and **RSG**.

LEMMA 1.12. *Let $(G, \equiv_G, -1)$, $(H, \equiv_H, -1)$ be psg's and φ, ψ be n forms on G. Then*

a) *A map $G \xrightarrow{f} H$ is a SG-morphism iff it is a group homomorphism such that $f(-1) = -1$ and satisfies*

[D] $\qquad \forall\ a \in G,\ f(D_G(1, a)) \subseteq D_H(1, f(a))$.

b) *A map $G \xrightarrow{\sigma} \mathbb{Z}_2$ is a morphism of psg's iff it is a group homomorphism, taking -1 to -1 and satisfying*

[ker] $\qquad \forall\ a \in G,\ a \in \ker \sigma\ \text{ implies }\ D_G(1, a) \subseteq \ker \sigma$.

Moreover, if $G \xrightarrow{f} H$ is a morphism of psg's and $\sigma : H \longrightarrow \mathbb{Z}_2$ is a group homomorphism satisfying [ker], the same is true of $\sigma \circ f : G \longrightarrow \mathbb{Z}_2$.

c) *If $G \xrightarrow{f} H$ is a morphism of special groups and φ, ψ are forms on G of the same dimension, then $\varphi \equiv_G \psi$ implies $f \star \varphi \equiv_H f \star \psi$.*

PROOF. a) Suppose that f is a morphism and $b \in D_G(1, a)$. Then there is $u \in G$ such that $\langle b, u \rangle \equiv \langle 1, a \rangle$. Since f is a morphism, $\langle f(b), f(u) \rangle \equiv_H \langle 1, f(a) \rangle$, and so $f(b) \in D_G(1, f(a))$. Conversely, assume that f is a group homomorphism, taking -1 to -1 and satisfying [D]. Let a, b, c, d be elements of G such that $\langle a, b \rangle \equiv \langle c, d \rangle$. Then, $ab = cd$ and $ac \in D_G(1, cd)$. Since $f(ab) = f(a)f(b) = f(c)f(d) = f(cd)$, to prove that f is a morphism of special groups, it is enough to verify (by 1.5.(a)) that $f(ac) \in D_H(1, f(cd))$. But this comes directly from [D].

b) It is straightforward that in this case, condition [D] is equivalent to [ker].

c) Straightforward induction on the dimension of φ and ψ, the result being true for forms of dimension 2 by definition. We omit the details. □

REMARK 1.13. Let $(G, \equiv_G, -1) \xrightarrow{f} (H, \equiv_H, -1)$ be a group homomorphism such that $f(-1) = -1$. It follows from the discussion in Examples 1.7 and 1.9 that f will be a morphism of psg whenever G is a fan (for any H), or if H carries the trivial special group structure (for any G). In particular, $\mathbb{Z}_2$ is the initial object in the categories **SG** and **RSG**. ◇

EXAMPLE 1.14. (The special subgroup of basic elements)

As an illustration of the notions introduced above we shall show that any special group, reduced or not, arises in a canonical way as a (possibly trivial) extension of a certain special subgroup of itself; if $G = G'[\Delta]$ is a proper extension of G' —i.e., Δ has at least two elements— the construction below canonically produces the smallest special subgroup of which G is an extension.

DEFINITION 1.15. Let G be a special group $\neq \mathbb{Z}_2$. An element $a \in G$ is called **basic** if either $D_G(1, a) \neq \{1, a\}$ or $D_G(1, -a) \neq \{1, -a\}$. The set of basic elements of G will be denoted by $Ba(G)$. We set $Ba(\mathbb{Z}_2) = \mathbb{Z}_2$ by convention. (In any case, $1, -1 \in Ba(G)$.) ◇

The proof of the following important result can be found in Marshall [Ma1], Thm. 5.18, p. 114, where further references are given.

Theorem (Berman). *$Ba(G)$ is a subgroup of G.* ◇

Since the axioms [SG0]–[SG5] are universal, $Ba(G)$ with isometry induced by that of G is a pre-special subgroup. We prove:

PROPOSITION 1.16. *$Ba(G)$ is a special subgroup of G.*

PROOF. Only axiom [SG6] is to be checked. Since, by assumption, G satifies it, the Proposition follows from Lemma 1.18 below. □

LEMMA 1.17. *Let G be a pre-special group, and $x, a, b \in G$.*
(a) *If $a \in Ba(G)$, $a \neq -1$, and $x \in D_G(1, a)$, then $x \in Ba(G)$.*
(b) *If $a, b \in Ba(G)$, $a \neq -b$, and $x \in D_G(a, b)$, then $x \in Ba(G)$.*

PROOF. (b) follows easily from (a).
(a) If the conclusion fails, then $D_G(1, x) = \{1, x\}$ and $D_G(1, -x) = \{1, -x\}$. From $x \in D_G(1, a)$ we get $-a \in D_G(1, -x)$ ([SG4]), whence $-a \in \{1, -x\}$. Then, either $a = -1$ or $a = x \in Ba(G)$, contrary to assumption. □

LEMMA 1.18. *For $i = 1, 2, 3$, let $a_i, b_i \in Ba(G)$. Then*
$$\langle a_1, a_2, a_3 \rangle \equiv_G \langle b_1, b_2, b_3 \rangle \Rightarrow \langle a_1, a_2, a_3 \rangle \equiv_{Ba(G)} \langle b_1, b_2, b_3 \rangle.$$

PROOF. By assumption there are $x, y, z \in G$ so that:
(1) $\langle a_1, x \rangle \equiv_G \langle b_1, y \rangle$, (2) $\langle a_2, a_3 \rangle \equiv_G \langle x, z \rangle$, (3) $\langle b_2, b_3 \rangle \equiv_G \langle y, z \rangle$.
The isometries (2) and (3), together with [SG3] imply that $xy = a_2 a_3$ and $yz = b_2 b_3$ are in $Ba(G)$. Since $Ba(G)$ is a subgroup, it suffices to prove that one of x, y or z is in $Ba(G)$ to conclude that all are, and prove the Lemma.

Since $x \in D_G(a_2, a_3)$ ((2)), if $a_2 \neq -a_3$, then Lemma 1.17.(a) implies $x \in Ba(G)$. Likewise, if $b_2 \neq -b_3$, (3) implies $y \in Ba(G)$. Thus, we may assume $a_2 = -a_3$ and $b_2 = -b_3$.

Our assumption and Lemma 1.5.(b) imply:
$$d(\langle a_1, a_2, a_3 \rangle) = a_1 a_2 a_3 = -a_1 = d(\langle b_1, b_2, b_3 \rangle) = b_1 b_2 b_3 = -b_1;$$

hence $a_1 = b_1$. Then, $\langle a_1, a_2, a_3 \rangle = \langle a_1, a_2, -a_2 \rangle$ and $\langle b_1, b_2, b_3 \rangle = \langle a_1, b_2, -b_2 \rangle$. Picking $x_0 = y_0 = -z_0 \in Ba(G)$, from [SG0] and [SG2] we get $\langle a_1, x_0 \rangle \equiv_{Ba(G)} \langle a_1, y_0 \rangle$, $\langle a_2, -a_2 \rangle \equiv_{Ba(G)} \langle x_0, z_0 \rangle$ and $\langle b_2, b_3 \rangle \equiv_{Ba(G)} \langle y_0, z_0 \rangle$, which proves $\langle a_1, a_2, a_3 \rangle \equiv_{Ba(G)} \langle b_1, b_2, b_3 \rangle$. □

Next we show that any special group G is an extension of $Ba(G)$.

PROPOSITION 1.19. *Let G be a special group and Δ be an orthogonal complement of $Ba(G)$ in G. Then, $Ba(G)[\Delta] \simeq G$.*

PROOF. By assumption, every $g \in G$ is uniquely expressed as $g = b \cdot d$ with $b \in Ba(G)$, $d \in \Delta$ (note Δ may be trivial). Let $f : G \to Ba(G) \times \Delta$ be the map $f(g) = (b, d)$, for $g \in G$. It is obvious that f is a bijective group homomorphism. Further, since $-1 \in Ba(G)$, f sends -1 in G to $-1 = (-1, 1)$ in $Ba(G)[\Delta]$. We finally prove that, for $g, g' \in G$:

$$g' \in D_G(1, g) \Leftrightarrow f(g') \in D_{Ba(G)[\Delta]}((1, 1), f(g)).$$

Let $f(g) = (b, d)$, $f(g') = (b', d')$, with $b, b' \in Ba(G)$, $d, d' \in \Delta$.

($\Rightarrow$) If $d = 1$, then $g = b \in Ba(G)$. If $b = -1$, the conclusion follows from the definition of representation in $Ba(G)[\Delta]$ (1.10). If $b \neq -1$, Lemma 1.17.(a) imples $g' \in Ba(G)$, whence $d' = 1$, $g' = b'$, and hence $b' \in D_G(1, b) \cap Ba(G) = D_{Ba(G)}(1, b)$.

Assume $d \neq 1$, i.e., $g \notin Ba(G)$; hence, $D_G(1, g) = \{1, g\}$ and $D_G(1, -g) = \{1, -g\}$. Then, the assumption $g' \in D_G(1, g)$ implies that either $g' = 1$, and hence $b' = 1$, $d' = 1$, or $g' = g$, whence $b' = b$, $d' = d$; this proves the conclusion.

($\Leftarrow$) Assume $(b', d') \in D_{Ba(G)[\Delta]}((1, 1), (b, d))$. If $d = 1$, from 1.10 we have either:

(i) $b \neq -1$, $d' = 1$ and $b' \in D_{Ba(G)}(1, b) \subseteq D_G(1, b)$, whence $g' \in D_G(1, g)$, or

(ii) $b = -1$, whence $g = b \cdot d = -1$, and $g' \in D_G(1, g)$ holds.

If $d \neq 1$, then (by 1.10) either $(b', d') = (1, 1)$, whence $g' = 1$, or $(b', d') = (b, d)$, whence $g' = g$; in both cases, $g' \in D_G(1, g)$. □

Remark. A finer analysis of the preceding arguments shows that Proposition 1.19 implies Proposition 1.16, Lemma 1.18, and has even stronger consequences (see Example 5.18, Chapter 5). For this it suffices to observe that, (1) The notion of extension (1.10) makes sense even for pre-special groups, and (2) If G' is a pre-special group, and $G'[\Delta]$ is a special group, then G' is also a special group (see Example 5.18). ◇

PROPOSITION 1.20. *Let G be a special group and Δ a group of exponent 2. Then $Ba(G[\Delta]) = Ba(G) \times \{1\}$; hence, modulo obvious identification,*

$Ba(G[\Delta]) \subseteq G$. In particular, $Ba(G)$ is the smallest special subgroup G' of G such that G is a (possibly trivial) extension of G'.

PROOF. This is straightforward from the definition of extension (1.10). As illustration we prove $g \in Ba(G) \Rightarrow (g, 1) \in Ba(G[\Delta])$. We assume $G \neq \mathbb{Z}_2$. Let $(g, 1) \notin Ba(G[\Delta])$, i.e., $D_{G[\Delta]}((1, 1), (g, 1)) = \{(1, 1), (g, 1)\}$ and $D_{G[\Delta]}((1, 1), (-g, 1)) = \{(1, 1), (-g, 1)\}$. Since $h \in D_G(1, \pm g)$ implies $(h, 1) \in D_{G[\Delta]}((1, 1), (\pm g, 1))$, we conclude that $h = 1$ or $h = \pm g$, which proves that $g \notin Ba(G)$. □

2. Characterizations of Special Groups

In this section, we present a useful set of equivalent conditions for a pre-special group to be a special group. The main result is Theorem 1.23, below. Its condition (4) is related to Witt's celebrated chain-equivalence Theorem (see Theorem 5.2, [L1], p. 21 ff).

Note that Proposition 1.6.(a) shows that **direct sum and tensor product preserve isometry in any pre-special group.**

If G is a group of exponent 2, $\varphi = \langle a_1, \ldots, a_n \rangle$ is a n-form over G and $\sigma \in S_n$ (the group of permutations of $\{1, \ldots, n\}$), write φ^σ for the n-form $\langle a_{\sigma(1)}, \ldots, a_{\sigma(n)} \rangle$.

LEMMA 1.21. *Let $(G, \equiv, -1)$ be a pre-special group. Let a, b, c, x, y be elements of G and φ, ψ be forms over G. Assume that $\langle a, b \rangle \equiv \langle x, y \rangle$. Then*

a) $\varphi \equiv \psi \oplus \langle a, b \rangle \Rightarrow \varphi \equiv \psi \oplus \langle x, y \rangle$.

b) $\forall \sigma \in S_3$, $\langle a, b, c \rangle \equiv \langle x, y, c \rangle^\sigma$.

PROOF. a) By induction on $dim(\psi)$. Write $\varphi = \langle z \rangle \oplus \varphi_1$. If $dim(\psi) = 1$, then $\psi = \langle \alpha \rangle$ and we must show that

$$\varphi \equiv \langle \alpha, a, b \rangle \Rightarrow \varphi \equiv \langle \alpha, x, y \rangle.$$

Then, there are γ, δ, $\mu \in G$, such that

$$\langle z, \gamma \rangle \equiv \langle \alpha, \delta \rangle, \quad \varphi_1 \equiv \langle \gamma, \mu \rangle \quad \text{and} \quad \langle a, b \rangle \equiv \langle \delta, \mu \rangle. \tag{1}$$

The third isometry in (1) and $\langle a, b \rangle \equiv \langle x, y \rangle$ yield $\langle x, y \rangle \equiv \langle \delta, \mu \rangle$ which, together with the first two isometries in (1), implies $\varphi \equiv \langle \alpha, x, y \rangle$.

Assume the result true for $dim(\psi) = n - 1$ and that $\psi = \langle y_1, \ldots, y_n \rangle$. Then $\varphi \equiv \psi \oplus \langle a, b \rangle$ means that there are γ, δ, μ_3, $\ldots$, μ_{n+2} in G, such that

$$\begin{cases} \langle z, \gamma \rangle \equiv \langle y_1, \delta \rangle; \\ \varphi_1 \equiv \langle \gamma, \mu_3, \ldots, \mu_{n+2} \rangle; \\ \langle y_2, \ldots, y_n \rangle \oplus \langle a, b \rangle \equiv \langle \delta, \mu_3, \ldots, \mu_{n+2} \rangle. \end{cases} \quad (2)$$

By the induction hypothesis, $\langle \delta, \mu_3, \ldots, \mu_{n+2} \rangle \equiv \langle y_2, \ldots, y_n \rangle \oplus \langle x, y \rangle$. But this isometry, and the first two in (2), show that $\varphi \equiv \psi \oplus \langle x, y \rangle$.

b) By (a) it is sufficient to verify that $\theta = \langle a, b, c \rangle$ is isometric to $\langle x, y, c \rangle$, $\langle y, x, c \rangle$ and $\langle c, x, y \rangle$. That the first two are isometric to θ follows directly from the preservation of $\equiv$ by sums and the hypothesis that $\langle a, b \rangle \equiv \langle x, y \rangle$ (and $\langle a, b \rangle \equiv \langle y, x \rangle$, by [SG1]). For the remaining permutation, observe that the isometries

$$\langle a, c \rangle \equiv \langle c, a \rangle, \quad \langle b, c \rangle \equiv \langle c, b \rangle \quad \text{and} \quad \langle x, y \rangle \equiv \langle a, b \rangle,$$

show that $\theta \equiv \langle c, x, y \rangle$. $\square$

We can readily adapt the classical notions of simple and chain-equivalence to our context (see [L1], Chapter I, §5, p. 20 ff).

Two forms $\varphi = \langle a_1, \ldots, a_n \rangle$, $\psi = \langle b_1, \ldots, b_n \rangle$, over the psg G are said to be **simply equivalent**, if there are i, j such that

i) $\langle a_i, a_j \rangle \equiv \langle b_i, b_j \rangle$;   ii) $a_k = b_k$, whenever k is distinct from i and j.

We say that φ, ψ are **chain-equivalent**, written $\varphi \approx \psi$, if there is a sequence of n-forms $\varphi_0, \varphi_1, \ldots, \varphi_m$, such that $\varphi_0 = \varphi$, $\varphi_m = \psi$, and φ_i is simply equivalent to φ_{i+1}, for $0 \leq i \leq m-1$.

LEMMA 1.22. *Chain-equivalence is an equivalence relation. Moreover, if φ, ψ are n-forms over G and c is an element of G, then*

a) $\varphi \approx \psi$ iff $\forall \sigma \in S_n, \varphi \approx \psi^\sigma$.

b) *If $\varphi \approx \psi$ implies $\langle c \rangle \oplus \varphi \approx \langle c \rangle \oplus \psi$.*

c) $\varphi \equiv \psi$ *implies* $\varphi \approx \psi$.

PROOF. The fact that $\approx$ is an equivalence relation is straightforward. Note that a form ψ is simply equivalent to ψ^τ, where τ is a transposition in S_n. Since $\approx$ is transitive and S_n is generated by transpositions, it is clear that the statements in (a) are equivalent. For (b), note that if φ is simply equivalent to ψ the same is true of $\langle c \rangle \oplus \varphi$ and $\langle c \rangle \oplus \psi$. So, any chain connecting φ and ψ becomes, adding $\langle c \rangle$ to each term, a chain connecting $\langle c \rangle \oplus \varphi$ to $\langle c \rangle \oplus \psi$.

c) By induction on the dimension of φ, noting that for 2-forms there is nothing to prove. So suppose the result true for forms of dimension n and let $\varphi = \langle a \rangle \oplus \theta_0$ and $\psi = \langle b \rangle \oplus \theta_1$, where $dim(\theta_i) = n$. If $\varphi \equiv \psi$, there are $x, y, \vec{z} = (z_1, \ldots, z_n)$ in G such that

$$\langle a, x \rangle \equiv \langle b, y \rangle, \quad \theta_0 \equiv \langle x, \vec{z} \rangle \quad \text{and} \quad \theta_1 \equiv \langle y, \vec{z} \rangle. \quad (1)$$

By the induction hypothesis, the last two isometries in (1) yield $\theta_0 \approx \langle x, \vec{z} \rangle$ and $\theta_1 \approx \langle y, \vec{z} \rangle$. By item (b),
$$\varphi = \langle a \rangle \oplus \theta_0 \approx \langle a, x, \vec{z} \rangle \quad \text{and} \quad \psi = \langle b \rangle \oplus \theta_1 \approx \langle b, y, \vec{z} \rangle,$$
with $\langle a, x, \vec{z} \rangle$ simply equivalent to $\langle b, y, \vec{z} \rangle$, because of the first isometry in (1). Since $\approx$ is an equivalence relation, we conclude $\varphi \approx \psi$, as desired. □

The following result is very useful in verifying that a psg is a special group. It will be applied in the next section, when it will be shown that classical quadratic form theory over fields can be stated in terms of special groups. Condition (5) in the statement originates in the work of M. Marshall ([Ma1]); the fact that it implies that a psg is a special group appears in [LP3], with a different proof. It is interesting to know that it can be presented at an early stage in the development of the theory of special groups.

THEOREM 1.23. *Let $(G, \equiv, -1)$ be a pre-special group. The following are equivalent :*

(1) $\equiv$ *is 3-transitive (i.e., transitive for 3-forms, and hence G is a special group).*

(2) $\equiv$ *is transitive (i.e., transitive for n-forms, for all $n \geq 2$).*

(3) *For all $n \geq 2$, for all n-forms φ, ψ over G and all $\sigma \in S_n$,*
$$\varphi \equiv \psi \quad \text{implies} \quad \varphi \equiv \psi^\sigma.$$

(4) *For all $n \geq 2$ and for all n-forms φ, ψ over G,*
$$\varphi \equiv \psi \quad \text{iff} \quad \varphi \approx \psi.$$

(5) *For all 3-forms φ and all $b_1, b_2, b_3 \in G$,*
$$\varphi \equiv \langle b_1, b_2, b_3 \rangle \quad \text{implies} \quad \varphi \equiv \langle b_2, b_1, b_3 \rangle.$$

PROOF. (1) $\Rightarrow$ (2). By induction on the dimension, which, when 2 or 3, are taken care of by assumption. Assume that $\langle a_1, \ldots, a_n \rangle \equiv \langle b_1, \ldots, b_n \rangle = \psi$ and $\psi \equiv \langle c_1, \ldots, c_n \rangle$, and that $\equiv$ is transitive on forms of dimension $n - 1 \geq 3$. The hypotheses yield $\alpha, \beta, \gamma, \delta, y_i, z_i \in G$, $3 \leq i \leq n$, such that (I) and (II) below hold true

$$\langle a_1, \alpha \rangle \equiv \langle b_1, \beta \rangle, \langle a_2, \ldots, a_n \rangle \equiv \langle \alpha, \vec{y} \rangle \text{ and } \langle b_2, \ldots, b_n \rangle \equiv \langle \beta, \vec{y} \rangle; \quad \text{(I)}$$

$$\langle b_1, \gamma \rangle \equiv \langle c_1, \delta \rangle, \langle b_2, \ldots, b_n \rangle \equiv \langle \gamma, \vec{z} \rangle \text{ and } \langle c_2, \ldots, c_n \rangle \equiv \langle \delta, \vec{z} \rangle, \quad \text{(II)}$$

where $\vec{y} = \langle y_3, \ldots, y_n \rangle$ and $\vec{z} = \langle z_3, \ldots, z_n \rangle$. By induction, $\equiv$ is transitive on $(n-1)$-forms, and so, $\langle \beta, \vec{y} \rangle \equiv \langle \gamma, \vec{z} \rangle$, since both are isometric to $\langle b_2, \ldots, b_n \rangle$. Thus, there are $x, y, \vec{t} = \{t_4, \ldots, t_n\}$ in G such that

$$\langle \beta, x \rangle \equiv \langle \gamma, y \rangle, \quad \langle \vec{y} \rangle \equiv \langle x, \vec{t} \rangle \quad \text{and} \quad \langle \vec{z} \rangle \equiv \langle y, \vec{t} \rangle. \quad \text{(III)}$$

Now, by the preservation of isometry by sum, the first isometry in (I), (II) and (III), as well as 3-transitivity, we may write

2. CHARACTERIZATIONS OF SPECIAL GROUPS

$$\langle a_1, \alpha, x \rangle = \langle a_1, \alpha \rangle \oplus \langle x \rangle \equiv \langle b_1, \beta \rangle \oplus \langle x \rangle \equiv \langle b_1 \rangle \oplus \langle \beta, x \rangle$$
$$\equiv \langle b_1 \rangle \oplus \langle \gamma, y \rangle \equiv \langle b_1, \gamma \rangle \oplus \langle y \rangle \equiv \langle c_1, \delta \rangle \oplus \langle y \rangle = \langle c_1, \delta, y \rangle.$$

Therefore, there are $u, v, w \in G$ such that

$$\langle a_1, u \rangle \equiv \langle c_1, v \rangle, \quad \langle \alpha, x \rangle \equiv \langle u, w \rangle \quad \text{and} \quad \langle \delta, y \rangle \equiv \langle v, w \rangle. \qquad \text{(IV)}$$

The preservation of isometry by sum, the transitivity of $\equiv$ for $(n-1)$-forms, the second and the third isometry in (I) and (II), respectively, together with the last two in (III) and (IV), yield

i) $\langle a_2, \ldots, a_n \rangle \equiv \langle \alpha, \vec{y} \rangle \equiv \langle \alpha, x, \vec{t} \rangle \equiv \langle u, w, \vec{t} \rangle$

and

ii) $\langle c_2, \ldots, c_n \rangle \equiv \langle \delta, \vec{z} \rangle \equiv \langle \delta, y, \vec{t} \rangle \equiv \langle v, w, \vec{t} \rangle$,

isometries which, together with the first one in (IV), prove that $\langle a_1, \ldots, a_n \rangle \equiv \langle c_1, \ldots, c_n \rangle$.

(2) $\Rightarrow$ (3). By induction on dimension; for 2-forms, the conclusion follows from [SG1]. Let $\sigma \in S_n$, $\varphi = \langle a \rangle \oplus \varphi_1$ and $\psi = \langle b_1, \ldots, b_n \rangle$.

Case A. $\sigma(1) = 1$.

We may write $\psi^\sigma = \langle b_1 \rangle \oplus \langle b_2, \ldots, b_n \rangle^\sigma$; moreover from $\varphi \equiv \psi$ we get $\alpha, \beta, \vec{y} = \langle y_3, \ldots, y_n \rangle \in G$ such that

$$\langle a, \alpha \rangle \equiv \langle b_1, \beta \rangle, \quad \varphi_1 \equiv \langle \alpha, \vec{y} \rangle \quad \text{and} \quad \langle b_2, \ldots, b_n \rangle \equiv \langle \beta, \vec{y} \rangle. \qquad \text{(V)}$$

By induction, $\langle b_2, \ldots, b_n \rangle^\sigma \equiv \langle \beta, \vec{y} \rangle$ and this, together with the first two isometries in (V), yields $\varphi \equiv \langle b_1 \rangle \oplus \langle b_2, \ldots, b_n \rangle^\sigma = \psi^\sigma$.

Case B. σ is a 2-cycle $(1, i)$, for some $i \geq 2$.

From $\varphi \equiv \psi$ we get the isometries in (V). Let $\vec{b} = \{b_k : k \neq 1, i\}$. By the induction hypothesis and the third isometry in (V), $\langle \beta, \vec{y} \rangle \equiv \langle b_i, \vec{b} \rangle$, and so it follows that $\langle \beta, \vec{y}, b_1 \rangle \equiv \langle b_i, \vec{b}, b_1 \rangle$.

Case A and the preservation of isometry by sum yield the following sequence of isometries:

$$\langle \beta, \vec{y}, b_1 \rangle \equiv \langle \beta, b_1, \vec{y} \rangle = \langle \beta, b_1 \rangle \oplus \langle \vec{y} \rangle \equiv \langle b_1, \beta \rangle \oplus \langle \vec{y} \rangle$$
$$\equiv \langle a, \alpha \rangle \oplus \langle \vec{y} \rangle = \langle a \rangle \oplus \langle \alpha, \vec{y} \rangle \equiv \varphi.$$

Since $\equiv$ is transitive, we get $\varphi \equiv \langle \beta, \vec{y}, b_1 \rangle$, and thus, $\varphi \equiv \langle b_i, \vec{b}, b_1 \rangle$. We may apply Case A once more to put b_1 in its desired place, preserving isometry, getting $\langle b_i, \vec{b}, b_1 \rangle \equiv \psi^\sigma$. The transitivity of $\equiv$ now yields $\varphi \equiv \psi^\sigma$, concluding the proof of Case B.

Cases A and B show that $\varphi \equiv \psi$ implies $\varphi \equiv \psi^\sigma$, for any transposition $\sigma \in S_n$. Since $\equiv$ is assumed transitive and S_n is generated by transpositions, we conclude the desired implication for all $\sigma \in S_n$.

(3) $\Rightarrow$ (4). By Lemma 1.22.(c) it is enough to verify that $\varphi \approx \psi$ implies $\varphi \equiv \psi$.

We first verify that simple equivalence implies isometry. If φ is simply equivalent to ψ, then there are a, b, x, y, $\vec{z}$ in G and permutations σ, $\tau \in S_n$, such that $\varphi^\sigma = \langle \vec{z}, a, b \rangle$ and $\psi^\tau = \langle \vec{z}, x, y \rangle$, with $\langle a, b \rangle \equiv \langle x, y \rangle$. By Lemma 1.21.(a), $\varphi^\sigma \equiv \psi^\tau$, and so (3) guarantees that $\varphi \equiv \psi$, because $(\varphi^\sigma)^{\sigma^{-1}} = \varphi$, for all $\sigma \in S_n$ and all forms φ over G.

We use induction on the length l of chains φ_i, $0 \leq i \leq l$, which witness $\varphi \approx \psi$. If $l = 1$, φ is simply equivalent to ψ and we have already remarked that (with (3)), $\varphi \equiv \psi$. Suppose the result true for chains of length l and that φ_i, $0 \leq i \leq l+1$, is a chain connecting $\varphi = \varphi_0$ and $\psi = \varphi_{l+1}$. By induction, $\varphi \equiv \varphi_l$, with φ_l simply equivalent to ψ. Thus, just as above, there are σ, $\tau \in S_n$ and a, b, x, y, $\vec{z}$ in G such that

$$\varphi_l^\sigma = \langle \vec{z} \rangle \oplus \langle a, b \rangle \quad \psi^\tau = \langle \vec{z} \rangle \oplus \langle x, y \rangle,$$

with $\langle a, b \rangle \equiv \langle x, y \rangle$. By (3), $\varphi \equiv \varphi_l^\sigma = \langle \vec{z} \rangle \oplus \langle a, b \rangle$. By Lemma 1.21.(a), $\varphi \equiv \langle \vec{z} \rangle \oplus \langle x, y \rangle = \psi^\tau$. Another application of (3) gives $\varphi \equiv \psi$, as desired.

(4) $\Rightarrow$ (5). This is a special case of Lemma 1.21.(b).

(5) $\Rightarrow$ (1). We show that if φ, ψ are 3-forms over G, then

$$\forall\, \sigma \in S_3,\ \varphi \equiv \psi \text{ implies } \varphi \equiv \psi^\sigma.$$

Once this is proven, then, exactly as in the proof of (3) $\Rightarrow$ (4), we have that for all 3-forms φ, ψ

$$\varphi \equiv \psi \text{ iff } \varphi \approx \psi.$$

Since $\approx$ is transitive, the same will be true of $\equiv$.

Let $\psi = \langle b_1, b_2, b_3 \rangle$, $\sigma \in S_3$, and assume that $\varphi \equiv \psi$.

i) $\sigma(1) = 1$. Since $\langle b_2, b_3 \rangle \equiv \langle b_3, b_2 \rangle$ ([SG1]), it follows from Lemma 1.21.(a) that $\varphi \equiv \psi$ implies $\varphi \equiv \psi^\sigma$.

ii) $\sigma(1) = 2$. In this case we have $\psi^\sigma = \langle b_2, b_i, b_j \rangle$, with $\{i, j\} = \{1, 3\}$. If $i = 1$, the desired isometry follows directly from (5). If $i = 3$, using (5) and (i) in succession, we get

$$\varphi \equiv \psi \text{ implies } \varphi \equiv \langle b_2, b_1, b_3 \rangle \text{ implies } \varphi \equiv \langle b_2, b_3, b_1 \rangle,$$

as needed.

iii) $\sigma(1) = 3$. By (i) above, we have $\varphi \equiv \langle b_1, b_3, b_2 \rangle$; by (5), we can exchange b_1 and b_3 to get $\varphi \equiv \langle b_3, b_1, b_2 \rangle$. Now, case (i) can be applied again, to get $\varphi \equiv \psi^\sigma$. $\square$

The statement of the next result should be compared to that of part (a) of Lemma 1.21. Condition (c) in the statement below appears to be new.

COROLLARY 1.24. *Let $(G, \equiv, -1)$ be a pre-special group. Let φ and ψ be forms over G and $a, b, x, y \in G$. The following are equivalent :*

a) G is a special group.

b) For all forms φ, ψ over G and all $a, b, x, y \in G$

$$\varphi \equiv \langle a, b \rangle \oplus \psi \text{ and } \langle a, b \rangle \equiv \langle x, y \rangle \quad \Rightarrow \quad \varphi \equiv \langle x, y \rangle \oplus \psi.$$

c) For all 3-forms φ over G and all $a, b, c, x, y \in G$,

$$\varphi \equiv \langle a, b, c \rangle \text{ and } \langle a, b \rangle \equiv \langle x, y \rangle \quad \Rightarrow \quad \varphi \equiv \langle x, y, c \rangle.$$

PROOF. That (a) implies (b) follows from Lemma 1.21.(a) and the fact that G satisfies condition (3) in Theorem 1.23. Clearly, (b) implies (c). It remains to prove that (c) implies (a). We verify that, in fact, (c) implies condition (5) of Theorem 1.23.

Assume that $\langle u, v, w \rangle \equiv \langle a, b, c \rangle$. Hence, there are α, β, γ in G such that

$$\langle u, \alpha \rangle \equiv \langle a, \beta \rangle, \quad \langle v, w \rangle \equiv \langle \alpha, \gamma \rangle \quad \text{and} \quad \langle b, c \rangle \equiv \langle \beta, \gamma \rangle. \tag{*}$$

By 1.21.(b), from $\langle u, \alpha \rangle \equiv \langle a, \beta \rangle$, we get $\langle \gamma, a, \beta \rangle \equiv \langle u, \alpha, \gamma \rangle$. Lemma 1.21.(a) (with $\langle \gamma, a, \beta \rangle = \varphi$ and $\langle u \rangle = \psi$) and the second isometry in (*) also yield $\langle \gamma, a, \beta \rangle \equiv \langle u, v, w \rangle$. Since $\langle a, \beta \rangle \equiv \langle \beta, a \rangle$, 1.21.(a) once again (this time with $\langle u, v, w \rangle = \varphi$ and $\langle \gamma \rangle = \psi$), implies $\langle u, v, w \rangle \equiv \langle \gamma, \beta, a \rangle$. Now, (c) and the third isometry in (*) yield $\langle u, v, w \rangle \equiv \langle b, c, a \rangle$, and yet another application of 1.21.(a) gives $\langle u, v, w \rangle \equiv \langle b, a, c \rangle$, as desired. □

1.25. The Witt ring of a special group. The usual construction of the Witt ring of a field can be carried out, in almost identical terms, for special groups as well. For the reader's convenience, and for later reference, we recall the main points; see [D] for more information. For the field case, see [L1], Chapter II, [L2], Chapter 1, or [Ma1], Chapters 2 and 3.

Let $\langle G, \equiv_G, -1 \rangle$ be a special group (not necessarily reduced). Two forms φ, ψ over G, φ are called **Witt-equivalent** (over G), written $\varphi \approx_G \psi$, if there are integers $n, m \geq 0$ such that

$$\varphi \oplus n \langle 1, -1 \rangle \equiv_G \psi \oplus m \langle 1-1 \rangle.$$

It is easily verified that $\approx_G$ is an equivalence relation on forms over G, compatible with (and, obviously, coarser than) the isometry relation $\equiv_G$.

We denote by $W(G)$ the set of equivalence classes of forms over G under Witt-equivalence, and by $\overline{\varphi}$ the Witt-equivalence class of the form φ. The following Fact summarizes the basic properties of this construction.

FACT 1.26. *Let G be a special group and let φ, ψ be forms over G.*

a) Witt-equivalence is a congruence with respect to sum and product of forms.

b) With the operations $\overline{\varphi} + \overline{\psi} = \overline{\varphi \oplus \psi}$ and $\overline{\varphi}\,\overline{\psi} = \overline{\varphi \otimes \psi}$, $W(G)$ is a commutative ring having as zero the class of hyperbolic forms and $\overline{\langle 1 \rangle}$ as multiplicative identity.

c) The set $I(G)$ of (classes of) even dimensional forms is a maximal ideal in $W(G)$ (called the **fundamental ideal** of $W(G)$). Moreover, $W(G)/I(G)$ is the two element field.

d) For $n \geq 1$, the nth power of $I(G)$, denoted $I^n(G)$, is generated, as an abelian group, by the multiples of Pfister forms of degree n, that is, every element of $I^n(G)$ is Witt-equivalent to a linear combination $\bigoplus_{i=1}^{k} a_i \varphi_i$ of Pfister forms φ_i of degree n, with coefficients $a_i \in G$.

e) For a field F, $W(G(F))$ is just the Witt ring of F, written $W(F)$. ◇

Comments. (1) Item (b) above depends in a crucial way on Witt's cancellation theorem 1.6.(b).

(2) Another, perhaps more intuitive description of the Witt ring stems from the following observations:

  i) For every form φ over G there is an anisotropic form φ_{an} over G — unique up to isometry in G — and a (uniquely determined) integer $r \geq 0$ such that $\varphi \equiv_G \varphi_{an} \oplus r\langle 1, -1 \rangle$.

  ii) $\varphi \approx_G \psi$ iff $\varphi_{an} \equiv_G \psi_{an}$.

These facts show that $W(G) \setminus \{0\}$ is the class of forms anisotropic over G modulo isometry. ◇

3. Fields and Special Groups

In this section, we shall present a proof that the usual quadratic form theories over fields of characteristic distinct from 2 – reduced and not necessarily reduced – yield special groups (Theorem 1.32). These examples are, of course, at the root of the concept of special group. The notion of special group is also tightly connected to that of abstract order space. This theme is discussed in Chapter 3. Our method will be quite general, covering all the known instances of quadratic form theory over fields, as well as being an illustration of the usefulness of condition (5) in Theorem 1.23. The section ends with the special group version of the Baer-Krull Theorem (Theorem 1.33), which is, in fact, a generalization of this classical result.

Let F be a field of characteristic $\neq 2$, which will remain fixed in what follows.

NOTATION 1.27. Let $\dot{F} = \{x \in F : x \neq 0\}$ be the multiplicative group of units in F.

1. $\dot{F}^2 = \{x^2 : x \in \dot{F}\}$. **2.** $\Sigma \dot{F}^2 = \{\sum_{i \in I} x_i^2 : I \text{ is finite and } x_i \in \dot{F}\}$.

Clearly, $\dot{F}^2$ is a subgroup of $\dot{F}$. We set $G(F) = \dot{F}/\dot{F}^2$. If $p = \Sigma_i\, y_i^2 \neq 0$ is in $\Sigma\dot{F}^2$, then $1/p = \Sigma_i\, (y_i/p)^2 \in \Sigma\dot{F}^2$. Thus, if F is formally real, then $\Sigma\dot{F}^2$ is a subgroup of $\dot{F}$. In this case we define $G_{red}(F) = \dot{F}/\Sigma\dot{F}^2$. Clearly, both $G(F)$ and $G_{red}(F)$ are groups of exponent 2. $\diamond$

We wish to show that the usual notion of isometry in $G(F)$ and in $G_{red}(F)$ yield special groups, the latter always reduced. To this end, we introduce the following

DEFINITION 1.28. *Let T be a subset of $\dot{F}$ and write $T^* = T \cup \{0\}$.*

a) If $a, b \in \dot{F}$,

$$D_T(a,b) = \{t \in \dot{F} : \exists\, p,\, q \in T^* \text{ such that } t = a\,p + b\,q\},$$

is the **set of elements represented** *by $\langle a, b \rangle$ over T. Clearly, $\{a, b\} \subseteq D_T(a,b)$.*

b) *T is a* **SG-subgroup** *of F iff it satisfies the following conditions :*

(i) *T is a proper subgroup of the multiplicative group $\dot{F}$;*

(ii) *$\dot{F}^2 \subseteq T$;*

(iii) *For all $a \in \dot{F}$, $D_T(1,a)$ is a subgroup of $\dot{F}$.*

Since T is a subgroup of $\dot{F}$, for all $p \in \dot{F}$, $p \in T$ iff $1/p \in T$. $\diamond$

We now show that squares, sums of squares and pre-orders are examples of SG-subgroups.

LEMMA 1.29. *With notation as above, let T be a subgroup of $\dot{F}$, containing $\dot{F}^2$ and satisfying :*

(CS) $\quad \forall\, p,\, q,\, u,\, v \in T$ and $\forall\, a \in \dot{F},\, \exists\, x \in F$ *such that*
$\quad (pua^2 + qv - xa) \in T^*$ *and* $(pv + qu + x) \in T^*$.

Then, T is a SG-subgroup of F. In particular,

— *$\dot{F}^2$ is a SG-subgroup of F;*

— *If F is formally real, then $\Sigma\dot{F}^2$ and pre-orders are SG-subgroups of F.*

PROOF. a) For $s, t \in D_T(1, a)$, we must show that $1/s$ and st are in $D_T(1, a)$. We may write

$$s = pa + q \quad \text{and} \quad t = ua + v, \qquad (1)$$

with $p, q, u, v \in T^*$. Dividing the first equation by $\dfrac{1}{s^2} \in T$ shows that $1/s \in D_T(1, a)$. To verify that $st \in D_T(1, a)$, consider the product of the equations in (1), namely

$$st = pua^2 + qv + (pv + qu)a. \tag{2}$$

Clearly, (2) implies that if any one of p, q, u, v is zero, then $st \in D_T(1, a)$. Assume then, that all these coefficients are in T. By (CS), there is $x \in F$ such that

$$st = pua^2 + qv - xa + xa + (pv + qu)a =$$
$$= \underbrace{(pua^2 + qv - xa)}_{\alpha} + \underbrace{(pv + qu + x)}_{\beta} a,$$

with $\alpha, \beta \in T^*$, and hence $D_T(1, a)$ is a subgroup of $\dot{F}$.

If T is closed under sums (as is the case of a pre-order or of $\Sigma \dot{F}^2$), then it satisfies (CS) with $x = 0$, for all $a \in \dot{F}$. If $T = \dot{F}^2$, then $p = p_1^2$, $q = q_1^2$, $u = u_1^2$ and $v = v_1^2$; we take $x = 2(p_1 q_1 u_1 v_1)$, to prove (CS) for all $a \in \dot{F}$. For instance, $pua^2 + qv - xa = (p_1 u_1 a - q_1 v_1)^2$. □

Let T be a fixed (but otherwise arbitrary) SG-subgroup of F. Let $G_T(F) = \dot{F}/T$ be the exponent-2 quotient of $\dot{F}$ by T; write a_T for the class of $a \in \dot{F}$ in $G_T(F)$. For $a, b \in \dot{F}$ we have

(*) $\qquad a_T = b_T \quad \text{iff} \quad ab \in T \quad \text{iff} \quad \exists\, p \in T \text{ such that } b = ap.$

LEMMA 1.30. *With notation as above, let a, b, c, d, t be elements of $\dot{F}$. Then,*

a) $t \in D_T(a, b) \;\Rightarrow\; t_T \subseteq D_T(a, b)$.

b) $t\, D_T(a, b) = D_T(ta, tb)$.

c) $a_T = c_T$ and $b_T = d_T \;\Rightarrow\; D_T(a, b) = D_T(c, d)$.

PROOF. Straightforward from definitions, elementary computations and (*) above. □

We now define a relation $\equiv$ on $G_T(F) \times G_T(F)$ by

$$\langle a_T, b_T \rangle \equiv \langle c_T, d_T \rangle \quad \text{iff} \quad (ab)_T = (cd)_T \text{ and } D_T(a, b) = D_T(c, d).$$

It follows from Lemma 1.30.(c) that $\equiv$ is well defined. Clearly, $\equiv$ is an equivalence relation on $G_T(F) \times G_T(F)$.

It is easily seen that when T is $\dot{F}^2$ or a pre-order, this relation is precisely the isometry of 2-forms in the non-reduced or reduced theory of quadratic forms, respectively. Our next result is a version of Proposition 5.1 in [L1] (p. 20) and will be crucial in the sequel.

PROPOSITION 1.31. *If $a, b, c, d, t \in \dot{F}$, then*

a) $t \in D_T(a, b) \quad \text{iff} \quad D_T(t, abt) = D_T(a, b)$
$\qquad\qquad \text{iff} \quad \langle t_T, (abt)_T \rangle \equiv \langle a_T, b_T \rangle.$

b) $\langle a_T, b_T \rangle \equiv \langle c_T, d_T \rangle$ iff $\begin{cases} (ab)_T = (cd)_T \\ \text{and} \\ D_T(a,b) \cap D_T(c,d) \neq \emptyset. \end{cases}$

PROOF. One should keep in mind that $\dot{F}^2 \subseteq T$. To prove (a), it is clearly sufficient to verify that

$$t \in D_T(a,b) \quad \text{implies} \quad D_T(t, abt) = D_T(a,b),$$

the other implications coming directly from the definition of $\equiv$. We first note the following

Fact. If $x \in D_T(1,y)$, then $D_T(x, xy) = x\, D_T(1,y) = D_T(1,y)$.

Since $D_T(1,y)$ is a subgroup of $\dot{F}$, if $x \in D_T(1,y)$, then $1/x \in D_T(1,y)$ and we have $xD_T(1,y) \subseteq D_T(1,y)$ and $1/x\, D_T(1,y) \subseteq D_T(1,y)$, relations which, together with Lemma 1.30.(b), prove $xD_T(1,y) = D_T(1,y)$, verifying the Fact.

If $t \in D_T(a,b)$, then $at \in D_T(1, ab)$ and so, by the Fact, $D_T(ta, (tab)a) = D_T(1, ab)$. Thus, we have

$$D_T(a,b) = D_T(a, ba^2) = aD_T(1, ab) = aD_T(ta, (tab)a)$$
$$= D_T(ta^2, (tab)a^2) = D_T(t, abt),$$

which proves (a).

b) If $t \in D_T(a,b) \cap D_T(c,d)$, by (a) we have

$$D_T(a,b) = D_T(t, abt) \quad \text{and} \quad D_T(t, cdt) = D_T(c,d).$$

Since $(ab)_T = (cd)_T$, we get $(abt)_T = (cdt)_T$ and so, by Lemma 1.30.(c), $D_T(t, abt) = D_T(t, cdt)$, proving that $D_T(a,b) = D_T(c,d)$ and that $\langle a_T, b_T \rangle \equiv \langle c_T, d_T \rangle$. The converse is clear, ending the proof. □

We take as distinguished element -1 in $G_T(F)$ the class of $-1 \in \dot{F}$, $(-1)_T$. We now prove

THEOREM 1.32. *If T is a SG-subgroup of a field F of characteristic $\neq 2$, then $(G_T(F), \equiv, -1)$ is a special group, which is reduced iff T is closed under sums.*

PROOF. We have to verify conditions [SG i] in Definition 1.2. Both [SG0] and [SG1] are straightforward. The validity of [SG3] is required in the very definition of $\equiv$, while [SG5] follows from Lemma 1.30.(b). It remains to verify that [SG2], [SG4] and [SG6]. Although [SG2] is a consequence of [SG4], the former will be used in the proof of the latter.

Verification of [SG2]. Note that if $a \in \dot{F}$, then there are $x, y \in F$ such that $a = x^2 - y^2$: just take $x = \dfrac{1+a}{2}$ and $y = \dfrac{1-a}{2}$. This shows that

$a \in D_T(1, -1) \cap D_T(a, -a)$. Since the discriminant of $\langle a, -a \rangle$ is the same as that of $\langle 1, -1 \rangle$ modulo T, 1.31.(c) yields $\langle a, -a \rangle \equiv \langle 1, -1 \rangle$.

Verification of [SG4]. By hypothesis, we have $a_T\, b_T = c_T\, d_T$ and $D_T(a, b) = D_T(c, d)$. Since the discriminant equation implies $a_T\,(-c_T) = (-b_T)\,d_T$, it is sufficient to verify, by Proposition 1.31.(c), that $D_T(a, -c)$ has non-empty intersection with $D_T(-b, d)$.

First observe that if $b_T = d_T$, then $a_T = c_T$, and [SG2] yields the desired conclusion. We assume, therefore, that $b_T \neq d_T$. Since $b \in D_T(c, d)$, there are $p, q \in T^*$ such that
$$b = c\, p + d\, q;$$
note that $p \neq 0$, otherwise, $b_T = d_T$. But then we may write
$$-c = d\,(q/p) - b(1/p),$$
and $-c \in D_T(-b, d)$. Since $-c$ is also in $D_T(a, -c)$, we have verified [SG4] and thus, that $(G_T(F), \equiv, -1)$ is a pre-special group.

Verification of [SG6]. By condition (5) in Theorem 1.23, it is enough to show that
$$\langle a_T, b_T, c_T \rangle \equiv \langle x_T, y_T, z_T \rangle \quad \text{implies} \quad \langle a_T, b_T, c_T \rangle \equiv \langle y_T, x_T, z_T \rangle.$$

The antecedent of the above implication means that there are $\alpha, \beta, \gamma \in \dot{F}$ such that
$$\langle a_T, \alpha_T \rangle \equiv \langle x_T, \beta_T \rangle, \langle b_T, c_T \rangle \equiv \langle \alpha_T, \gamma_T \rangle \text{ and } \langle y_T, z_T \rangle \equiv \langle \beta_T, \gamma_T \rangle. \quad (1)$$

From the first isometry in (1) we get $a \in D_T(x, \beta)$, while the last one implies $\beta \in D_T(y, z)$. Thus, there are $p_a, q_a, p_\beta, q_\beta$ in T^* such that equations (2) and (3) below hold true:
$$a = x\, p_a + \beta\, q_a \tag{2}$$
$$\beta = y\, p_\beta + z\, q_\beta \tag{3}$$

Substituting equation (3) in (2), we arrive at
$$a = x\, p_a + \beta\, q_a = x\, p_a + q_a(y\, p_\beta + z\, q_\beta)$$
$$= x\, p_a + y\, p_\beta q_a + z\, q_\beta q_a = y\, p_\beta q_a + (x\, p_a + z\, q_\beta q_a).$$

Now define
$$v = x\, p_a + z\, q_\beta q_a. \tag{4}$$

Then,
$$a = y\, p_\beta q_a + v. \tag{5}$$

We discuss two cases.

3. FIELDS AND SPECIAL GROUPS

<u>Case I</u> : $v = 0$. Then, from (5), we have $a_T = y_T$. Consequently, the third isometry in (1) can be written as $\langle a_T, z_T \rangle \equiv \langle \beta_T, \gamma_T \rangle$. This isometry, the first one in (1) and [SG4] yield

$$\langle x_T, -\alpha_T \rangle \equiv \langle a_T, -\beta_T \rangle \equiv \langle -z_T, \gamma_T \rangle,$$

and so, $\langle x_T, -\alpha_T \rangle \equiv \langle -z_T, \gamma_T \rangle$. Another application of [SG4] yields $\langle x_T, z_T \rangle \equiv \langle \alpha_T, \gamma_T \rangle$, which together with the second isometry in (1), gives $\langle x_T, z_T \rangle \equiv \langle b_T, c_T \rangle$. Then, we have

$$\langle a_T, x_T \rangle \equiv \langle a_T, x_T \rangle, \quad \langle b_T, c_T \rangle \equiv \langle x_T, z_T \rangle \quad \text{and} \quad \langle x_T, z_T \rangle \equiv \langle x_T, z_T \rangle,$$

which shows that $\langle a_T, b_T, c_T \rangle \equiv \langle a_T, x_T, z_T \rangle$, as required.

<u>Case II</u> : $v \neq 0$. Equation (5) implies $a \in D_T(y, v)$, while (4) yields $v \in D_T(x, z)$. Therefore, Proposition 1.31.(a) gives

$$\langle a_T, (vay)_T \rangle \equiv \langle y_T, v_T \rangle \quad \text{and} \quad \langle v_T, (vxz)_T \rangle \equiv \langle x_T, z_T \rangle.$$

These isometries imply that, to prove $\langle a_T, b_T, c_T \rangle \equiv \langle y_T, x_T, z_T \rangle$, it is enough to verify that $\langle (vay)_T, (vxz)_T \rangle \equiv \langle b_T, c_T \rangle$. Since the discriminant of these forms in $G_T(F)$ are the same, by Proposition 1.31, they are isometric iff $D_T(vay, vxz) = D_T(b, c)$. From the isometries in (1) we get $\alpha_T = (ax\beta)_T$, $\gamma_T = (yz\beta)_T$ and $D_T(b, c) = D_T(\alpha, \gamma)$. By Lemma 1.30.(c), we conclude $D_T(b, c) = D_T(ax\beta, yz\beta)$.

Hence, what is needed is equivalent to $D_T(ax\beta, yz\beta) = D_T(vay, vxz)$. Since the discriminants are the same, it is enough to prove $ax\beta \in D_T(vay, vxz)$. Multiplying this relation through by axv, we arrive at yet another equivalent condition, namely

$$v\beta \in D_T(xy, az),$$

which we shall now verify. Equations (3), (4) and (2) yield, with $t = z\, q_\beta$,

$$v\,\beta = (x\, p_a + t\, q_a)(y\, p_\beta + t) = xy\, p_a p_\beta + t\, x\, p_a + t\, y\, p_\beta q_a + t^2\, q_a =$$
$$= xy\, p_a p_\beta + t\, (x\, p_a + y\, p_\beta q_a + t\, q_a) =$$
$$= xy\, p_a p_\beta + t\, (x\, p_a + q_a\, (y\, p_\beta + t)) = xy\, p_a p_\beta + t\, (x\, p_a + \beta\, q_a) =$$
$$= xy\, p_a p_\beta + t\, a = xy\, p_a p_\beta + az\, q_\beta,$$

showing that $v\beta \in D_T(xy, az)$ and concluding the verification of [SG6].

Regarding reduction, note that $\langle a_T, a_T \rangle \equiv \langle 1, 1 \rangle$ iff a is a sum of elements of T; thus, requiring that $a_T = 1_T$ is tantamount to stipulating that T be closed under sums, finishing the proof. $\square$

As a final thought on the theme, we mention that the above can be generalized to semi-local rings which are semi-simple and where 2 is invertible. What is needed is the Chinese Remainder Theorem, together with what has been proven above, applied to the residue field modulo each maximal ideal of the ring in question.

Our next topic is to give a special group version of the classical Baer-Krull Theorem. This will help to make clear the relationship between special groups of the form $G_T(F)$, where T is a pre-order of a formally real field F, and the corresponding group $G_{T_v}(F_v)$ of the residue field F_v, under a valuation v (fully) compatible with T. It will also give a natural example of the extension operation introduced in 1.10.

In the sequel, F denotes a formally real field, T a pre-order of F, v a valuation fully compatible with T, so that the push-down T_v of T is a pre-order of the residue field F_v (see 1.35.(6) below). Let

$$\dot{T} = T - \{0\} = T \cap \dot{F}.$$

We shall write Γ_v for the value group of v, and set $\Gamma'_v = \Gamma_v/v[\dot{T}]$, a group of exponent 2 (see 1.35.(3) below). We shall prove:

THEOREM 1.33. *The reduced special group $G_T(F)$ is isomorphic (as a special group) to the extension of $G_{T_v}(F_v)$ by the group Γ'_v.*

REMARK 1.34. The original version of the Baer-Krull theorem gave a complete and explicit description of all the liftings of a (total) order in F_v to F along a valuation v; see [L2; Chapter 3] for references. The version given in [L2; Thm. 3.10] asserts the existence of a homeomorphism between the space $\chi(F,T)$ of (total) orders of F extending T and the product of the residual space of orders $\chi(F_v,T_v)$ with the space $\chi(\Gamma'_v)$ of $\{\pm 1\}$-characters of the group Γ'_v. The version above is more general insofar it gives an isomorphism of abstract order spaces —not merely a homeomorphism— between these spaces of orders. The (dual) terminology of abstract order spaces is introduced in Chapter 3 below, where the Duality Theorem 3.19 that gives the interpretation just mentioned is proved. The proof of the foregoing result is essentially that of [L2; Thm. 3.10] conveniently adapted to the technology of special groups.

NOTATION AND PRELIMINARIES 1.35.

(1) Following standard usage in valuation theory, the symbols A_v, M_v, U_v, F_v, Γ_v, $v(a)$ denote, respectively, the ring, the maximal ideal, the (multiplicative) group of elements of value 0, the residue field, the value group, and the value of an element $a \in \dot{F}$, of a valuation v of F. We shall write $\bar{b}$ for the image b/M_v of an element $b \in A_v$ in $F_v = A_v/M_v$.

(2) We recall our earlier notation (section 1.3) under which a_T stands for the image $a/\dot{T}$ of an element $a \in \dot{F}$ in the group $G_T(F) = \dot{F}/\dot{T}$.

(3) For $a \in \dot{F}$ we set $v'(a) = v(a)/v[\dot{T}]$. The group $\Gamma'_v = \Gamma_v/v[\dot{T}]$ has exponent 2 ($\Gamma_v \subseteq v[\dot{T}]$, since $\dot{F}^2 \subseteq T$); its operation will be denoted additively.

We shall fix, once and for all, a set $\{a_i | i \in I\} \subseteq \dot{F}$ such that $\{v'(a_i)|i \in I\}$ is a basis of Γ'_v as an $\mathbb{F}_2$-vector space. Thus, every $a \in \dot{F}$ determines a unique finite set $I(a) \subseteq I$ so that $v'(a) = \sum_{i \in I(a)} v'(a_i) = v'(\prod_{i \in I(a)} a_i)$. Hence, there exists an element $t \in \dot{T}$ so that $v(a) = v(\prod_{i \in I(a)} a_i) + v(t)$; in turn, this equality entails the existence of an element $c \in F$, $v(c) = 0$, such that

$$a = t \, c \prod_{i \in I(a)} a_i.$$

Note that, $v'(a) = 0$ iff $I(a) = \emptyset$ iff $a = tc$ for some $t \in \dot{T}$ and $v(c) = 0$.

As for uniqueness of the representation above, if $a = t c \prod_{i \in I(a)} a_i = t' c' \prod_{i \in I(a)} a_i$ are two such representations of an element $a \in \dot{F}$, then $t\, t'^{-1} = c\, c^{-1}$, and it follows that $v(t) = v(t')$ and $c\, c' \in \dot{T}$, i.e., $c_T = c'_T$. Thus, the representation is unique upon fixing:

- The finite subset $I(a)$ of I.
- The value $v(t)$ of t.
- The class c_T of c in $G_T(F)$.

Furthermore, we have:

FACT 1.36. Let $a, b \in \dot{F}$ be represented as $a = t c \prod_{i \in I(a)} a_i$, $b = t' c' \prod_{i \in I(b)} a_i$, with $I(a), I(b) \subseteq I$ finite, $t, t' \in \dot{T}$, and $v(c) = v(c') = 0$. Then, $a_T = b_T$ iff $I(a) = I(b)$ and $c\, c' \in \dot{T}$. ◇

(4) Replacing, if necessary, a_i by a_i^{-1}, the elements $\{a_i \mid i \in I\}$ can be chosen so that $v(a_i) > 0$ for all $i \in I$. The same trick shows that every $x \in G_T(F)$ can be represented in the form $x = a/\dot{T} = a_T$ for some $a \in A_v$.

(5) Every group of exponent 2, Γ, is a $\mathbb{F}_2$-vector space, and hence isomorphic to the group $\langle P_{fin}(B), \triangle, \emptyset \rangle$ of finite subsets of a $\mathbb{F}_2$-base B of Γ, with the operation of symmetric difference. Thus, every element of Γ gets identified with a finite subset of B. We shall adopt this notation for $\Gamma = \Gamma'_v$, which gets identified with $P_{fin}(I)$; thus, for $a \in \dot{F}$, the element $v'(a)$ will be identified with the finite set $I(a) \subseteq I$.

(6) Recall from [L2],§3, that a pre-order T of F is called <u>compatible</u> (resp., <u>fully compatible</u>) with a valuation v (or vice-versa) if $(1 + M_v) \cap -T = \emptyset$ (resp. $(1 + M_v) \subseteq T$). For (total) orders both notions are identical, and v is fully compatible with T iff it is compatible with every order containing T. If T and v are compatible, the set $T_v = \{a/M_v \mid a \in T\}$ is a (proper) pre-order of the residue field F_v. Routine checking shows that "1" can be replaced in these definitions by any element in T of value zero:

(a) $(T \cap U_v) + M_v \subseteq \dot{T}(1 + M_v)$.
(b) T is compatible with v iff $((T \cap U_v) + M_v) \cap -T = \emptyset$.
(c) T is fully compatible with v iff $(T \cap U_v) + M_v \subseteq T$. ◇

Proof of Theorem 1.33.

I. Construction of the homomorphism κ.

We start by setting up a map $G_{T_v}(F_v) \times \Gamma'_v \xrightarrow{\kappa} G_T(F)$ defined by:
$$(\bar{c}_{T_v}, J) \xrightarrow{\kappa} (c \prod_{i \in J} a_i)_T,$$
where $c \in \dot{F}$, $v(c) = 0$, and $J \in P_{fin}(I)$.

(1) κ is well defined, i.e.,
$$\bar{c}_{T_v} = \bar{c'}_{T_v} \Rightarrow (c \prod_{i \in J} a_i)_T = (c' \prod_{i \in J} a_i)_T.$$

Our assumption means that $\overline{c c'} \in \dot{T}_v$, i.e., $\overline{c c'} = \bar{t}$ for some $t \in \dot{T}$, $v(t) = 0$, i.e., $t - c c' \in M_v$. By full compatibility, $t - (t - c c') = c c' \in (T \cap U_v) + M_v \subseteq T$ (see 1.35.(6.b)), and we get $c (\prod_{i \in J} a_i) \cdot c' (\prod_{i \in J} a_i) = c c' (\prod_{i \in J} a_i)^2 \in \dot{T}$, as required.

(2) κ is a group homomorphism.

Let $c_1, c_2 \in F$, $v(c_1) = v(c_2) = 0$, and let J_1, J_2 be finite subsets of I. We have $\kappa(((\bar{c_i})_{T_v}, J_i)) = (c_i \prod_{j \in J_i} a_j)_T$, for i = 1,2, and
$$((\bar{c_1})_{T_v}, J_1) \cdot ((\bar{c_2})_{T_v}, J_2) = ((\overline{c_1 c_2})_{T_v}, J_1 \triangle J_2).$$

Hence,
$$\kappa(((\overline{c_1 c_2})_{T_v}, J_1 \triangle J_2)) = (c_1 c_2 \prod_{j \in J_1 \triangle J_2} a_j)_T = (c_1 \prod_{j \in J_1} a_j)_T \cdot (c_2 \prod_{j \in J_2} a_j)_T$$

$$= \kappa(((\bar{c_1})_{T_v}, J_1)) \cdot \kappa(((\bar{c_2})_{T_v}, J_2)).$$

Also, the group unit of $G_{T_v}(F_v) \times \Gamma'_v$ is $(1, 0) = (\bar{1}_{T_v}, \emptyset)$. Then:
$$\kappa((1, 0)) = (1 \cdot \prod_{j \in \emptyset} a_j)_T = 1_T = 1_{G_T(F)}.$$

(3) κ is injective.

Assume $\kappa((\bar{c}_{T_v}, J)) = (c \prod_{j \in J} a_j)_T = 1_{G_T(F)}$, i.e., $c \prod_{j \in J} a_j \in \dot{T}$. Then, $v(c \prod_{j \in J} a_j) = v(\prod_{j \in J} a_j) \in v[\dot{T}]$, i.e., $\sum_{j \in J} v'(a_i) = 0$. By $\mathbb{F}_2$-linear independence, $v'(a_j) = 0$ for $j \in J$. After our notational conventions (cf. 1.35.(5)) it follows that $J = \emptyset$. Then $c \in \dot{T}$, whence $\bar{c}_{T_v} = 1_{G_{T_v}(F_v)}$, and $(\bar{c}_{T_v}, J) = (1_{G_{T_v}(F_v)}, \emptyset) =$ the unit of $G_{T_v}(F_v) \times \Gamma'_v$.

(4) κ is surjective.

Let $a \in \dot{T}$. The canonical representation of 1.35.(3) gives $a = t c \prod_{i \in I(a)} a_i$. Hence $\kappa((\bar{c}_{T_v}, I(a))) = (c \prod_{i \in I(a)} a_i)_T = a_T$.

Now we consider the group $G_{T_v}(F_v) \times \Gamma'_v$ endowed with the SG-structure defined by the extension of $G_{T_v}(F_v)$ by Γ'_v. The distinguished element -1

of $G_{T_v}(F_v)[\Gamma'_v]$ is $(-1_{G_{T_v}(F_v)}, \emptyset) = (\overline{-1}_{T_v}, \emptyset)$. The definition of κ yields at once that its image is $-1_T = -1_{G_T(F)}$.

(5) κ preserves representation.

With notation as above, assume that

$$((\overline{c_1})_{T_v}, J_1) \in D_{G_{T_v}(F_v)[\Gamma'_v]}((1_{G_{T_v}(F_v)}, \emptyset), ((\overline{c_2})_{T_v}, J_2)),$$

and show:

(*) $\qquad \kappa(((\overline{c_1})_{T_v}, J_1)) \in D_{G_T(F)}(1_T, \kappa(((\overline{c_2})_{T_v}, J_2))).$

We consider several cases according to the definition of representation in $G_{T_v}(F_v)[\Gamma'_v]$ (see Example 1.10).

(a) $J_2 = \emptyset$.

Then, $J_1 = \emptyset$ and $(\overline{c_1})_{T_v} \in D_{G_{T_v}(F_v)}(\overline{1}_{T_v}, (\overline{c_2})_{T_v})$. By the definition of representation in $G_{T_v}(F_v)$ there are $r, p \in T \cap A_v$ such that $\overline{c_1} = \overline{r} + \overline{c_2 p}$, i.e., $c_1 - (r + c_2 p) = m \in M_v$. Since $\overline{c_1} \neq 0$, at least one of $\overline{r}$ or $\overline{p}$ is $\neq 0$.

(i) $\overline{r} \neq 0$. In this case, $r \in T \cap U_v$, and we have $c_1 = (r + m) + c_2 p$. By full compatibility, $r + m \in (T \cap U_v) + M_v \subseteq T$ (cf. 1.35(6.c)), which shows that $(c_1)_T \in D_{G_T(F)}(1, (c_2)_T)$.

(ii) $\overline{r} = 0$. Then, $\overline{p} \neq 0$, and $p \in T \cap U_v$. We also have $c_1 = r + c_2(p + \dfrac{m}{c_2})$. Since $m/c_2 \in M_v$, as in (i) we have $p + \dfrac{m}{c_2} \in T$, showing that $(c_1)_T \in D_{G_T(F)}(1, (c_2)_T)$.

(b) $J_2 \neq \emptyset$.

Then, $((\overline{c_1})_{T_v}, J_1) = (1_{G_{T_v}(F_v)}, \emptyset)$ or $((\overline{c_1})_{T_v}, J_1) = ((\overline{c_2})_{T_v}, J_2)$. In both cases it is clear that (*) holds.

II. κ is a SG-isomorphism.

To establish this it suffices to prove, for arbitrary elements $a, b \in \dot{F}$, represented in the form $a = tc \prod_{i \in I(a)} a_i$, $b = t'c' \prod_{i \in I(b)} a_i$, that the following holds:

$$a_T \in D_{G_T(F)}(1, b_T)$$

(**) $\qquad\qquad$ implies

$$(\overline{c}_{T_v}, I(a)) \in D_{G_{T_v}(F_v)[\Gamma'_v]}((1_{G_{T_v}(F_v)}, \emptyset), ((\overline{c'})_{T_v}, I(b)))$$

Our assumption is:

(+) $\quad a = r + bp$, with $r, p \in T$.

We consider two cases, according to the definition of representation in $G_{T_v}(F_v)[\Gamma'_v]$, cf. 1.10.

Case 1. $I(b) \neq \emptyset$.

We show that either

(a) $I(a) = \emptyset$ and $\bar{c}_{T_v} = 1_{G_{T_v}(F_v)}$ (i.e., $\bar{c} \in \dot{T}_v$),

or

(b) $I(a) = I(b)$ and $\bar{c}_{T_v} = \overline{c'}_{T_v}$ (i.e., $\overline{cc'} \in \dot{T}_v$).

Assume first $r = 0$. Then, (+) yields $a = tc \prod_{i \in I(a)} a_i = bp = t'pc' \prod_{i \in I(b)} a_i$ which, by $\mathbb{F}_2$-linear independence, gives $I(a) = I(b)$ and $cc' = \frac{t'p}{t}(c')^2 \in T$; hence, $\overline{cc'} \in \dot{T}_v$.

Next, if $p = 0$, then (+) yields $a = r = tc \prod_{i \in I(a)} a_i$; $\mathbb{F}_2$-linear independence gives $I(a) = \emptyset$ and $c = r/t \in T$, whence $\bar{c} \in \dot{T}_v$.

From now on we assume $r, p \neq 0$. Clearly, $v(r) \in v[\dot{T}]$, whence $v'(r) = 0$ and $v'(bp) = (v(p)/v[\dot{T}]) + v'(b) = \sum_{i \in I(b)} v'(a_i) \neq 0$, as $p \in \dot{T}$, and $I(b) \neq \emptyset$. It follows that $v(r) \neq v(bp)$, and (+) yields $v(a) = \min\{v(r), v(bp)\}$.

We consider two cases:

(a') $v(r) < v(bp)$.

Then, $v(a) = v(r) \in v[\dot{T}]$; hence $v'(a) = 0$ and $I(a) = \emptyset$. Also, $a = rd$, with $v(d) = 0$, and (+) gives $a = r(1 + \frac{p}{r}b) = rd$; we get $d = 1 + \frac{p}{r}b$, and $v(1 + \frac{p}{r}b) = 0$. Since we also have $a = tc$, we get $c = \frac{r}{t}(1 + \frac{p}{r}b)$, and $v(a) = v(t) = v(r)$, which proves $v(r/t) = 0$. Likewise, $v(t) = v(r) < v(bp)$ gives $v(\frac{bp}{r}) > 0$, which implies $\overline{1 + \frac{bp}{r}} = \bar{1}$. Finally, since $r/t \in T$, we conclude that

$$\bar{c} = \overline{\frac{r}{t} \cdot (1 + \frac{bp}{r})} = \overline{r/t} \in \dot{T}_v.$$

This proves (a).

(b') $v(bp) < v(r)$.

In this case $v(a) = \min\{v(r), v(bp)\} = v(p) + v(b)$; this gives $v'(a) = v'(b)$, and hence $I(a) = I(b)$. We also get $a = bpc_0$, with $v(c_0) = 0$. Using the representation of a, b, we obtain:

$$a = tc \prod_{i \in I(a)} a_i = bpc_0 = t'pc'c_0 \prod_{i \in I(b)} a_i,$$

which yields $cc' = (\frac{t'p}{t}(c')^2)c_0$. Since the expression in parentheses is in $\dot{T}$, proving $\overline{c_0} \in \dot{T}_v$ would give $\overline{cc'} \in \dot{T}_v$, showing that (b) holds.

From $bpc_0 = a = r + bp$ we get $r = bp(c_0 - 1)$, and $v(r/p) = v(b) + v(c_0 - 1)$, which yields

$$\sum_{i \in I(b)} v(a_i) + v(c_0 - 1) \in v[\dot{T}].$$

Since the first summand is not in $v[\dot{T}]$ ($I(b) \neq \emptyset$), it follows that $v(c_0 - 1) \neq 0$, hence $v(c_0 - 1) > 0$, showing that $\overline{c_0} = \overline{1} \in \dot{T}_v$, as required.

Case 2. $I(b) = \emptyset$.

Assume $\overline{c}'_{T_v} = \overline{-1}_{T_v}$, i.e., $\overline{-c'} \in \dot{T}_v$. Then, $((\overline{c'})_{T_v}, I(b)) = ((\overline{-1})_{T_v}, \emptyset) = -1_{G_{T_v}(F_v)}$, and the conclusion in (**) holds trivially, as the form $\langle 1, -1 \rangle$ represents the whole group $G_{T_v}(F_v)[\Gamma'_v]$. Thus, we may assume $\overline{-c'} \notin \dot{T}_v$, and will show:

(c) $I(a) = \emptyset$ and $(\overline{c})_{T_v} \in D_{G_{T_v}(F_v)}(1_{G_{T_v}(F_v)}, (\overline{c'}_{T_v}))$,

which proves (**) in Case 2.

Under the present assumptions, (+) takes the form

(++) $a = tc \prod_{i \in I(a)} a_i = r + t'pc'$.

First we prove:

(d) $v(r + t'pc') \in v[\dot{T}]$.

Obviously, this implies $I(a) = \emptyset$.

Proof of (d).   If $v(r) \neq v(t'pc') = v(t'p)$, then $v(r + t'pc')$ equals one of $v(r)$ or $v(t'p)$, and hence is in $v[\dot{T}]$. So, we may assume $v(r) = v(t'p)$.

Obviously, $r + t'pc' = r(1 + \frac{t'p}{r}c')$. We analyze the term $1 + \frac{t'p}{r}c'$. Since both $v(\frac{t'p}{r})$ and $v(c')$ are zero, then $v(1 + \frac{t'p}{r}c') \geq 0$. If this value is > 0, then $1 + \frac{t'p}{r}c' \in M_v$, and full compatibility yields

$$1 - (1 + \frac{t'p}{r}c') = (-c')\frac{t'p}{r} \in 1 + M_v \subseteq T;$$

this clearly implies $-c' \in T$, and $\overline{-c'} \in \dot{T}_v$, contrary to our assumption.

Hence, $v(1 + \frac{t'p}{r}c') = 0$, and $v(r + t'pc') = v(r) \in v[\dot{T}]$, as asserted. □

The proof of (d) also shows:

(e) $v(1 + \frac{t'p}{r}c') = 0$.

Since $I(a) = \emptyset$, the identity (++) yields:

(+++) $c = \frac{r}{t} + \frac{t'p}{t}c'$.

Below we show that both the coefficients $r' = \dfrac{r}{t}$, and $p' = \dfrac{t'p}{t}$ have non-negative value. Hence $r', p' \in T \cap A_v$, and $(+++)$ entails $\bar{c} = \overline{r'} + \overline{p'}\,\overline{c'}$, which proves the representation statement in (c).

Thus, to complete the proof we observe:

- If $v(r) > v(t'p)$, then $v(a) = v(t) = v(t'p)$ (cf. $(++)$), and $v(\dfrac{r}{t}) > v(\dfrac{t'p}{t}) = 0$.
- If $v(r) < v(t'p)$, then $v(a) = v(t) = v(r)$ (cf. $(++)$), and $v(\dfrac{t'p}{t}) > v(\dfrac{r}{t}) = 0$.
- If $v(r) = v(t'p)$, then, using $(++)$ and (e) we have
$$v(a) = v(t) = v(r + t'pc') = v(r) + v(1 + \dfrac{t'p}{r}c') = v(r).$$
Hence, $v(\dfrac{r}{t}) = v(\dfrac{t'p}{t}) = 0$.

This ends the proof of the Baer-Krull theorem. □

CHAPTER 2

Pfister Forms, Saturated Subgroups and Quotients

The main themes treated in this section are an algebraic characterization of the kernels of SG-morphisms and the construction of quotients of special groups. Here the crucial notion is that of **saturated subgroup**. We also give a particularly simple proof of a version of the Hahn-Banach theorem in the context of reduced special groups (the separation theorem, Theorem 2.11), as well as of a version of Pfister's local-global principle (Theorem 2.30). These results pave the way to the duality theory which is presented in the next Chapter. In view of the important role played here by the so-called Pfister forms, we begin by briefly reviewing some of their basic properties, used implicitly or explicitly in the sequel.

1. Pfister Forms

DEFINITION 2.1. Let G be special group (not necessarily reduced). A **Pfister form** over G is a quadratic form φ of the type $\bigotimes_{i=1}^{n} \langle 1, a_i \rangle$, where $n \geq 1$ and $a_1, \ldots, a_n \in G$, or the form $\langle 1 \rangle$. In the first case, the integer n is called the **degree of** φ and written $deg(\varphi)$; also, degree $(\langle 1 \rangle) = 0$. If the coefficients of φ happen to belong to a subgroup Δ of G, we say that φ is **Pfister over** Δ.

Since a Pfister form φ contains 1 as a coefficient, we may write φ as $\langle 1 \rangle \oplus \varphi'$; φ' is called the **pure subform** of φ.

We denote the Pfister form $\bigotimes_{i=1}^{n} \langle 1, a_i \rangle$ by $\langle\langle a_1, \ldots, a_n \rangle\rangle$, and the form $\langle\langle 1, \ldots, 1 \rangle\rangle$ by 2^n. If ψ is a form over G and $n \geq 1$ is an integer, we write $2^n \otimes \psi$ as $2^n \cdot \psi$ or, simply, $2^n \psi$. $\diamond$

In the next Proposition we state, mostly without proof, a number of the basic algebraic properties of Pfister forms. Some of these properties are proved in [L1] (Chapter X) for the case of fields; his proofs can be adapted to the present context of special groups. In fact, the properties below are proven by induction on the degree, using, if necessary previous statements in the list and/or Proposition 1.6.

PROPOSITION 2.2. (Basic properties of Pfister forms)
Let G be a special group, $\varphi = \langle\langle a_1, \ldots, a_n \rangle\rangle$ a Pfister form over G of degree $n \geq 1$ and $b \in G$. Recall that φ' is the pure sub-form of φ. Then

a) $b \in D_G(1, a_1) \Rightarrow \langle\langle a_1, a_2 \rangle\rangle \equiv_G \langle\langle a_1, a_2 b \rangle\rangle$.

b) $b \in D_G(a_1, a_2) \Rightarrow \langle\langle a_1, a_2 \rangle\rangle \equiv_G \langle\langle b, a_1 a_2 \rangle\rangle$.

c) $\langle\langle a_1 b, \ldots, a_n b \rangle\rangle \equiv_G \langle 1, a_1 b \rangle \otimes \langle\langle a_1 a_2, \ldots, a_1 a_n \rangle\rangle$.

d) If $b \in D_G(\varphi')$, then $\varphi \equiv_G \langle\langle b, b_2, \ldots, b_n \rangle\rangle$, with $b_2, \ldots, b_n \in G$.

e) An isotropic Pfister form is hyperbolic.

f) $D_G(\varphi) = \{x \in G : x\varphi \equiv_G \varphi\}$. Hence, $D_G(\varphi)$ is a subgroup of G. If ψ is a Pfister form over G, then $D_G(\varphi) D_G(\psi) \subseteq D_G(\varphi \otimes \psi)$.

g) If $a \in D_G(\varphi)$, then $\langle\langle a_1, \ldots, a_n, b \rangle\rangle \equiv_G \langle\langle a_1, \ldots, a_n, ab \rangle\rangle$.

h) $a \in D_G(\varphi) \Rightarrow \langle 1, a \rangle \otimes \varphi \equiv_G 2 \otimes \varphi$ and $\langle 1, -a \rangle \otimes \varphi$ is hyperbolic.

i) $a \in D_G(\varphi)$ and $b \in D_G(1, a) \Rightarrow b \in D_G(2 \otimes \varphi)$.

j) $\langle 1, a \rangle \otimes \varphi \equiv_G 2 \otimes \varphi \Rightarrow a \in D_G(\varphi)$.

k) $\langle 1, -a \rangle \otimes \varphi$ hyperbolic $\Rightarrow a \in D_G(\varphi)$.

l) The following are equivalent :

 (1) G is a reduced special group.

 (2) For every Pfister form φ over G and $a \in G$,
$$a, -a \in D_G(\varphi) \Rightarrow \varphi \text{ hyperbolic.}$$

 (3) For every Pfister form φ over G and $a \in G$,
$$a \in D_G(\langle 1, -a \rangle \otimes \varphi) \Rightarrow a \in D_G(\varphi).$$

PROOF. We prove only item (d). Proceed by induction on the degree n of φ; if $n = 1$, there is nothing to prove. Assume that $\varphi = \langle 1, a \rangle \otimes \psi$, where ψ is Pfister form of degree n. Since $\varphi' = \psi' \oplus a\psi$, the hypothesis $b \in D_G(\varphi')$ and 1.6.(c) yield $x \in D_G(\psi')$ and $y \in D_G(\psi)$, such that
$$b \in D_G(x, ay), \text{ that is, } \langle b, baxy \rangle \equiv \langle x, ay \rangle. \tag{I}$$
By the induction hypothesis, there are $z_2, \ldots, z_n \in G$, such that,
$$\psi \equiv \langle\langle x, z_2, \ldots, z_n \rangle\rangle. \tag{II}$$
Then, (II), (I) and $y\psi \equiv \psi$, yield, with $\alpha = \langle\langle z_2, \ldots, z_n \rangle\rangle$,
$$\langle 1, b \rangle \otimes \langle 1, axy \rangle \otimes \alpha = \langle 1, b, axy, abxy \rangle \otimes \alpha =$$
$$= (\langle 1, axy \rangle \oplus \langle b, abxy \rangle) \otimes \alpha$$
$$= (\langle 1, axy \rangle \oplus \langle x, ay \rangle \otimes \alpha \equiv (\langle 1, x \rangle \oplus ay \langle 1, x \rangle) \otimes \alpha$$
$$= (\langle 1, x \rangle \otimes \alpha) \oplus ay(\langle 1, x \rangle \otimes \alpha) = \psi \oplus ay\psi = \psi \oplus a\psi = \varphi,$$
completing the induction step. □

2. Saturated Subgroups

As a motivation for the notion introduced below (Definition 2.3), we note the following property of the kernels of SG-morphisms $G \xrightarrow{f} H$, **with H reduced**, related to [D] and [ker] in Lemma 1.12 :
$$\forall\, a \in G, \quad a \in \ker f \quad \Rightarrow \quad D_G(1,a) \subseteq \ker f.$$
Indeed, if $b \in D_G(1,a)$, we have $\langle\, b, ba\,\rangle \equiv_G \langle\, 1, a\,\rangle$, and then $\langle\, f(b), f(b)\,\rangle \equiv_H \langle\, 1, 1\,\rangle$. Since H is reduced, $f(b) = 1$. We now give a name to the subgroups of G enjoying this property.

DEFINITION 2.3. *Let G be a (not necessarily reduced) special group and let Δ be a subgroup of G. We say that Δ is **saturated** if for all $a \in G$,*

[sat] $\qquad\qquad a \in \Delta \;\Rightarrow\; D_G(1,a) \subseteq \Delta.$

Note that if, in addition, $-1 \in \Delta$, then $\Delta = G$. Thus, we will reserve the noun saturated for those subgroups satisfying [sat] such that $-1 \notin \Delta$, while G will be called the improper saturated subgroup of itself. ◇

Along with a few preservation properties of saturation by intersections and directed unions, we show that the binary form $\langle 1, a \rangle$ in the above definition can be replaced by an arbitrary Pfister form over Δ.

LEMMA 2.4. *Let G be special group and Δ a subgroup of G.*

a) The intersection of any family of saturated subgroups is saturated. The union of an upward directed family of saturated subgroups is saturated.

b) The following are equivalent :

(1) *Δ is saturated.*

(2) *For any Pfister forms φ, ψ over Δ and any $b, c \in \Delta$*
$$D_G(\varphi), D_G(\psi) \subseteq \Delta \quad \Rightarrow \quad D_G(b\varphi \oplus c\psi) \subseteq \Delta.$$

(3) *For any Pfister form φ over Δ, $D_G(\varphi) \subseteq \Delta$.*

PROOF. (a) is straightforward. In (b), it is clear that (3) implies (1).

(1) $\Rightarrow$ (2). If $a \in D_G(b\varphi \oplus c\psi)$, there are $x \in D_G(\varphi)$ and $y \in D_G(\psi)$ such that $a \in D_G(bx, cy)$, which implies $abx \in D_G(1, bcxy)$. Since $D_G(\varphi)$ and $D_G(\psi)$ are contained in Δ, we have $x, y \in \Delta$; hence $bcxy \in \Delta$ and, by (1), $abx \in \Delta$. Since $bx \in \Delta$, we get $a \in \Delta$.

(2) $\Rightarrow$ (3). By induction on the $deg(\varphi) = n$. The case $n = 0$ is obvious. For the induction step, φ can be written
$$\varphi = \langle\, 1, a\,\rangle \otimes \psi \equiv_G \psi \oplus a\psi,$$
with $a \in \Delta$ and ψ a Pfister form over Δ of degree $n - 1$. Hence, $D_G(\psi) \subseteq \Delta$ and the conclusion follows from (2) for the values $b = 1$ and $c = a$. □

In the sequel it will be shown that saturated subgroups exist in profusion.

LEMMA 2.5. *Let G be a special group and let Δ be a subgroup of G.*

a) The family $\mathcal{P}_\Delta$ of all Pfister forms over Δ contains the form $2 = \langle 1, 1 \rangle$ and is closed under tensor products.

b) The following are equivalent :

(1) Δ *is proper saturated subgroup of G.*

(2) *There is a family $\mathcal{S}$ of anisotropic Pfister forms over G containing 2, closed under tensor products and such that $\Delta = \bigcup \{D_G(\varphi) : \varphi \in \mathcal{S}\}$.*

PROOF. (a) is clear. For (b), if Δ is saturated, consider $\mathcal{S} = \mathcal{P}_\Delta$. Lemma 2.4.c implies that $\Delta = \bigcup \{D_G(\varphi) : \varphi \in \mathcal{S}\}$. If some $\varphi \in \mathcal{S}$ is isotropic, then $-1 \in D_G(\varphi) \subseteq \Delta$, and so $\Delta = G$. This proves (1) $\Rightarrow$ (2).

Conversely, suppose that $\mathcal{S}$ is as in (2), and $\Delta = \bigcup \{D_G(\varphi) : \varphi \in \mathcal{S}\}$. Let $a \in \Delta$ and $b \in D_G(1, a)$. Thus, $a \in D_G(\varphi)$ for some $\varphi \in \mathcal{S}$ and by 2.2.(i), we have $b \in D_G(2 \otimes \varphi)$. Since this form is in $\mathcal{S}$, we conclude that $b \in \Delta$. If $-1 \in D_G(\varphi)$ for some $\varphi \in \mathcal{S}$, then by 2.2.(f), $-\varphi \equiv_G \varphi$. But this means that $\varphi \oplus \varphi \equiv_G 2 \otimes \varphi$ is an isotropic form in $\mathcal{S}$. □

Lemma 2.5 yields at once

PROPOSITION 2.6. *Let G be a special group and let Δ be a subgroup of G. Then,*

[saturation] $$\overline{\Delta} = \bigcup \{D_G(\varphi) : \varphi \in \mathcal{P}_\Delta\}$$

is the smallest saturated subgroup of G containing Δ. In particular, if Δ is saturated and φ is a Pfister form over Δ, then $D_G(\varphi) \subseteq \Delta$.

PROOF. Items (a) and (b) of Lemma 2.5 show that $\overline{\Delta}$ is saturated (possibly improper). Since $\langle 1, a \rangle$ is in $\mathcal{P}_\Delta$, for all $a \in \Delta$, we have $\Delta \subseteq \overline{\Delta}$. If Γ is a saturated subgroup containing Δ and $\varphi \in \mathcal{P}_\Delta$, then φ is Pfister over Γ and item 3 in 2.4.(b) shows that $D_G(\varphi) \subseteq \Gamma$. Hence, $\overline{\Delta} \subseteq \Gamma$. □

DEFINITION 2.7. We call $\overline{\Delta}$ the **saturation** of Δ. If $\Delta = \{1\}$, we write $\overline{\{1\}} = Sat(G)$. ◇

REMARKS. (i) $\overline{\Delta}$ may be improper even if -1 is not in Δ.

(ii) $Sat(G) = \bigcup \{D_G(n\langle 1 \rangle) : n \in \omega\} = \bigcup \{D_G(2^k \langle 1 \rangle) : k \in \omega\}$. $Sat(G)$ is the smallest saturated subgroup of G.

(iii) $Sat(G) = \{1\}$ if and only if G is reduced. (Routine checking). ◇

In case G is a **reduced** special group, there are further examples of saturated subgroups.

2. SATURATED SUBGROUPS

COROLLARY 2.8. *Let G be a reduced special group. Then:*

a) For any Pfister form φ on G, $D_G(\varphi)$ is a saturated subgroup of G. In fact, if $\varphi = \langle\langle a_1, \ldots, a_n \rangle\rangle$ then
$$D_G(\varphi) = \overline{D_G(1, a_1) D_G(1, a_2) \cdots D_G(1, a_n)}.$$

b) For any form ψ on G, the set $\{a \in G : a\psi \equiv_G \psi\}$ is a saturated subgroup of G.

PROOF. a) Proposition 2.2.(i) shows that $a \in D_G(\varphi)$ and $b \in D_G(1, a)$ imply $b \in D_G(2 \otimes \varphi)$. Now use 1.6(e.2) to conclude that $b \in D_G(\varphi)$.

If $\varphi = \langle\langle a_1, \ldots, a_n \rangle\rangle$, since $D_G(\varphi)$ is saturated and $a_i \in D_G(\varphi)$, it contains $D_G(1, a_i)$, $1 \leq i \leq n$. Thus, being a subgroup, $D_G(\varphi)$ contains the product of the $D_G(1, a_i)$.

If Γ is a saturated subgroup of G containing $D_G(1, a_i)$, $1 \leq i \leq n$, then it follows from 2.6 that Γ contains $D_G(\varphi)$, since φ is a Pfister form over Γ. Hence, $D_G(\varphi)$ is in fact the saturation of the product of the $D_G(1, a_i)$.

b) Clearly, the set of the statement is a subgroup of G. Assume that $a \in D_G(1, b)$, where $a\psi \equiv_G \psi$. Then, $\langle a, ab \rangle \equiv_G \langle 1, b \rangle$; tensoring both sides of this isometry with ψ yields $a\psi \oplus ab\psi \equiv_G \psi \oplus b\psi$, and hence $a\psi \oplus a\psi \equiv_G \psi \oplus \psi$. Now 1.6(e.4) gives $a\psi \equiv_G \psi$, as required. □

Remark. As an exercise, the reader may check that each of the properties in Corollary 2.8 is, in fact, equivalent to the group G being reduced. ◇

In the remainder of this section we prove two important properties of maximal saturated subgroups. The next lemma will be used in both proofs.

If S is a subset of a group G, let $[S]$ denote the subgroup generated by S in G.

LEMMA 2.9. *Let G be a special group, Δ a saturated subgroup of G and $x \in G$. Then,*
$$\overline{[\Delta \cup \{x\}]} = G \quad \text{iff} \quad -x \in \Delta.$$

PROOF. It is clear that if $-x \in \Delta$, then the subgroup generated by Δ and x will have -1 in it, and so its saturation is G.

Now assume that the saturation Γ of $[\Delta \cup \{x\}] = \Delta \cup x\Delta$ is equal to G. Thus, $-x \in \Gamma$, and by the definition of saturation (Proposition 2.6) there is a Pfister form φ over $\Delta \cup x\Delta$ such that $-x \in D_G(\varphi)$. We may write φ in the form

(*) $$\varphi = \langle\langle a_1, \ldots, a_n, b_1 x, \ldots, b_m x \rangle\rangle,$$

with $a_1, \ldots, a_n, b_1, \ldots, b_m \in \Delta$. By Proposition 2.2.(c)
$$\langle\langle b_1 x, \ldots, b_m x \rangle\rangle \equiv_G \langle\langle b_1 b_2, \ldots, b_1 b_m \rangle\rangle \otimes \langle 1, b_1 x \rangle.$$

Substituting this isometry in (*) we get

(**) $$\varphi \equiv_G \psi \otimes \langle 1, b_1 x \rangle \equiv_G \psi \oplus b_1 x \psi,$$

where $\psi = \langle\!\langle a_1, \ldots, a_n, b_1 b_2, \ldots, b_1 b_m \rangle\!\rangle$, a Pfister form over Δ. Since $-x \in D_G(\varphi)$, (**) implies the existence of $y, z \in D_G(\psi)$ such that $-x \in D_G(y, b_1 xz)$, i.e.

$$\langle -x, -b_1 yz \rangle \equiv_G \langle y, b_1 xz \rangle.$$

It follows that

$$\langle -y, -b_1 yz \rangle \equiv_G \langle x, b_1 xz \rangle,$$

and hence $-xy \in D_G(1, b_1 z)$. Since Δ is saturated, Lemma 2.4.(b.3) yields $y, z \in \Delta$ and so $b_1 z \in \Delta$. By saturatedness again, we get $-xy \in \Delta$ and hence $-x = (-xy)y \in \Delta$. $\square$

PROPOSITION 2.10. *Let Δ be a saturated subgroup of a special group G. Then,*

Δ is a maximal saturated subgroup iff $\forall\, x \in G,\ x \in \Delta$ or $-x \in \Delta$.

PROOF. If $x \notin \Delta$ and Δ is maximal, the saturation of the subgroup generated by Δ and x must be G. By Lemma 2.9, we conclude $-x \in \Delta$.

Conversely, suppose Δ is saturated and such that either x or $-x$ is in Δ, $\forall\, x \in G$. Then any proper saturated extension Γ of Δ will contain z and $-z$, for some $z \in G$, whence $\Gamma = G$, by Proposition 2.6. Thus, Δ is a maximal saturated subgroup of G. $\square$

THEOREM 2.11. (The separation theorem) *Let G be a special group, Δ a saturated subgroup of G and a an element of G such that $a \notin \Delta$. Then there is a maximal saturated subgroup Γ of G such that $\Delta \subseteq \Gamma$ and $a \notin \Gamma$.*

PROOF. Let $\Sigma = \overline{[\Delta \cup \{-a\}]}$; Lemma 2.9 implies that Σ is a proper subgroup of G, otherwise a would be in Δ. In particular, $a \notin \Sigma$. Now consider

$$\mathcal{V} = \{\Lambda : \Lambda \text{ is a proper saturated subgroup of } G,\ \Sigma \subseteq \Lambda \text{ and } a \notin \Lambda\},$$

ordered by inclusion. Clearly $\Sigma \in \mathcal{V}$. Since an upward directed family of saturated subgroups is again saturated (Lemma 2.4.(a)), Zorn's Lemma furnishes a maximal element Γ in $\mathcal{V}$. To see that Γ is indeed a maximal saturated subgroup of G, let Θ be a saturated subgroup of G properly containing Γ. Then Θ is not in $\mathcal{V}$ and so $a \in \Theta$. Since Θ contains Σ, we have both a and $-a$ in Θ, which implies that $\Theta = G$. $\square$

3. Pfister quotients

We begin by an example which shows that a completely general theory of quotients does not exist in the realm of special groups.

EXAMPLE 2.12. Let $(G, \equiv_G, -1)$ be any special group where $dim_{\mathbf{F}_2}(G) \geq 2$. As in Example 1.9, let $G_t = (G, \equiv_t, -1)$ denote the trivial structure on G, with the same distinguished element -1. G_t is a non-reduced special group. Since $(G, \equiv_G, -1)$ satisfies the discriminant axiom [SG3], the identity map $id : G \longrightarrow G_t$ is a surjective special group morphism. However, unless the special relation $\equiv_G$ is also trivial, G_t cannot be obtained as a quotient of $(G, \equiv_G, -1)$ since $ker(id) = \{1\}$. This shows that the homomorphic images of special groups may not be quotients. Note that $(G, \equiv_G, -1)$ may even be reduced (eg., by taking the <u>fan</u> special relation on G), which shows in addition that the reduction axiom is not preserved under surjective special group morphisms. $\diamond$

We shall see below that there is an important class of subgroups for which a satisfactory theory of quotients exists. In Proposition 2.23 below we give a simple criterion for a reduced homomorphic image of a special group to be a quotient.

We begin by the following

DEFINITION 2.13. Let Δ be a subgroup of a special group $(G, \equiv_G, -1)$. We define a quaternary relation on the quotient group G/Δ as follows :

$$\langle a/\Delta, b/\Delta \rangle \equiv^*_{G/\Delta} \langle c/\Delta, d/\Delta \rangle \text{ iff } \begin{cases} \exists\ a', b', c', d' \in G \text{ such that} \\ aa', bb', cc', dd' \in \Delta \text{ and} \\ \langle a', b' \rangle \equiv_G \langle c', d' \rangle. \end{cases}$$

Remark that no conditions are imposed on Δ. $\diamond$

The relation $\equiv^*_{G/\Delta}$ has the following properties :

PROPOSITION 2.14. *With notation as in Definition 2.13, we have*

a) *The relation $\equiv^*_{G/\Delta}$ is well defined.*

b) *$(G/\Delta, \equiv^*_{G/\Delta}, -1/\Delta)$ verifies the axioms [SG1] – [SG5] of special groups. The relation $\equiv^*_{G/\Delta}$ (on $G/\Delta \times G/\Delta$) is reflexive and symmetric, but not transitive in general. The canonical quotient map $\pi : G \longrightarrow G/\Delta$ satisfies, $\forall\ a, b, c, d \in G$*

[quo] $\qquad \langle a, b \rangle \equiv_G \langle c, d \rangle \ \Rightarrow\ \langle \pi(a), \pi(b) \rangle \equiv^*_{G/\Delta} \langle \pi(c), \pi(d) \rangle.$

c) *$\equiv^*_{G/\Delta}$ is the smallest binary relation $\equiv$ on $G/\Delta \times G/\Delta$ satisfying condition [quo], for all $a, b, c, d \in G$.*

PROOF. Left to the reader. $\square$

The relation $\equiv^*_{G/\Delta}$ may not be transitive or 3-transitive. This state of affairs suggests, for an arbitrary special group G, some natural questions :

(i) For which subgroups Δ is $(G/\Delta, \equiv^*_{G/\Delta}, -1/\Delta)$ a special group ?

(ii) Are there other natural relations $\equiv$ on G/Δ such that $(G/\Delta, \equiv, -1/\Delta)$ is a special group and the canonical map $\pi : G \longrightarrow G/\Delta$ is a special group morphism ?

The only class of subgroups, known so far, for which there is a complete answer to question (i) are the **Pfister subgroups** of G, defined below.

DEFINITION 2.15. Let G be a special group. A collection $\mathcal{S}$ of Pfister forms is said to be (upward) **directed** iff for every $\varphi, \psi \in \mathcal{S}$, there is $\theta \in \mathcal{S}$ such that $D_G(\varphi), D_G(\psi) \subseteq D_G(\theta)$.

A subgroup Δ of G is a **Pfister subgroup** iff there is a directed family $\mathcal{S}$ of Pfister forms over G such that $\Delta = \bigcup \{D_G(\varphi) : \varphi \in \mathcal{S}\}$. ◇

REMARK 2.16. Lemma 2.5.b proves that any saturated subgroup is Pfister, since a family of Pfister forms closed under tensor products is directed (Proposition 2.2.(f)). Note that the subgroups $D_G(\varphi)$, φ a Pfister form, are Pfister ($\mathcal{S} = \{\varphi\}$ is directed). ◇

The class of Pfister and saturated subgroups are not identical, except in the case of reduced groups – and only in that case – as shown by the following

PROPOSITION 2.17. *Let G be a special group such that $1 \neq -1$. Then G is reduced if and only if every Pfister subgroup of G is saturated.*

PROOF. Suppose G is reduced and Δ is a Pfister subgroup, say $\Delta = \bigcup \{D_G(\varphi) : \varphi \in \mathcal{S}\}$, $\mathcal{S}$ a directed family of Pfister forms. By Corollary 2.8, $D_G(\varphi)$ is saturated, for each $\varphi \in \mathcal{S}$. Thus, Δ is the directed union of saturated subgroups, and so itself saturated. The converse follows from item (iii) of the Remarks after Definition 2.7 : since $\langle 1 \rangle$ is a Pfister form such that $\{1\} = D_G(\langle 1 \rangle)$, $\{1\}$ is saturated. But this is equivalent to G being reduced. □

Thus, $\{1\}$ is a non-saturated Pfister subgroup in any non-reduced special group. In general there are many other such subgroups. For a more sophisticated example, see Example 2.32 below.

In the case of a Pfister subgroup Δ, the relation $\equiv^*_{G/\Delta}$ admits a manageable characterization described in

PROPOSITION 2.18. *Let G be a special Group and Δ a Pfister subgroup of G, $\Delta = \bigcup \{D_G(\varphi) : \varphi \in \mathcal{S}\}$, $\mathcal{S}$ a directed family of Pfister forms. For $a, b, c, d \in G$, the following are equivalent :*

(1) $\langle a/\Delta, b/\Delta \rangle \equiv^*_{G/\Delta} \langle c/\Delta, d/\Delta \rangle$.

(2) There is a $\varphi \in \mathcal{S}$ such that $\langle a, b \rangle \otimes \varphi \equiv_G \langle c, d \rangle \otimes \varphi$.

Before proving this result we will deal with the particular case where $\mathcal{S}$ consists of a single form, a result that will be used several times afterwards.

LEMMA 2.19. *Let G be a special group and φ a Pfister form over G.*

a) *For $a, b, c, d \in G$, the following are equivalent:*

(i) $\langle a, b \rangle \otimes \varphi \equiv_G \langle c, d \rangle \otimes \varphi$.

(ii) *There are $a', b', c', d' \in G$ such that $aa', \ldots, dd' \in D_G(\varphi)$ and $\langle a', b' \rangle \equiv_G \langle c', d' \rangle$.*

b) *Conditions (i) or (ii) imply $abcd \in D_G(\varphi)$.*

PROOF. a) (i) $\Rightarrow$ (ii) : By assumption, $a\varphi \oplus b\varphi \equiv_G c\varphi \oplus d\varphi$; in particular, $a \in D_G(c\varphi \oplus d\varphi)$. By Proposition 1.6, there are $x, y \in D_G(\varphi)$ such that $a \in D_G(cx, dy)$, i.e.

(*) $\qquad\qquad\qquad \langle a, acdxy \rangle \equiv_G \langle cx, dy \rangle$.

Setting $a' = a$, $b' = acdxy$, $c' = cx$ and $d' = dy$, we have $\langle a', b' \rangle \equiv_G \langle c', d' \rangle$. Further, $aa' = 1$, $cc' = x$ and $dd' = y$ are in $D_G(\varphi)$. Tensoring (*) with φ gives

$$a\varphi \oplus b'\varphi \equiv_G cx\varphi \oplus dy\varphi \equiv_G c\varphi \oplus d\varphi \equiv_G a\varphi \oplus b\varphi.$$

Cancelling $a\varphi$ on both sides (1.6) yields $b\varphi \equiv_G b'\varphi$, that is, $bb' \in D_G(\varphi)$.

(ii) $\Rightarrow$ (i) : Assume $\langle a', b' \rangle \equiv_G \langle c', d' \rangle$ with $aa', \ldots, dd' \in D_G(\varphi)$. Then,

1. $a'\varphi \oplus b'\varphi = \langle a', b' \rangle \otimes \varphi \equiv_G \langle c', d' \rangle \otimes \varphi = c'\varphi \oplus d'\varphi$.

2. $a'\varphi \equiv_G a\varphi, \ldots, d\varphi \equiv_G d'\varphi$.

Substituting the isometries in (2) for the corresponding terms in (1), we get

$$a\varphi \oplus b\varphi \equiv_G c\varphi \oplus d\varphi, \text{ i.e., } \langle a, b \rangle \otimes \varphi \equiv_G \langle c, d \rangle \otimes \varphi,$$

as required.

b) Using (ii) we obtain $a'b'c'd' = 1$. Since $aa', \ldots, dd' \in D_G(\varphi)$, which is a subgroup, we conclude $abcda'b'c'd' = abcd \in D_G(\varphi)$. $\square$

Proof of Proposition 2.18. (1) $\Rightarrow$ (2). By assumption (1) there are elements $a', \ldots, d'$ in G such that $aa', \ldots, dd' \in \Delta$ and $\langle a', b' \rangle \equiv_G \langle c', d' \rangle$. Since Δ is Pfister, there are forms $\varphi_1, \ldots, \varphi_4$ in $\mathcal{S}$ such that $aa' \in D_G(\varphi_1)$, $\ldots$, $dd' \in D_G(\varphi_4)$. Using the directedness of $\mathcal{S}$, pick a form $\varphi \in \mathcal{S}$ so that $D_G(\varphi_i) \subseteq D_G(\varphi)$, $1 \leq i \leq 4$. Now (2) follows from (ii) $\Rightarrow$ (i) of Lemma 2.19(a) applied to φ.

(2) $\Rightarrow$ (1) is immediate from (i) $\Rightarrow$ (ii) in Lemma 2.19(a). $\square$

LEMMA 2.20. *Let G be a special group and let φ_1, φ_2 be anisotropic Pfister forms over G, such that $D_G(\varphi_1) \subseteq D_G(\varphi_2)$. Then, for all forms ψ, θ over G,*

$$\psi \otimes \varphi_1 \equiv_G \theta \otimes \varphi_1 \Rightarrow \psi \otimes \varphi_2 \equiv_G \theta \otimes \varphi_2.$$

PROOF. By induction on $n = dim(\psi) = dim(\theta)$. For $n = 1$, the conclusion is clear, while for $n = 2$ it is a consequence of Lemma 2.19(a). Assume the result is true for $n \geq 2$, and that $\psi = \langle a \rangle \oplus \lambda$, where $dim(\lambda) = n$. Write $\theta = \langle b_1, \ldots, b_n, b_{n+1} \rangle$; thus,

$$(\langle a \rangle \oplus \lambda) \otimes \varphi_1 \equiv \theta \otimes \varphi_1 = \bigoplus_{j=1}^{n+1} b_j \varphi_1,$$

and so, by Proposition 1.6.(c), there are $x_j \in D_G(\varphi_1)$, $1 \leq j \leq (n+1)$, such that

$$a \in D_G(b_1 x_1, \ldots, b_{n+1} x_{n+1}),$$

or equivalently, there are $c_2, \ldots, c_{n+1} \in G$, such that

$$\langle a, c_2, \ldots, c_{n+1} \rangle \equiv \langle b_1 x_1, \ldots, b_{n+1} x_{n+1} \rangle. \tag{I}$$

Multiplying (I) by φ_1 yields

$$a \varphi_1 \oplus \bigoplus_{k=2}^{n+1} c_k \varphi_1 \equiv \bigoplus_{j=1}^{n+1} b_j x_j \varphi_1 = \bigoplus_{j=1}^{n+1} b_j \varphi_1 \equiv a\varphi_1 \oplus (\lambda \otimes \varphi_1). \tag{II}$$

Cancelling $a\varphi_1$ on both sides of (II) gives

$$\langle c_2, \ldots, c_{n+1} \rangle \otimes \varphi_1 = \bigoplus_{k=2}^{n+1} c_k \varphi_1 \equiv \lambda \otimes \varphi_1. \tag{III}$$

From the induction hypothesis, we get

$$\langle c_2, \ldots, c_{n+1} \rangle \otimes \varphi_2 = \bigoplus_{k=2}^{n+1} c_k \varphi_1 \equiv \lambda \otimes \varphi_2. \tag{IV}$$

Tensoring (I) with φ_2, yields, recalling that $D_G(\varphi_1) \subseteq D_G(\varphi_2)$,

$$a\varphi_2 \oplus \bigoplus_{k=2}^{n+1} c_k \varphi_2 = \bigoplus_{j=1}^{n+1} b_j x_j \varphi_2 = \bigoplus_{j=1}^{n+1} b_j \varphi_2 = \theta \otimes \varphi_2.$$

The substitution of (IV) in this last isometry shows that $\psi \otimes \varphi_2 \equiv \theta \otimes \varphi_2$, completing the induction step and the proof. $\square$

Proposition 2.18, together with Lemmas 2.19 and 2.20, yield

PROPOSITION 2.21. *Let G be a special group and Δ a Pfister subgroup of G, determined by the directed family $\mathcal{S}$ of Pfister forms over G. Then, $(G/\Delta, \equiv^*_{G/\Delta}, -1/\Delta)$ is a special group, and the quotient map $G \xrightarrow{\pi} G/\Delta$ is a morphism of special groups. Further, $1 \neq -1$ in G/Δ iff $-1 \notin \Delta$. Moreover, in this situation we have*

a) If φ, ψ are n-forms in G, then

$$\pi \star \varphi \equiv^*_{G/\Delta} \pi \star \psi \quad \text{iff} \quad \begin{cases} \text{There is a Pfister form } \mathcal{P} \text{ in } \mathcal{S} \\ \text{such that } \varphi \otimes \mathcal{P} \equiv_G \psi \otimes \mathcal{P}. \end{cases}$$

b) If $(G, \equiv_G, -1) \xrightarrow{f} (H, \equiv_H, -1)$ is a morphism of special groups satisfying $\Delta \subseteq \ker f$, then there is a unique **SG**-morphism
$$\widehat{f} : (G/\Delta, \equiv^*_{G/\Delta}, -1) \longrightarrow (H, \equiv_H, -1),$$
such that $f = \widehat{f} \circ \pi$.

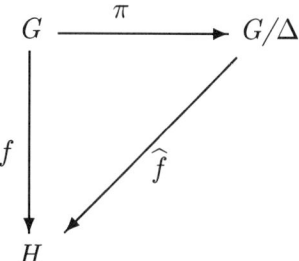

PROOF. a) Transitivity of the relation $\equiv^*_{G/\Delta}$ follows easily from Lemma 2.20, using Proposition 2.18. Likewise, axiom [SG6] is an immediate consequence of (a), which is proven by induction on n, using 2.20 and 2.18. For (b), it is straightforward to verify that, setting $\widehat{f}(\pi(a)) = f(a)$ one gets a well defined morphism of special groups. □

The next result gives a criterion for a (reduced) homomorphic image to be a quotient.

DEFINITION 2.22. A morphism of psg's $f : G \longrightarrow H$, is **regular** if for every $a, b \in G$ such that $f(a) \in D_H(1, f(b))$, there are $x, y \in G$, such that $f(a) = f(x)$, $f(y) = f(b)$ and $x \in D_G(1, y)$. ◇

PROPOSITION 2.23. (Quotient criterion)

Let $G \xrightarrow{f} H$ be a surjective morphism of psg's, with H reduced. Let $\widehat{f} : G/\ker f \longrightarrow H$ be the group homomorphism induced by f (defined by $\widehat{f} \circ \pi = f$, with $\pi : G \longrightarrow G/\ker f$, the canonical quotient map). The following are equivalent:

1) f is regular.

2) $\widehat{f}$ is an isomorphism of psg's.

In particular, if f is regular and G is a special group, then H is a special group as well.

PROOF. (1) ⇒ (2). We know that $\widehat{f}$ is a well-defined surjective morphism of psg's of $(G/\ker f, \equiv^*_{G/\ker f}, -1/\ker f)$ onto $(H, \equiv_H, -1)$. It only remains to be checked that, for $a, b \in G$,
$$\widehat{f}(\pi(a)) \in D_H(1, \widehat{f}(\pi(b))) \quad \Rightarrow \quad \pi(a) \in D_{G/\ker f}(1, \pi(b)).$$

44   2. PFISTER FORMS, SATURATED SUBGROUPS AND QUOTIENTS

Owing to the definitional identity $\widehat{f} \circ \pi = f$, the antecedent amounts to $f(a) \in D_H(1, f(b))$. By regularity, there are $x, y \in G$ such that $f(x) = f(a)$, $f(y) = f(b)$ and $x \in D_G(1, y)$. Since π is a morphism of psg's and $\pi(x) = \pi(a)$, $\pi(y) = \pi(b)$, we conclude that $\pi(a) \in D_{G/\ker f}(1, \pi(b))$.

(2) $\Rightarrow$ (1). Assume $f(a) \in D_H(1, f(b))$; then $\widehat{f}(\pi(a)) \in D_H(1, \widehat{f}(\pi(b)))$. Since $\widehat{f}$ is an isomorphism, $\pi(a) \in D_{G/\ker f}(1, \pi(b))$. Definition 2.13 gives elements $a', b' \in G$ such that $aa', bb' \in \ker f$ (i.e., $f(a) = f(a')$, $f(b) = f(b')$) and $a' \in D_G(1, b')$; hence, f is regular. $\square$

EXAMPLE 2.24. As an application of the preceding Proposition we give a short argument to deduce that $G_T(F)$ is a special group from the fact that $G(F)$ is one, whenever F is a formally real field and T is a preorder of F. (A proof that $G(F)$ is a special group, different from that given in section 1.3, can be found in sections 4 and 5 of Chapter 1 in [Ma1]).

It is quite obvious that: (1) $G_T(F)$ is a reduced psg, and, from the definition of representation 1.28, (2) the canonical map $f_T : G(F) \longrightarrow G_T(F)$, given by $f_T(a/\dot{F}^2) = a/\dot{T}$ $(a \in \dot{F})$, is a surjective morphism of psg's. Then, it suffices to show that f_T is regular.

Assuming that $f_T(a) \in D_{G_T(F)}(1, f_T(b))$, there are $t, s \in T$ so that $a = s + tb$. If $s = 0$, let $a' = a$ and $b' = tb$. If $s \neq 0$, we get $\frac{1}{s}a = 1 + \frac{t}{s}b$; setting $a' = \frac{a}{s}$, $b' = \frac{t}{s}b$, we have $a' = 1 + b'$. In both cases, $a' \in D_{G(F)}(1, b')$, $a'/\dot{T} = a/\dot{T}$ and $b'/\dot{T} = b/\dot{T}$, whence $f(a') = f(a)$ and $f(b') = f(b)$, as required. In case $T = \sum F^2$, we have $\ker f_T = Sat(G(F))$; hence, $G_{red}(F) \cong G(F)/Sat(G(F))$. $\diamond$

DEFINITION 2.25. If G is a special group, set $G_{red} = G/Sat(G)$. $\diamond$

By Proposition 2.21, G_{red} is a special group. When $Sat(G) \neq G$ (i.e., G is formally real, see Definition 3.3), Proposition 2.23 implies that G_{red} is the *largest reduced quotient of G*. As shown by Example 2.24, this notation is consistent with that of section 1.3 for fields.

The preceding results hold, in particular, for saturated subgroups (Remark 2.16). However, in this case, we can be more precise. First, we define, for an arbitrary subgroup of G :

DEFINITION 2.26. Let G be a special group and Δ a subgroup of G. Define a relation $\equiv_{G/\Delta}$ on G/Δ by the following clause : for $a, b, c, d \in G$

$$\langle a/\Delta, b/\Delta \rangle \equiv_{G/\Delta} \langle c/\Delta, d/\Delta \rangle \text{ iff } \begin{cases} \text{There is a Pfister form } \varphi \text{ over } \Delta \\ \text{such that} \\ \langle a, b \rangle \otimes \varphi \equiv_G \langle c, d \rangle \otimes \varphi. \end{cases} \diamond$$

The relation $\equiv_{G/\Delta}$ has the following properties :

FACT 2.27. *Let G be a special group and Δ a subgroup of G. Then*

1. $\equiv_{G/\Delta}$ is well defined.

2. $(G/\Delta, \equiv_{G/\Delta}, -1/\Delta)$ verifies the axioms for special groups, except, possibly, [SG3] (the discriminant axiom).

3. The canonical quotient map $\pi : G \longrightarrow G/\Delta$ verifies

$$\langle a,b \rangle \equiv_G \langle c,d \rangle \;\Rightarrow\; \langle \pi(a), \pi(b) \rangle \equiv_{G/\Delta} \langle \pi(c), \pi(d) \rangle.$$

Hence, by Proposition 2.14.3

4. $\equiv^*_{G/\Delta}$ is contained in $\equiv_{G/\Delta}$.

PROOF. (1) If $aa', \ldots, dd' \in \Delta$ and $\langle a,b \rangle \otimes \varphi \equiv_G \langle c,d \rangle \otimes \varphi$, φ Pfister over Δ, let $\psi = \varphi \otimes \langle 1, aa' \rangle \otimes \langle 1, bb' \rangle \otimes \langle 1, cc' \rangle \otimes \langle 1, dd' \rangle$. It is straightforward to verify that $\langle a', b' \rangle \otimes \psi \equiv_G \langle c', d' \rangle \otimes \psi$. Items (2) and (3) are routine. $\square$

Now we show

PROPOSITION 2.28. *For a special group G and Δ a proper subgroup of G, the following are equivalent :*

(1) $\equiv^*_{G/\Delta} \;=\; \equiv_{G/\Delta}$.

(2) $(G/\Delta, \equiv_{G/\Delta}, -1/\Delta)$ *satisfies [SG3] (and hence it is a special group).*

(3) Δ *is saturated.*

(4) $(G/\Delta, \equiv^*_{G/\Delta}, -1/\Delta)$ *is a reduced special group.*

REMARK. Obviously (2) can be replaced by

(2') $(G/\Delta, \equiv_{G/\Delta}, -1/\Delta)$ is a reduced special group.

But Proposition 2.21 and the fact that G may have Pfister subgroups which are not saturated, show that condition (4) <u>cannot</u> be weakened to

"$(G/\Delta, \equiv^*_{G/\Delta}, -1/\Delta)$ is a special group". $\diamond$

PROOF. (1) $\Rightarrow$ (3). Let φ Pfister over Δ and let $x \in D_G(\varphi)$. Then $x\varphi \equiv_G \varphi$; hence $x\varphi \oplus \varphi \equiv_G \varphi \oplus \varphi$, i.e. $\langle x, 1 \rangle \otimes \varphi \equiv_G \langle 1, 1 \rangle \otimes \varphi$ and $\langle x/\Delta, 1 \rangle \equiv_{G/\Delta} \langle 1, 1 \rangle$. By (1), $\langle x/\Delta, 1 \rangle \equiv^*_{G/\Delta} \langle 1, 1 \rangle$, that is, there are $b, c, d \in \Delta$ and $x' \in G$ such that $xx' \in \Delta$ and $\langle x', b \rangle \equiv_G \langle c, d \rangle$. This implies, by [SG3], that $x' = bcd \in \Delta$, and so $x \in \Delta$. Thus, $D_G(\varphi) \subseteq \Delta$, showing that Δ is saturated.

(3) $\Rightarrow$ (2). Lemma 2.19.(a) shows that $\langle a,b \rangle \otimes \varphi \equiv_G \langle c,d \rangle \otimes \varphi$ for φ Pfister over Δ —i.e., $\langle a/\Delta, b/\Delta \rangle \equiv_{G/\Delta} \langle c/\Delta, d/\Delta \rangle$— implies $abcd \in D_G(\varphi)$. By (3), $D_G(\varphi) \subseteq \Delta$ (Lemma 2.4.b), and so $abcd/\Delta = 1$.

(2) $\Rightarrow$ (1). Only the inclusion $\equiv_{G/\Delta} \subseteq \equiv^*_{G/\Delta}$ requires proof (Fact 2.27.(4)). Lemma 2.19(a) and the assumption $\langle a,b \rangle \otimes \varphi \equiv_G \langle c,d \rangle \otimes \varphi$ yield $\langle a',b' \rangle \equiv_G \langle c', d' \rangle$, for $a',b',c',d' \in G$ such that $aa', \ldots, dd' \in D_G(\varphi)$. It suffices to show that (2) entails

(*) $\qquad$ For a Pfister form φ over Δ, $D_G(\varphi) \subseteq \Delta$.

Let $x \in D_G(\varphi)$, i.e. $x\varphi \equiv_G \varphi$; then $\langle x, 1 \rangle \otimes \varphi \equiv_G \langle 1, 1 \rangle \otimes \varphi$, and so $\langle x/\Delta, 1 \rangle \equiv_{G/\Delta} \langle 1, 1 \rangle$. By 2, $x/\Delta = 1$, i.e., $x \in \Delta$.

(3) $\Rightarrow$ (4). Since $(G/\Delta, \equiv^*_{G/\Delta}, -1/\Delta)$ satisfies [SG1] - [SG5] (2.14.(b)) and $(G/\Delta, \equiv_{G/\Delta}, -1/\Delta)$ satisfies [SG0] and [SG6] (Fact 2.27.2), (1) tells us that $(G/\Delta, \equiv^*_{G/\Delta}, -1/\Delta)$ is a special group (note that (3) $\Rightarrow$ (1)).

For the reduction axiom, let $\langle a/\Delta, a/\Delta \rangle \equiv^*_{G/\Delta} \langle 1, 1 \rangle$, that is $\langle a', a'' \rangle \equiv_G \langle b, c \rangle$, where $b, c, aa', aa'' \in \Delta$. Then, $a' \in D_G(b,c)$, whence $ba' \in D_G(1, bc)$. Since $bc \in \Delta$, (3) yields $ba' \in \Delta$, and so $a' \in \Delta$, which gives $a \in \Delta$.

(4) $\Rightarrow$ (3). If $a \in \Delta$ and $x \in D_G(1, a)$, then $\langle x, xa \rangle \equiv_G \langle 1, a \rangle$ and so $\langle x/\Delta, x/\Delta \rangle \equiv^*_{G/\Delta} \langle 1, 1 \rangle$. By (4), $x/\Delta = 1$, i.e., $x \in \Delta$. $\square$

Remark. The first three equivalences in Proposition 2.28 appear as Proposition 26 in [LP1], and also in [LP3].

COROLLARY 2.29. *Let G be a special group and Δ a subgroup of G.*

*a) The proper saturated subgroups of G are exactly the kernels of the SG-morphisms from G into **reduced** special groups.*

b) The maximal saturated subgroups of G are exactly the kernels of special group characters of G (i.e., the SG-morphisms of G into $\mathbb{Z}_2$).

c) $\overline{\Delta} = \bigcap \{\ker \sigma : \sigma$ is a SG-character of G and $\Delta \subseteq \ker \sigma\}$.

PROOF. (a) follows from Proposition 2.28 and the remark at the beginning of section 2. Proposition 2.10 implies that G/Δ is isomorphic to $\mathbb{Z}_2$ for a maximal saturated subgroup Δ. The converse is clear. Item (c) follows directly from (b) and Theorem 2.11. $\square$

Remark. If the special group G is **reduced**, 2.29 follows from the duality to be presented in Chapter 3, instead of Propositions 2.18 – 2.28. $\diamond$

The theory of quotients presented above yields the following result which, in fact, is a version of Pfister's local-global principle (see Proposition 3.7).

THEOREM 2.30. *For $a_1, \ldots, a_n$, $b_1, \ldots, b_n$ in a reduced special group G, the following are equivalent:*

(1) $\langle a_1, \ldots, a_n \rangle \equiv_G \langle b_1, \ldots, b_n \rangle$.

(2) $\langle a_1/\Delta, \ldots, a_n/\Delta \rangle \equiv_{G/\Delta} \langle b_1/\Delta, \ldots, b_n/\Delta \rangle$, for all maximal saturated subgroups Δ of G.

Before proving this result, we show that it implies a more general version, holding in all special groups, reduced or not.

PROPOSITION 2.31. *Let G be a special group such that $-1 \neq 1$ and $a_1, \ldots, a_n, b_1, \ldots, b_n \in G$. The following are equivalent :*

(1) *For some integer $k \geq 0$, $2^k \cdot \langle a_1, \ldots, a_n \rangle \equiv_G 2^k \cdot \langle b_1, \ldots, b_n \rangle$.*

(2) $\langle a_1/\Delta, \ldots, a_n/\Delta \rangle \equiv_{G/\Delta} \langle b_1/\Delta, \ldots, b_n/\Delta \rangle$, *for every maximal saturated subgroup Δ of G.*

PROOF. (1) $\Rightarrow$ (2) is clear in view of Proposition 1.6.(e) (G/Δ is reduced, Proposition 2.28) and that the quotient map $G \longrightarrow G/\Delta$ is a SG-morphism.

(2) $\Rightarrow$ (1). We apply Theorem 2.30 to $G/Sat(G)$. Assume $2^k \cdot \langle a_1, \ldots, a_n \rangle \not\equiv_G 2^k \cdot \langle b_1, \ldots, b_n \rangle$ for every $k \geq 0$. Proposition 2.21.(a) applied to the family $\{2^n : n \in \omega\}$ of Pfister forms yields :

(*) $\quad \langle a_1/Sat(G), \ldots, a_n/Sat(G) \rangle \not\equiv_{G/SatG} \langle b_1/Sat(G), \ldots, b_n/Sat(G) \rangle$.

By the preceding theorem, $G/Sat(G)$ contains a maximal saturated subgroup, Γ, such that (*) holds modulo Γ, i.e., in $G/Sat(G)/\Gamma$. Let $\Delta = \pi^{-1}[\Gamma]$, where π is the canonical map from G to $G/Sat(G)$. Using Proposition 2.14.(c), $\ker(\pi) \subseteq \Delta$, and the surjectivity of π, it is easily checked that Δ is a maximal saturated subgroup of G. Since $\Gamma = \Delta/Sat(G)$ and G/Δ is isomorphic to $G/Sat(G)/\Gamma$ (as special groups), we obtain

$$\langle a_1/\Delta, \ldots, a_n/\Delta \rangle \not\equiv_{G/\Delta} \langle b_1/\Delta, \ldots, b_n/\Delta \rangle,$$

contrary to (2). $\square$

Our proof of Theorem 2.30 uses a technique similar to that of Marshall's proof for a part of Pfister's local-global principle for abstract Witt rings (see [Ma1], Theorem 4.12, p.76ff).

Proof of Theorem 2.30. Lemma 1.12.(c) gives (1) $\Rightarrow$ (2). To prove the converse, assume G is reduced and $\langle a_1, \ldots, a_n \rangle \not\equiv_G \langle b_1, \ldots, b_n \rangle$. Then (Proposition 1.6.(e))

$$2^k \cdot \langle a_1, \ldots, a_n \rangle \not\equiv_G 2^k \cdot \langle b_1, \ldots, b_n \rangle,$$

for all $k \geq 0$. Hence, every Pfister form φ in the family $\{2^n : n \in \omega\}$ has the property

(*) $\quad \langle a_1, \ldots, a_n \rangle \otimes \varphi \not\equiv_G \langle b_1, \ldots, b_n \rangle \otimes \varphi;$

further, it is closed under tensor products. By Zorn's Lemma, there is a maximal family $\mathcal{L}$ of Pfister forms over G containing $\langle 1, 1 \rangle$, closed under tensor products, and such that every $\varphi \in \mathcal{L}$ verifies (*). Note that every Pfister form φ verifying (*) is anisotropic; otherwise, φ would be hyperbolic (Proposition 2.2.(e)), and we know that for $l \geq 1$

(**) $\qquad \langle a_1, \ldots, a_n \rangle \otimes l \cdot \langle 1, -1 \rangle \equiv_G \langle b_1, \ldots, b_n \rangle \otimes l \cdot \langle 1, -1 \rangle,$

as the coefficients on each side occur in pairs $c, -c$.

Let $\Delta = \bigcup \{D_G(\varphi) : \varphi \in \mathcal{L}\}$. By Lemma 2.5.(b), Δ is a proper saturated subgroup of G. We show

Claim. Δ is maximal saturated.

<u>Proof of Claim.</u> By 2.10, it suffices to show that, for $x \in G$, either $x \in \Delta$ or $-x \in \Delta$.

For $y \in G$, let $\mathcal{L}_y = \mathcal{L} \cup \{\langle 1, y \rangle \otimes \varphi : \varphi \in \mathcal{L}\}$. $\mathcal{L}_y$ contains $\mathcal{L}$ and is closed under tensor products (because $\langle 1, y \rangle \otimes \langle 1, y \rangle \equiv_G \langle 1, 1 \rangle \otimes \langle 1, y \rangle$). If $y \notin \Delta$, by the maximality of $\mathcal{L}$, the isometry

(***) $\qquad \langle a_1, \ldots, a_n \rangle \otimes \langle 1, y \rangle \otimes \varphi \equiv_G \langle b_1, \ldots, b_n \rangle \otimes \langle 1, y \rangle \otimes \varphi$

holds for some form $\varphi \in \mathcal{L}$. Note that if we have $y, z \notin \Delta$ and φ_1, φ_2 are the forms in $\mathcal{L}$ satisfying (***) in relation to y and z, respectively, then $\varphi = \varphi_1 \otimes \varphi_2$ is in $\mathcal{L}$ and satisfies (***) with respect to y and z, simultaneously. Thus, we may assume that (***) holds as stated for x and $-x$.

Assume $x, -x \notin \Delta$. Applying (***) to x and $-x$, and adding up the instances of (***) thus obtained, yields :

$\langle a_1, \ldots, a_n \rangle \otimes \varphi \otimes (2 \oplus \langle 1, -1 \rangle) \equiv_G \langle b_1, \ldots, b_n \rangle \otimes \varphi \otimes (2 \oplus \langle 1, -1 \rangle),$

recalling that $\langle 1, x, 1, -x \rangle \equiv_G \langle 1, 1 \rangle \oplus \langle 1, -1 \rangle$. Cancelling out the terms $\langle a_1, \ldots, a_n \rangle \otimes \varphi \otimes \langle 1, -1 \rangle \equiv_G \langle b_1, \ldots, b_n \rangle \otimes \varphi \otimes \langle 1, -1 \rangle$, (cf. (**)), we get

$\qquad \qquad \langle a_1, \ldots, a_n \rangle \otimes 2\varphi \equiv_G \langle b_1, \ldots, b_n \rangle \otimes 2\varphi,$

in contradiction to (*), since $2\varphi \in \mathcal{L}$. □

Next we exhibit a more interesting example of a non-saturated Pfister group than that given by the proof of Proposition 2.17.

EXAMPLE 2.32. We consider the extension $G[\Delta]$ of a special group G by Δ (Example 1.10); recall that the extension is reduced iff G is reduced.

It is easily verified that the projection $pr : G[\Delta] \longrightarrow G$ onto the first coordinate is a special group morphism, whose kernel is $\{1\} \times \Delta$ (we identify this subgroup of $G[\Delta]$ with Δ). The map $\widehat{pr} : G[\Delta]/\Delta \longrightarrow G$ induced by pr is an isomorphism of special groups, where the group $G[\Delta]/\Delta$ is endowed with the relation $\equiv^*_{G[\Delta]/\Delta}$ given in Definition 2.13.

Thus, the quotient structure structure $G[\Delta]/\Delta$ is a special group; further, it is reduced iff G is reduced. In view of Proposition 2.28, this shows that Δ is not a saturated subgroup of $G[\Delta]$, unless G is reduced. However, the following Proposition, whose proof is omitted, shows that Δ is a Pfister subgroup of $G[\Delta]$.

PROPOSITION 2.33. *Let G be a sg and Δ a group of exponent 2.*

a) Let $\{d_1, \ldots, d_n\} \subseteq \Delta$ ($n \geq 1$) be a $\mathbb{F}_2$-independent subset of Δ. Then,
$$D_{G[\Delta]}(\langle\langle 1 \cdot d_1, \ldots, 1 \cdot d_n \rangle\rangle) = \{1 \cdot d : d \in [d_1, \ldots, d_n]\},$$
where $[d_1, \ldots, d_n]$ is the subgroup generated by the d_i's.

b) Let $\mathcal{S}$ denote the family of Pfister forms over $G[\Delta]$, consisting of $\langle \mathbb{1} \rangle$ and all forms $\langle\langle 1 \cdot d_1, \ldots, 1 \cdot d_n \rangle\rangle$ where $\{d_1, \ldots, d_n\}$ is an $\mathbb{F}_2$-independent subset of Δ ($n \geq 1$). Then, $\mathcal{S}$ is directed and $\Delta = \bigcup \{D_{G[\Delta]}(\varphi) : \varphi \in \mathcal{S}\}$. ◇

The results proven suggest the following

PROBLEM 2.34. Give conditions on a special group G and a subgroup Δ under which if $(G/\Delta, \equiv^*_{G/\Delta}, -1/\Delta)$ is a special group, then Δ is a Pfister subgroup of G. ◇

Note that under the hypothesis that G/Δ is reduced, the solution to this problem is given by Proposition 2.28.

We close the chapter with an example showing that $(G/\Delta, \equiv^*_{G/\Delta}, -1/\Delta)$ may be a special group without Δ being a Pfister subgroup.

EXAMPLE 2.35. Given a group G of exponent 2 and cardinality ≥ 4, recall (Example 1.9) that G_t denotes the trivial special group structure on G, with $-1 \neq 1$. It follows from the discussion presented therein, that $D_{G_t}(\varphi) = G$, for all Pfister forms of degree ≥ 1, and, no proper subgroup of G_t is Pfister.

FACT 2.36. For every (proper) subgroup Δ of G_t, $(G_t/\Delta, \equiv^*_{G_t/\Delta}, -1/\Delta)$ is a special group. Further, $-1 \neq 1$ in G_t/Δ iff $-1 \notin \Delta$. More generally, $(G_t/\Delta, \equiv, -1)$ is a special group for any quaternary relation $\equiv$ on G_t/Δ that verifies [SG3], and such that the map $\pi : G_t \longrightarrow G_t/\Delta$ satisfies, for all $a, b, c, d \in G_t$:

(*) $\quad \langle a, b \rangle \equiv_{G_t} \langle c, d \rangle \quad \Rightarrow \quad \langle \pi(a), \pi(b) \rangle \equiv \langle \pi(c), \pi(d) \rangle$

PROOF. It suffices to show that (*) implies that the relation $\equiv$ endows G_t/Δ with the trivial group structure. Let us prove :
$$ab/\Delta = cd/\Delta \quad \Rightarrow \quad \langle \pi(a), \pi(b) \rangle \equiv \langle \pi(c), \pi(d) \rangle.$$

Note that the converse is the assumption that $(G_t/\Delta, \equiv, -1)$ satisfies [SG3]. We know that $x = abcd \in \Delta$. Let $d' = dx$; then $ab = cd'$, i.e. $\langle a, b \rangle \equiv_{G_t} \langle c, d' \rangle$. Since $d/\Delta = d'/\Delta$, (*) entails that $\langle \pi(a), \pi(b) \rangle \equiv \langle \pi(c), \pi(d) \rangle$. □

CHAPTER 3

The Space of Orders of a Reduced Group. Duality

We begin this Chapter by constructing the space of orders of a reduced special group, and proving that it is an abstract order space in the sense of Marshall ([Ma2]). Conversely, we show that each abstract order space naturally gives rise to a reduced special group.

The main result proved here is the Duality Theorem (Theorem 3.19) which asserts that the correspondence associating to each special group its space of orders can be extended to a functor which defines a contravariant isomorphism of the respective categories.

As it will be apparent from results in later sections, the Duality Theorem is a fundamental tool for the study of reduced special groups. The Theorem appears in Lira's thesis [LP3]; it was previously reported in [LP1]. Our presentation is similar to Lira's, with the important exception that we avoid use of the Witt ring machinery (Pfister's local-global principle, required for the proof, follows from Theorem 2.30 above).

DEFINITION 3.1. Let G be a special group. The **space of orders** of G is the set X_G of all SG-morphisms of G into $\mathbb{Z}_2$, endowed with the topology induced by the product $\{\pm 1\}^G$. ◇

LEMMA 3.2. *Let G be a special group.*

a) $X_G \neq \emptyset$ *if and only if* $Sat(G) \neq G$ *if and only if* $-1 \notin Sat(G)$.

Furthermore,

b) *If $\sigma \in X_G$, then $Sat(G) \subseteq ker\sigma$. Thus, σ factors through π to give a character $\hat{\sigma} \in X_{G/Sat(G)}$ satisfying $\hat{\sigma} \circ \pi = \sigma$, where $G \xrightarrow{\pi} G/Sat(G)$ is the canonical quotient map.*

c) *The map $X_{G/Sat(G)} \mapsto X_G$ given by $\tau \mapsto \tau \circ \pi$, is a homeomorphism.*

PROOF. Left to the reader (see Lemma 2.5 in [DM1]). □

DEFINITION 3.3. Any special group verifying the equivalent conditions in 3.2.(a) will be called **formally real**. ◇

It is immediate that a field F is formally real iff the group $G(F)$ is formally real.

The space of orders has the following properties :

FACT 3.4. *Let G be a rsg. Then,*

a) X_G is closed in $\{\pm 1\}^G$ (and in $\chi(G)$).

Hence :

b) X_G is a Boolean space (= compact, Hausdorff, totally disconnected).

c) (Separation property) $\bigcap_{\sigma \in X_G} \ker \sigma = \{1\}$.

PROOF. (a) and (b) are straightforward.

(c) follows from the Separation Theorem 2.11 : since G is reduced, $\{1\}$ is saturated; given $a \neq 1$, there is a maximal saturated subgroup Δ of G such that $a \notin \Delta$. But we have $\Delta = \ker \sigma$, for some $\sigma \in X_G$ (Corollary 2.29). □

NOTATION 3.5. For a form $\varphi = \langle a_1, \ldots, a_n \rangle$ over G and a map $G \xrightarrow{\sigma} \{\pm 1\}$, we set

$$sgn_\sigma(\varphi) = \sum_{i=1}^n \sigma(a_i) \qquad \text{(addition in } \mathbb{Z})$$

$sgn_\sigma(\varphi)$ is called the **signature** of φ at σ. ◇

REMARK 3.6. With notation as in 3.2, if φ is a form over a formally real special group G, then for all $\sigma \in X_G$,

$$sgn_\sigma(\varphi) = sgn_{\hat{\sigma}}(\pi \star \varphi).$$

◇

Our next result, a reformulation of Theorem 2.30, is well known in both the theory of quadratic forms over fields ([L1], Chapter X) and in its abstract versions ([Ma1], Thm. 4.12). It will be of crucial importance in the sequel.

PROPOSITION 3.7. (Pfister's local-global principle)

For $a_1, \ldots, a_n, b_1, \ldots, b_n$ in a reduced special group G, the following are equivalent :

(a) $\langle a_1, \ldots, a_n \rangle \equiv_G \langle b_1, \ldots, b_n \rangle$.

(b) For every $\sigma \in X_G$, $sgn_\sigma(\langle a_1, \ldots, a_n \rangle) = sgn_\sigma(\langle b_1, \ldots, b_n \rangle)$.

PROOF. Immediate from Theorem 2.30 and Corollary 2.29, recalling (Example 1.7) that for s, t forms over $\mathbb{Z}_2$, we have

$s \equiv_{fan} t$ iff $\sum_{i \leq n} s(i) = \sum_{i \leq n} t(i)$ (in $\mathbb{Z}$) iff $p_s = p_t$ and $n_s = n_t$,

where p_s and n_s are, respectively, the number of 1's and -1's in s. □

Remarks. a) The following is a more standard form of the local-global principle. It can be obtained from the above, for n even, by taking one-half of the b_i's to be 1, and the other half as -1.

Proposition 3.7*. *With notation as in 3.5, the following are equivalent :*

(1) $\langle a_1, \ldots, a_n \rangle$ is a hyperbolic form.

(2) For every $\sigma \in X_G$, $sgn_\sigma(\langle a_1, \ldots, a_n \rangle) = 0$. ◇

b) A similar statement, for a not necessarily reduced special group G, can be derived from Proposition 2.31. ◇

NOTATION 3.8. Let G be a group of exponent 2 with a distinguished element -1. Let $X \subseteq \chi(G)$.

a) The notion of **weak isometry modulo X**, $\equiv_X$, is defined as follows : for forms $\varphi = \langle a_1, \ldots, a_n \rangle$ and $\psi = \langle b_1, \ldots, b_n \rangle$ over G

$$\varphi \equiv_X \psi \quad \text{iff} \quad \forall \sigma \in X, (sgn_\sigma(\varphi) = sgn_\sigma(\psi)).$$

b) If φ is a form of dimension n over G, define

$$D_X(\varphi) = \{b \in G : \exists\, b_2, \ldots, b_n \in G \text{ such that } \varphi \equiv_X \langle b, b_2, \ldots, b_n \rangle\},$$

the set of elements **weakly represented by φ modulo X**.

c) We denote by $\equiv_X^*$ the extension to forms of arbitrary dimension of the weak isometry relation on *binary forms*, as constructed in Notation 1.1.(e). The relation $\equiv_X^*$ is referred to as **strong isometry modulo X**. ◇

We recall the notion of an abstract order space introduced by M. Marshall.

DEFINITION 3.9. With notation as in 3.8, the structure $(X, G, -1)$ is an **abstract order space, AOS**, if it verifies the following conditions :

[O1] X is closed in $\chi(G)$ (equivalently, in $\{\pm 1\}^G$).

[O2] $\sigma(-1) = -1$, \quad for all $\sigma \in X$.

[O3] $\bigcap_{\sigma \in X} \ker \sigma = \{1\}$.

[O4] If φ, ψ are forms over G, and $x \in G$

$$x \in D_X(\varphi \oplus \psi) \Rightarrow \begin{cases} \text{There are } y \in D_X(\varphi) \text{ and } z \in D_X(\psi) \\ \text{such that } x \in D_X(y, z). \end{cases} \text{◇}$$

The results proved above at once yield

PROPOSITION 3.10. *If $(G, \equiv_G, -1)$ is a rsg, then $(X_G, G, -1)$ is an AOS.*

PROOF. [O1] and [O3] come from Fact 3.4.(a) and (c); [O2] is contained in the definition of X_G; [O4] is an immediate consequence of Propositions 3.7 and 1.6.(c). □

Conversely, any abstract order space generates a reduced special group, as follows :

PROPOSITION 3.11. *If $(X, G, -1)$ be an AOS, then $(G, \equiv_X, -1)$ is a reduced special group.*

Before proving this result, we state a few simple properties of weak isometry.

FACT 3.12. *Let G be a group of exponent 2 and $X \subseteq \chi(G)$. Then,*

a) Weak isometry modulo X is transitive on forms of any dimension.

b) Strong isometry implies weak isometry (but not conversely, see Proposition A below).

c) Let φ, θ_1, θ_2 be forms over G; then

$$\varphi \oplus \theta_1 \equiv_X \varphi \oplus \theta_2 \iff \theta_1 \equiv_X \theta_2. \quad \diamond$$

Proof of Proposition 3.11. Checking that $(G, \equiv_X, -1)$ satisfies [SG0] – [SG5] and the reduction axiom [red] is routine. As for [SG6], since weak isometry modulo X is transitive, it would be sufficient to show that, under our assumptions, strong and weak isometry modulo X are identical on forms of dimension 3. Indeed, this follows from [O4], as we now prove : assume

(i) $\langle a_1, a_2, a_3 \rangle \equiv_X \langle b_1, b_2, b_3 \rangle$;

then, $b_1 \in D_X(\langle a_1, a_2, a_3 \rangle)$. By [O4], there is $x \in D_X(\langle a_2, a_3 \rangle)$ such that $b_1 \in D_X(\langle a_1, x \rangle)$, i.e. there are $y, z \in G$ so that

(ii) $\langle a_1, x \rangle \equiv_X \langle b_1, y \rangle$; (iii) $\langle a_2, a_3 \rangle \equiv_X \langle x, z \rangle$.

It only remains to show that

(iv) $\langle b_2, b_3 \rangle \equiv_X \langle y, z \rangle$.

The isometry (iii) and Fact 3.12.(c) give : $\langle a_1, a_2, a_3 \rangle \equiv_X \langle a_1, x, z \rangle$.
Similarly, (ii) and 3.12.(c) yield : $\langle a_1, x, z \rangle \equiv_X \langle b_1, y, z \rangle$.

Since $\equiv_X$ is transitive, (i) and the last two isometries prove $\langle b_1, b_2, b_3 \rangle \equiv_X \langle b_1, y, z \rangle$, which, using 3.12.(c) again, implies (iv). $\square$

In her thesis, Lira proves the following interesting result :

Proposition A ([LP1], [LP3]). *Let G be a group of exponent 2 with a distinguished element -1, and $X \subseteq \chi(G)$. If the structure $(X, G, -1)$ satisfies the axioms [O1], [O2] and [O3], the following are equivalent :*

(1) $(X, G, -1)$ *satisfies* [O4].

(2) *Strong and weak isometry modulo X are identical on forms of all dimensions.*

(3) *Strong isometry modulo X is transitive on forms of dimension 3, and $(X, G, -1)$ verifies the following maximality condition :*

[O5] *For every $\sigma \in \chi(G)$, such that $\sigma(-1) = -1$,*

if $\forall a \in G \, [a \in \ker \sigma \Rightarrow D_X(1, a) \subseteq \ker \sigma]$, **then** $\sigma \in X$. $\quad \diamond$

Summarizing, we have two correspondences between reduced special groups and abstract order spaces, as follows :

NOTATION 3.13.

$$\Phi : \mathbf{RSG} \longrightarrow \mathbf{AOS}, \qquad (G, \equiv_G, -1) \mapsto (X_G, G, -1),$$

and

$$\Psi : \mathbf{AOS} \longrightarrow \mathbf{RSG}, \qquad (X, G, -1) \mapsto (G, \equiv_X, -1). \qquad \diamond$$

We show next that these correspondences are reciprocal to each other.

PROPOSITION 3.14. *With notation as in 3.13, we have* $\Phi \circ \Psi = Id_{AOS}$ *and* $\Psi \circ \Phi = Id_{RSG}$. *In particular,* Φ *and* Ψ *are bijective.*

PROOF. (1) $\Psi \circ \Phi = Id_{RSG}$. Since

$$\Psi \circ \Phi((G, \equiv_G, -1)) = (G, \equiv_{X_G}, -1),$$

it suffices to show that the relations $\equiv_G$ and $\equiv_{X_G}$ are identical on binary forms. This is asserted by Proposition 3.7.

(2) $\Phi \circ \Psi = Id_{AOS}$. Let us denote $\Psi((X, G, -1)) = (G, \equiv_X, -1)$ by $G[X]$. Thus, $\Phi(G[X]) = (X_{G[X]}, G, -1)$, and we have to prove that $X = X_{G[X]}$.

The inclusion $X \subseteq X_{G[X]}$ is clear : by definition of the relation $\equiv_X$ (cf. 3.8.(a)), each $\sigma \in X$ is a SG-morphism from $G[X]$ into $\mathbb{Z}_2$. Conversely, since the special relation of $G[X]$ is $\equiv_X$, every $\sigma \in X_{G[X]}$ verifies the assumption of the maximality condition [O5] in Proposition A.(3) above; hence $\sigma \in X$.

The gist of the proof that any AOS verifies [O5] – due to Marshall ([Ma2]; Prop. 4.1, [Ma4]; Lemma 4.1) — is as follows :

a) Let $\sigma \in \chi(G)$ be a character verifying the assumptions of [O5]. Using [O4], it is proved by induction on n that if $\varphi = \bigotimes_{i=1}^n \langle 1, a_i \rangle$ is a Pfister form over G, then

(*) $\qquad\qquad a_1, \ldots, a_n \in \ker \sigma \Rightarrow D_X(\varphi) \subseteq \ker \sigma.$

b) Condition (*) implies that σ is in the closure of X in $\chi(G)$. Otherwise, (by $(\star)$ in Notation 1.1.(a)), there are $a_1, \ldots, a_n \in G$ such that $\bigcap_{i=1}^n [a_i = 1] \cap X = \emptyset$ and $\sigma \in \bigcap_{i=1}^n [a_i = 1]$. But this entails $-1 \in D_X(\varphi)$, whence, by (*), $-1 \in \ker \sigma$, a contradiction. The same conclusion can be obtained using Proposition 4.11 below.

Since X is closed in $\chi(G)$, we get $\sigma \in X$. $\qquad\square$

Our next step is to extend the correspondences Φ, Ψ, to functors so as to establish a *duality of categories*. The natural notion of morphism of AOS is as follows ([Ma3]; Def. 1.4) :

DEFINITION 3.15. Let $(X, G, -1)$ and $(Y, H, -1)$ be AOS's. A map $X \xrightarrow{\gamma} Y$ is a **morphism of AOS's** iff there is a continuous group homomorphism $\Gamma : \chi(G) \longrightarrow \chi(H)$, such that $\gamma = \Gamma_{|X}$. Abstract order spaces and their morphisms constitute a category, denoted by **AOS**. $\diamond$

We now show that every morphism of AOS's is the restriction of a **unique** continuous group homomorphism of $\chi(G)$ into $\chi(H)$. This is an immediate consequence of

PROPOSITION 3.16. *Let G be a group of exponent 2 and X a subset of $\chi(G)$ satisfying the separation axiom [O3] (cf.3.9). Then, the subgroup of $\chi(G)$ generated by X, $[X]$, is dense in $\chi(G)$.*

For the proof of this result we need the following

LEMMA 3.17. *Let K be a finite group of exponent 2, and let $\sigma_i \in \chi(K)$, $1 \leq i \leq n$. Then, $\{\sigma_1, \ldots, \sigma_n\}$ generates $\chi(K)$ iff $\bigcap_{i=1}^n \ker \sigma_i = \{1\}$.*

PROOF. ($\Rightarrow$) If the conclusion fails, consider $a \in \bigcap_{i=1}^n \ker \sigma_i$, $a \neq 1$, and any character σ such that $\sigma(a) = -1$ (such a σ exists because $\{a\}$ is an $\mathbb{F}_2$-linearly independent subset of K); then $\sigma \notin [\sigma_1, \ldots, \sigma_n]$.

($\Leftarrow$) We may assume that the character constantly equal to 1, $\mathbb{1}$, is not in $\{\sigma_1, \ldots, \sigma_n\}$. By induction on n we select an irredundant subset of $\{\sigma_1, \ldots, \sigma_n\}$, say $\{\sigma_1, \ldots, \sigma_m\}$, i.e. a subset with the following properties :

(i) $\bigcap_{i=1}^m \ker \sigma_i = \{1\}$. (ii) For $2 \leq j \leq m$, $\bigcap_{i=1}^{j-1} \ker \sigma_i \not\subseteq \ker \sigma_j$.

By induction on m we choose elements $b_1, \ldots, b_m \in K$ such that

(iii) $b_1 \notin \ker \sigma_1$. (iv) For $2 \leq j \leq m$, $b_j \in \bigcap_{i=1}^{j-1} \ker \sigma_i$, $b_j \notin \ker \sigma_j$.

Claim. $\{b_1, \ldots, b_m\}$ is a $\mathbb{F}_2$-basis of K.

Proof of Claim. Induction on m.

$m = 1$) By (i), $\ker \sigma_1 = \{1\}$, that is, σ_1 is an isomorphism between K and $\{\pm 1\}$. By (iii), $b_1 = -1$, a basis for K.

$m - 1 \to m$, $m \geq 2$) Consider the (proper) subgroup $K_1 = \ker \sigma_1$. For $i \geq 2$, let $\sigma'_i = \sigma_{i|K_1}$; then $\sigma'_2, \ldots, \sigma'_m$ are in $\chi(K_1)$, $b_2, \ldots, b_m$ are in K_1, and

(i') $\bigcap_{i=2}^m \ker \sigma'_i = \bigcap_{i=1}^m \ker \sigma_i = \{1\}$;

(ii') If $3 \leq j \leq m$, then $\bigcap_{i=2}^{j-1} \ker \sigma'_i = \bigcap_{i=1}^{j-1} \ker \sigma_i \not\subseteq \ker \sigma_j$;

whence $\bigcap_{i=2}^{j-1} \ker \sigma'_i \not\subseteq \ker \sigma_1 \cap \ker \sigma_j = \ker \sigma'_j$. By induction hypothesis, $\{b_2, \ldots, b_m\}$ is a basis for K_1. Since $b_1 \notin K_1$, it is easily checked that $\{b_1, \ldots, b_m\}$ is a basis of K, proving the Claim.

The claim implies that $\{\sigma_1, \ldots, \sigma_m\}$ is an $\mathbb{F}_2$-basis of $\chi(K)$. Observe first that $\{\sigma_1, \ldots, \sigma_m\}$ is $\mathbb{F}_2$-linearly independent : if $1 \leq j \leq m$, we have $\sigma_j(b_j) = -1$, while $\sigma_i(b_j) = 1$ for $1 \leq i < j$; hence $\sigma_j \notin [\sigma_1, \ldots, \sigma_{j-1}]$. Now, the well-known fact that the $\mathbb{F}_2$-dimension of K is equal to the $\mathbb{F}_2$-dimension of $\chi(K)$, shows that $\{\sigma_1, \ldots, \sigma_m\}$ is a basis for $\chi(K)$. □

Proof of Proposition 3.16. Let $U \neq \emptyset$ be a clopen in $\chi(G)$; we show that $[X] \cap U \neq \emptyset$. The set U is of the form (cf. Notation 1.1.(a))

$$U = \bigcap_{i=1}^{n} [a_i = \delta(i)],$$

for some $\{a_1, \ldots, a_n\} \subseteq G$, and $\{1, \ldots, n\} \xrightarrow{\delta} \{\pm 1\}$. Let $K = [a_1, \ldots, a_n]$; K is finite, and so is $\chi(K)$. Also, the finite set $X_{|K} = \{\sigma_{|K} : \sigma \in X\}$ separates points in K, i.e. $\bigcap_{\gamma \in X_{|K}} \ker \gamma = \{1\}$. Hence, there is a finite set $\{\sigma_1, \ldots, \sigma_n\} \subseteq X$ such that $\bigcap_{i=1}^{n} \ker (\sigma_{i|K}) = \{1\}$. Lemma 3.17 shows that $S = \{\sigma_{i|K} : i \leq n\}$ generates $\chi(K)$. Thus, if $\lambda \in U$, we have that $\lambda_{|K}$ is a linear combination of S, say $\lambda_{|K} = \prod_{j=1}^{r} \sigma_{i_j|K}$. Consequently, we have $\prod_{j=1}^{r} \sigma_{i_j} \in [X] \cap U$, as required. □

3.18. Construction of the functors Φ and Ψ.

(1) Let G, H be reduced groups and let $G \xrightarrow{f} H$ be a SG-morphism. The map $\Phi(f) : (X_H, H, -1) \longrightarrow (X_G, G, -1)$ is obtained by composition :

(*) $\qquad \Phi(f)(\sigma) = \sigma \circ f \qquad$ for $\sigma \in X_H$.

Clearly, $\sigma \circ f$ is a SG-morphism, hence it is in X_G. Also, $\Phi(f)$ is the restriction of the map $\chi(H) \longrightarrow \chi(G)$ given by (*), which is a continuous group homomorphism. Hence, $\Phi(f)$ is a morphism of AOS's.

(2) Extending Ψ to morphisms is a more delicate task which requires a simple case of Pontrjagin's duality Theorem, namely

Theorem (Pontrjagin [Pon]; Ch. 6, §36, 37)

*Let G be a group of exponent 2 and $\chi_c(\chi(G))$ be the group of **continuous** group characters of $\chi(G)$ into $\{\pm 1\}$. Let $ev : G \longrightarrow \chi_c(\chi(G))$ denote the **evaluation** map : for $g \in G$,*

$$ev(g) : \chi(G) \longrightarrow \{\pm 1\}, \qquad \sigma \mapsto \sigma(g).$$

Then, ev is a group isomorphism between G and $\chi_c(\chi(G))$. ◇

Let $\mu : (X_1, G_1, -1) \longrightarrow (X_2, G_2, -1)$ be a morphism of AOS's. By Proposition 3.16, μ is the restriction of a **unique** continuous homomorphism of $\chi(G_1)$ into $\chi(G_2)$, which we also denote by μ. By composition, μ induces a group homomorphism $\overline{\mu} : \chi_c(\chi(G_2)) \longrightarrow \chi_c(\chi(G_1))$:

$$\overline{\mu}(\gamma) = \gamma \circ \mu, \qquad \text{for } \gamma \in \chi_c(\chi(G_2)).$$

If $b \in G_2$, then $ev_2(b) \in \chi_c(\chi(G_2))$, and hence $\overline{\mu}(ev_2(b)) \in \chi_c(\chi(G_1))$. Since ev_1 is an isomorphism between G_1 and $\chi_c(\chi(G_1))$, there is a **unique** $a \in G_1$ such that

(+) $\quad ev_1(a) = ev_2(b) \circ \mu = \overline{\mu}(ev_2(b))$

$$\chi(G_1) \xrightarrow{\mu} \chi(G_2)$$

with $ev_1(a)$ and $ev_2(b)$ mapping down to $\{\pm 1\}$.

This is equivalent to :

[dual] $\qquad$ For all $\sigma \in \chi(G_1), \quad \sigma(a) = \mu(\sigma)(b)$.

We now define,
$$\Psi(\mu)(b) = a.$$

Thus, setting $\mu^* = \Psi(\mu)$, we have $\mu^* : G_2 \longrightarrow G_1$, while from [dual] comes

(**) $\qquad$ For every $\sigma \in \chi(G_1)$ and $b \in G_2, \quad \sigma(\mu^*(b)) = \mu(\sigma)(b)$.

Moreover, (+) and the fact that $ev_1(a)$ is an isomorphism yield

(++) $\qquad \mu^* = ev_1^{-1} \circ \overline{\mu} \circ ev_2.$ $\hfill \diamond$

Now we prove,

THEOREM 3.19. (Duality Theorem) *The correspondences Φ, Ψ are contravariant functors. Further, the compositions $\Phi \circ \Psi$ and $\Psi \circ \Phi$ are the identity functors, which shows that the pair (Φ, Ψ) establishes an isomorphism between the category* **RSG** *and* **AOS**^{op}, *the opposite category of* **AOS**.

PROOF. In view of Propositions 3.10, 3.11 and 3.14, only the assertions concerning morphisms require proof.

Straightforward checking shows that Φ, Ψ are contravariant functors (use the identity (++) of 3.18.(2) to check that Ψ reverses composition). The assertions of the statement are items (1), (2) and (3) below.

(1) $\quad$ The map $\mu^* = \Psi(\mu)$ is a SG-morphism of $(G_2, \equiv_{X_2}, -1)$ into $(G_1, \equiv_{X_1}, -1)$.

(i) μ^* is a group homomorphism.

Follows at once from 3.18 (++), since $ev_2, \overline{\mu}$ and ev_1^{-1} are group homomorphisms.

(ii) $\mu^*(-1) = -1$.

Let $a = \mu^*(-1)$; (**), with $b = -1$ yields, $\forall\, \sigma \in X_1$,
$$\sigma(a) = \mu(\sigma)(-1) = -1,$$
since $\mu(\sigma) \in X_2$ (axiom [O2]). But then, axiom [O3] guarantees that $a = -1$, since $\forall\, \sigma \in X_1$, $\sigma(-a) = 1$.

(iii) For $a, b, c, d \in G_2$,
$$\langle a, b\rangle \equiv_{X_2} \langle c, d\rangle \Rightarrow \langle \mu^*(a), \mu^*(b)\rangle \equiv_{X_1} \langle \mu^*(c), \mu^*(d)\rangle.$$

Let $\sigma \in X_1$; since $\mu(\sigma) \in X_2$, we have
$$\mu(\sigma)(a) + \mu(\sigma)(b) = \mu(\sigma)(c) + \mu(\sigma)(d).$$

From (**) we get: $\sigma(\mu^*(a)) + \sigma(\mu^*(b)) = \sigma(\mu^*(c)) + \sigma(\mu^*(d))$. Since this holds for arbitrary $\sigma \in X_1$, we have $\langle \mu^*(a), \mu^*(b)\rangle \equiv_{X_1} \langle \mu^*(c), \mu^*(a)\rangle$.

(2) $\Phi \circ \Psi(\mu) = \mu$, for any morphism $\mu : (X_1, G_1, -1) \longrightarrow (X_2, G_2, -1)$ of AOS's.

Writing $\mu' = \Phi(\mu^*)$, we have $\mu'(\sigma) = \sigma \circ \mu^*$, for $\sigma \in \chi(G_1)$ (see 3.18 (*)). The left-hand side of (**) gives $\sigma(\mu^*(a)) = \sigma \circ \mu^*(a) = \mu'(\sigma)(a)$; hence
$$\mu'(\sigma)(a) = \mu(\sigma)(a)$$
for arbitrary $a \in G_2$, $\sigma \in \chi(G_1)$. Fixing σ, this shows that $\mu'(\sigma) = \mu(\sigma)$, and hence $\mu' = \mu$.

(3) For every SG morphism $f : (G_2, \equiv_{G_2}, -1) \longrightarrow (G_1, \equiv_{G_1}, -1)$, $\Psi \circ \Phi(f) = f$.

Let $\Phi(f) = f'$ and $f^* = \Psi(f')$. From 3.18 (*) we have $f'(\sigma) = \sigma \circ f$, for $\sigma \in \chi(G_1)$. Computing the right-hand side of (**) for $\mu^* = f^*$ and $\mu = f'$, gives
$$\sigma(f^*(a)) = f'(\sigma)(a) = \sigma(f(a))$$
for arbitrary $a \in G_2$, $\sigma \in \chi(G_1)$. Fixing a, this shows that $f^*(a) = f(a)$, and so $f = f^*$. □

CHAPTER 4

Boolean Algebras and Reduced Special Groups

An abstract order space $(X, G, -1)$ is called a **SAP space** (SAP = Strong Approximation Property) if the canonical embedding of G into the multiplicative group $C(X) = C(X, \pm 1)$ of all continuous functions of X into $\{\pm 1\}$ (under pointwise operations, $\{\pm 1\}$ endowed with the discrete topology) given by evaluation at a point of G, is an isomorphism.

As is well known, we may view $C(X)$ as a Boolean algebra, under under pointwise operations and order $(1 < -1)$. This is just the Boolean algebra of clopens in X.

Setting these ideas into purely algebraic terms, in the first section of this chapter we endow every Boolean algebra with a suitable structure defined in terms of order and symmetric difference and prove that this is, indeed, a reduced special group.

It turns out that the space of orders of a Boolean algebra with this natural special group structure is just its Stone space. Thus, Boolean algebras are precisely the special group duals of the SAP spaces.

In the last section of this chapter we show how any reduced special group can be embedded in a canonically defined 'minimal' Boolean algebra, its **Boolean hull**. The correspondence that assigns to each rsg G its Boolean hull B_G is a functor. This observation reveals itself to be a powerful tool which shall be put to use in later chapters.

Among other things, the aforementioned functor is adjoint to the functor that assigns to each Boolean algebra its special group structure. Thus, the duality "reduced special groups/abstract order spaces" can be viewed as a composition of the adjunction we are presenting and classical Stone duality.

Convenient references on Boolean algebras are, for instance, [HBA] or [BD]. We use standard notation for Boolean algebras : $\leq$ is the order, $\vee$ denotes join, $\wedge$ denotes meet, and $-$ indicates complementation, while $\bot$, $\top$, stand, respectively, for the least and the largest element of a Boolean algebra.

We shall also use the standard notation for intervals. If $a \leq b$ in a Boolean algebra B, write $[a, b] = \{c \in B : a \leq c \leq b\}$ and $(a, b]$ for $\{c \in B : a < c \leq b\}$. The intervals (a, b) and $[a, b)$ are defined analogously.

The acronym BA will stand for (non-trivial) "Boolean algebra", i.e., $\bot \neq \top$, while the category of (non-trivial) Boolean algebras and Boolean algebra homomorphisms shall be denoted by **BA**.

1. Boolean Algebras as Reduced Special Groups

If B is a BA, define the operation of **symmetric difference** on B by

$$a \triangle b = (a \wedge -b) \vee (-a \wedge b) \qquad (a, b \in B)$$

It is well-known that $(B, \wedge, \triangle, \bot, \top)$ is a commutative ring with $0 = \bot$ and multiplicative identity $\top$. Moreover, $(B, \triangle, \bot)$ is a group of exponent 2.

A subgroup of a BA B is a subset of B containing $\top$ and closed under $\triangle$.

DEFINITION 4.1. For a BA, B, we define :

Product : Symmetric difference, $\triangle$;

Distinguished elements : $1 = \bot$ and $-1 = \top$.

Isometry : $\langle a, b \rangle \equiv_B \langle c, d \rangle$ iff $a \wedge b = c \wedge d$ and $a \triangle b = c \triangle d$.

Since, for all $a, b \in B$ we have

[dis] : $a \triangle b = a \vee b$ iff $a \wedge b = \bot$;

[sup] : $(a \triangle b) \vee (a \wedge b) = (a \triangle b) \triangle a \wedge b = a \vee b$,

it is straightforward to verify that $\langle a, b \rangle \equiv_B \langle c, d \rangle$ iff $a \wedge b = c \wedge d$ and $a \vee b = c \vee d$.

A BA, B, endowed with the structure defined above, will be denoted by $Sg(B) = (B, \equiv_B, -1)$. ◇

PROPOSITION 4.2. *If B is a BA, $Sg(B) = (B, \equiv_B, -1)$ is a reduced pre-special group.*

PROOF. The axioms [SG0], [SG1] and [SG3] (Definition 1.2) follow easily from the definition of isometry.

[SG2] : $\langle a, -a \rangle \equiv_B \langle 1, -1 \rangle$, since $a \wedge -a = \bot = \bot \wedge \top$ and $a \triangle -a = \bot \triangle \top = \top$.

[SG4] : $\langle a, b \rangle \equiv_B \langle c, d \rangle$ implies $\langle a, -c \rangle \equiv_B \langle -b, d \rangle$

From the definition of isometry we have

$(*)$ $\qquad a \wedge b = c \wedge d \qquad$ and $\qquad (**) \qquad a \triangle b = c \triangle d.$

$(**)$ immediately implies $a \triangle -c = -b \triangle d$.

On the other hand, we know that these equations imply that $a \vee b = c \vee d$. Thus,

$$(a \wedge -c) \wedge (b \vee -d) = (a \wedge -c \wedge b) \vee (a \wedge -c \wedge -d)$$
$$= \bot \vee (a \wedge -c \wedge -d) = a \wedge -c \wedge -d$$
$$\leq (a \vee b) \wedge -(c \vee d) = \bot$$

From this it follows that $a \wedge -c \leq -(b \vee -d) = -b \wedge d$. By symmetry, we may conclude that $a \wedge -c = -b \wedge d$, as required.

[SG5]: $\forall x \in B, \langle a, b \rangle \equiv_B \langle c, d \rangle$ implies $\langle x \triangle a, x \triangle b \rangle \equiv_B \langle x \triangle c, x \triangle d \rangle$. Since $x \triangle x = \bot$, we get $x \triangle a \triangle x \triangle b = a \triangle b = c \triangle d = x \triangle c \triangle x \triangle d$. Next, using distributivity of $\wedge$ over $\triangle$ together with (*) and (**), we get

$$(x \triangle a) \wedge (x \triangle b) = x \triangle (x \wedge b) \triangle (x \wedge a) \triangle (a \wedge b) =$$
$$= x \triangle (x \wedge (a \triangle b)) \triangle (a \wedge b) = x \triangle (x \wedge (c \triangle d)) \triangle (c \wedge d)$$
$$= (x \triangle c) \wedge (x \triangle d),$$

as required. It is immediate from (*) and $a \wedge a = a$, that the axiom of reduction is true in B. □

Our next task is the verification that, with the structure of pre-special group defined above, a Boolean algebra is a special group. We shall obtain this from the following

THEOREM 4.3. *Let G be a subgroup of a BA B, and $p, q, u, v \in G$. The following are equivalent:*

(1) $\langle p, q, p \triangle q \rangle \equiv_G \langle u, v, u \triangle v \rangle$.

(2) There is $\gamma \in G$ such that
$$p \vee \gamma = u \vee \gamma = p \vee q = u \vee v.$$

PROOF. To ease notation, in this proof meets will be written as products, $a \wedge b = ab$.

First observe that, since G is a subgroup of B, the following are equivalent:

(A) $\langle p, q, p \triangle q \rangle \equiv_G \langle u, v, u \triangle v \rangle$;

(B) There are $\alpha, \beta, \gamma \in G$ such that
$$\langle p, \alpha \rangle \equiv_B \langle u, \beta \rangle, \langle q, p \triangle q \rangle \equiv_B \langle \alpha, \gamma \rangle \text{ and } \langle v, u \triangle v \rangle \equiv_B \langle \beta, \gamma \rangle.$$

(C) There are $\alpha, \beta, \gamma \in G$ such that

(1a) $p\alpha = u\beta$ (1b) $p \triangle \alpha = u \triangle \beta$.

(2a) $q \triangle pq = \alpha\gamma$ (2b) $p = \alpha \triangle \gamma$.

(3a) $v \triangle uv = \beta\gamma$ (3b) $u = \beta \triangle \gamma$.

The equivalence of (A) and (B) comes from the definition of isometry of 3-forms, while (B) and (C) are equivalent because the isometry of 2-forms in G is that of B. Regarding equations (2a) and (3a), note that $p(p \triangle q) = p \triangle pq$, and similarly for u, v. So, it suffices to prove that (C) is equivalent to condition (2) of the statement.

(C) $\Rightarrow$ (2). First we verify that $u \vee v = p \vee q$. Formula [sup] in Definition 4.1 will be of constant use: $\quad a \triangle b \triangle ab = a \vee b. \quad (*)$

Taking $\triangle$ on both sides of (3a) and (3b) and using (*), yields $u \vee v = \beta \vee \gamma$. Similarly, from (2a) and (2b) comes $p \vee q = \alpha \vee \gamma$; so, we must show that $\alpha \vee \gamma = \beta \vee \gamma$. Taking meet with α on both sides of (2b) yields $p\alpha = \alpha \triangle \alpha\gamma$; a similar procedure in (3b) gives $u \beta = \beta \triangle \beta\gamma$. Thus, (1a) implies $\alpha \triangle \alpha\gamma = \beta \triangle \beta\gamma$. Taking $\triangle$ with γ on both sides of this last equality yields

$$\alpha \vee \gamma = \gamma \triangle \alpha \triangle \alpha\gamma = \gamma \triangle \beta \triangle \beta\gamma = \beta \vee \gamma,$$

as needed.

From (3b) comes $\beta = u \triangle \gamma$. Thus, using (3a) and (*) we get

$$u \vee v = u \triangle v \triangle uv = u \triangle \beta\gamma = u \triangle (u \triangle \gamma)\gamma = u \triangle u\gamma \triangle \gamma = u \vee \gamma.$$

On the other hand, (2b) yields $\alpha = p \triangle \gamma$; whence, from (2a) comes

$$p \vee q = p \triangle q \triangle pq = p \triangle \alpha\gamma = p \triangle (p \triangle \gamma)\gamma = p \vee \gamma.$$

It is now clear that we have all the equalities ascertained in (2).

(2) $\Rightarrow$ (C). Given γ satisfying the equations in (2), define $\alpha = p \triangle \gamma$ and $\beta = u \triangle \gamma$. Since G is closed under $\triangle$, we have α, β and γ in G. It is also clear that the equations (1b) - (3b) in (C) are all satisfied. It remains to show that the equations (1a), (2a), and (3a) are also verified.

For (1a) we have, using the equations in (2), formula (*), and the definitions of α and β :

$$p\alpha = p(p \triangle \gamma) = p \triangle p\gamma = (p \vee \gamma) \triangle \gamma = (u \vee \gamma) \triangle \gamma$$
$$= u \triangle u\gamma \triangle \gamma \triangle \gamma = u \triangle u\gamma = u(u \triangle \gamma) = u\beta.$$

For (2a), we have

$$\alpha\gamma = (p \triangle \gamma)\gamma = p\gamma \triangle \gamma = (p \vee \gamma) \triangle p = (p \vee q) \triangle p = q \triangle pq.$$

Finally, for (3a), a similar computation yields

$$\beta\gamma = (u \triangle \gamma)\gamma = (u \vee \gamma) \triangle u = (u \vee v) \triangle u = v \triangle uv,$$

ending the proof. $\square$

COROLLARY 4.4. *Let B be a Boolean algebra and p, q, u, v be elements in B. Then*

a) $\langle p, q, p \triangle q \rangle \equiv_B \langle u, v, u \triangle v \rangle$ *iff* $p \vee q = u \vee v$.

b) $Sg(B) = (B, \equiv_B, -1)$ *is a reduced special group.*

c) $p \in D_B(1, q)$ *iff* $p \leq q$ *(in B).*

1. BOOLEAN ALGEBRAS AS REDUCED SPECIAL GROUPS

PROOF. a) Theorem 4.3 proves the implication ($\Leftarrow$). Conversely, if $p \vee q = u \vee v$, just take $\gamma = p \vee q = u \vee v$, to satisfy all the equations in item (2) of Theorem 4.3.

b) First note that in **any** pre-special group G we have
$$\langle a,b,c \rangle \equiv_G \langle x,y,z \rangle \quad \text{iff} \quad \alpha \langle a,b,c \rangle \equiv_G \alpha \langle x,y,z \rangle,$$
where α is any element of G. This follows directly from the definition of isometry of 3-forms. Now, note that if two forms are isometric, then their discriminants are the same (Lemma 1.5.(b)). Thus, to prove that [SG6] holds in G it is sufficient to verify 3-transitivity for forms of discriminant 1, i.e., of type $\langle p, q, p \triangle q \rangle$, for some $p, q \in B$. This follows directly from (a).

c) If $p \in D_B(1,q)$, then $\langle p, p \triangle q \rangle \equiv_B \langle 1, q \rangle$. Thus, $p \wedge (p \triangle q) = 1 \wedge q = 1 (= \bot)$. It follows that $p \triangle (p \wedge q) = \bot$, which implies $p = p \wedge q$, i.e., $p \leq q$. Conversely, if $p \leq q$, then
$$q = p \triangle (p \triangle q) \quad \text{and} \quad p \wedge (p \triangle q) = 1 = q \wedge 1.$$
By the definition of $\equiv_B$, $\langle p, p \triangle q \rangle \equiv_B \langle 1, q \rangle$, whence, $p \in D_B(1,q)$. $\square$

Remark. The equivalence in (a) of the preceding Corollary is just a special instance of the Horn-Tarski conditions, to be studied in Chapter 7. $\diamond$

We now show that the morphisms of special groups between BA's are precisely the BA-morphisms.

PROPOSITION 4.5. *Let A, B be Boolean algebras and $f : |A| \longrightarrow |B|$ a map between their underlying sets. The following are equivalent :*

1. *f is a SG-morphism from $Sg(A)$ to $Sg(B)$.*

2. *f is a morphism of BA's.*

PROOF. (1) $\Rightarrow$ (2). If f is a SG-morphism, then f preserves $\triangle$ and sends $\top (= -1)$ to $\top$. Therefore, $f(-a) = f(\top \triangle a) = f(\top) \triangle f(a) = -f(a)$, showing that f preserves complements. To verify that f is a BA-morphism, it is now enough to check that it preserves inf's or sup's.

We first observe that f must be increasing. If $x \leq y$ in A, then (Corollary 4.4.(c)), $x \in D_A(1,y)$ and so (f is a SG-morphism), $f(x) \in D_B(1, f(y))$, whence, $f(x) \leq f(y)$ in B.

Now, for a, b in A, the isometry $\langle a, b \rangle \equiv_A \langle a \vee b, a \wedge b \rangle$ is verified in A; since f is a SG-morphism, we have $\langle f(a), f(b) \rangle \equiv_B \langle f(a \vee b), f(a \wedge b) \rangle$. Because f is increasing, we have $f(a \vee b) \geq f(a \wedge b)$, and so, from the definition of the isometry in B, we get

(i) $f(a) \vee f(b) = f(a \vee b) \vee f(a \wedge b) = f(a \vee b)$, and

(ii) $f(a) \wedge f(b) = f(a \vee b) \wedge f(a \wedge b) = f(a \wedge b)$,

as needed.

(2) ⇒ (1). If f is a BA-morphism, then it sends $\top$ to $\top$ and preserves $\triangle$ (it is defined using $\wedge$, $\vee$ and complements). By Lemma 1.12 we need only show that for all $a \in A$, $f[D_A(1,a)] \subseteq D_B(1, f(a))$. But, by Corollary 4.4, this amounts to f being increasing, which is the case for all BA-morphisms. Thus, f is a SG-morphism. □

COROLLARY 4.6. *The correspondence which assigns to each BA, B, its special group structure $Sg(B)$, and to every BA-morphism $A \xrightarrow{f} B$ the same mapping $Sg(A) \xrightarrow{f} Sg(B)$ (viewed as a SG-morphism), is a covariant functor $Sg :$ **BA** $\longrightarrow$ **RSG**, from the category of Boolean algebras to the category of reduced special groups.* ◇

Remark. In view of this, we shall not distinguish between the morphisms of BA's and the SG-morphisms of their underlying algebras. ◇

PROPOSITION 4.7. *Let B be a BA, and $\Delta \subseteq B$. Then,*

a) Δ is a saturated subgroup of $Sg(B)$ iff Δ is an ideal in B.

b) If Δ is a saturated subgroup of $Sg(B)$, then $Sg(B)/\Delta$ is naturally isomorphic (as a BA and as a special group) to $Sg(B/\Delta)$.

c) Any reduced SG-homomorphic image of B is (the special group of) a BA.

PROOF. If a is an element of B, write $a^{\leftarrow} = \{b \in B : b \leq a\}$ for the ideal generated by a in B. By Corollary 4.4.(c), a subgroup Δ of $Sg(B)$ is saturated iff it satisfies

(*) $\qquad \forall\, a \in B, \quad a \in \Delta \Rightarrow a^{\leftarrow} \subseteq \Delta.$

a) We must show that a subgroup of B satisfies (*) iff it is an ideal.

It is clear that an ideal Δ satisfies (*). If $a, b \in \Delta$, then $a \triangle b \leq a \vee b \in \Delta$, and so $a \triangle b \in \Delta$. Thus, Δ is a subgroup satisfying (*).

If Δ is a subgroup satisfying (*), we must show that $a, b \in \Delta$, implies $a \vee b \in \Delta$. By (*), $a \wedge b \in \Delta$, and so $a \vee b = (a \triangle b) \triangle (a \wedge b)$ is in Δ.

b) We start with

Fact. If $a, b \in B$, then the following are equivalent:

(1) $\exists\, c \in \Delta$, $a \vee c = b \vee c$ $\qquad$ (2) $a \triangle b \in \Delta$.

Proof. Suppose $c \in \Delta$ is as in (1). Then

(I) $\qquad a \vee c = a \triangle c \triangle (a \wedge c) = b \triangle c \triangle (b \wedge c) = b \vee c.$

It follows from (I) that $a \triangle b = c \wedge (a \triangle b)$, i.e. $a \triangle b \leq c$. Since Δ is an ideal, we conclude (2).

Conversely, suppose that $a \triangle b \in \Delta$. Note that $a \vee (a \triangle b) = b \vee (a \triangle b) = a \vee b$, to conclude that (1) holds. ◇

By the Fact, the equivalence relation induced by Δ is the same whether Δ is regarded as an ideal or as a saturated subgroup of B. Hence, the map $Sg(B)/\Delta \longrightarrow Sg(B/\Delta)$, $a/\Delta \mapsto a/\Delta$, is well defined and bijective group homomorphism. Let $\equiv$ denote the isometry in $Sg(B)/\Delta$ and $\approx$ the isometry in $Sg(B/\Delta)$. To prove that this map is an SG-isomorphism we must show that for all a_1, a_2, a_3, a_4 in $Sg(B)/\Delta$

$$\langle a_1/\Delta, a_2/\Delta \rangle \equiv \langle a_3/\Delta, a_4/\Delta \rangle \quad \text{iff} \quad \langle a_1/\Delta, a_2/\Delta \rangle \approx \langle a_3/\Delta, a_4/\Delta \rangle$$

So, suppose $\langle a_1/\Delta, a_2/\Delta \rangle \equiv \langle a_3/\Delta, a_4/\Delta \rangle$ holds in $Sg(B)/\Delta$; this means (Definition 2.13) that there are $b_i \in Sg(B)$ such that $\langle b_1, b_2 \rangle \equiv_B \langle b_3, b_4 \rangle$ and $a_i \triangle b_i \in \Delta$ ($1 \leq i \leq 4$). Since $Sg(B)$ is the special group of a Boolean algebra, we have

(II) $\qquad b_1 \vee b_2 = b_3 \vee b_4 \quad \text{and} \quad b_1 \wedge b_2 = b_3 \wedge b_4.$

Since the quotient by an ideal is a congruence, (II) yields

(III) $\quad b_1/\Delta \vee b_2/\Delta = b_3/\Delta \vee b_4/\Delta \quad \text{and} \quad b_1/\Delta \wedge b_2/\Delta = b_3/\Delta \wedge b_4/\Delta.$

Since $b_i/\Delta = a_i/\Delta$, (III) yields $\langle a_1/\Delta, a_2/\Delta \rangle \approx \langle a_3/\Delta, a_4/\Delta \rangle$.

Now suppose that $\langle a_1/\Delta, a_2/\Delta \rangle \approx \langle a_3/\Delta, a_4/\Delta \rangle$ in $Sg(B/\Delta)$. From the definition of isometry in a BA, taking into account the congruence generated by the ideal Δ, we get elements $y, z \in \Delta$, such that

$$a_1 \vee a_2 \vee y = a_3 \vee a_4 \vee y \quad \text{and} \quad (a_1 \wedge a_2) \vee z = (a_3 \wedge a_4) \vee z.$$

By taking $x = y \vee z$ we may write the above equations only in x. It is easily checked that $\langle a_1 \triangle x, a_2 \triangle x \rangle \equiv_B \langle a_3 \triangle x, a_4 \triangle x \rangle$, with $(a_i \triangle x)/\Delta = a_i/\Delta$, ($1 \leq i \leq 4$), by the Fact above. But this says that $\langle a_1/\Delta, a_2/\Delta \rangle \equiv \langle a_3/\Delta, a_4/\Delta \rangle$, completing the proof of 4.7.(b).

Item (c) is a direct consequence of (b) and Proposition 2.21. $\square$

We now determine the space of orders of a BA, construed as a special group.

PROPOSITION 4.8. *Let B be a BA. Then,*

a) The space of orders of $Sg(B)$, $X_{Sg(B)}$, is naturally homeomorphic to the Stone space of B, $S(B)$.

b) The AOS dual to $Sg(B)$ is (naturally isomorphic to) $(S(B), B, \top)$.

PROOF. a) As is well-known, there is a natural bijective correspondence

$$\{\sigma : \sigma : B \longrightarrow 2 \text{ a BA-morphism}\} \longrightarrow S(B) = \{\text{ultrafilters of } B\},$$

given by $\sigma \mapsto \{b \in B : \sigma(b) = \top\}$.

This correspondence is a homeomorphism, because the sub-basic clopen $[a = -1] = \{\sigma : \sigma(a) = \top\}$ is mapped bijectively onto the sub-basic clopen $S_a = \{F \in S(B) : a \in F\}$ ($a \in B$).

Upon noting that $Sg(2) = \mathbb{Z}_2$, Proposition 4.5 gives a natural bijective correspondence between the SG-morphisms from $Sg(B)$ into $\mathbb{Z}_2$, and the BA-morphisms from B to 2. Moreover, this correspondence takes the sub-basic clopen $[a = -1]$ (in $X_{Sg(B)}$) bijectively onto the sub-basic clopen $[a = -1]$ (in $S(B)$). Thus, these spaces are naturally homeomorphic.

Assertion (b) is a direct consequence of (a) and Theorem 3.19. □

We end this section with some comments on classical Stone duality which will be useful later. Notation and basic definitions can be found in [BD].

Let **BA** be the category of BA's and **BS** be the category of Boolean spaces (Hausdorff, compact, totally disconnected). Stone's construction establishes a duality (contravariant equivalence) $\mathbf{BA} \underset{S}{\overset{B}{\longleftrightarrows}} \mathbf{BS}$ as follows:

The functor S: If B is a BA, $S(B) = \{\mathcal{F} : \mathcal{F} \text{ is a proper ultrafilter in } B\}$ is its Stone space (with the Stone topology). If $A \xrightarrow{f} B$ is a BA-morphism, $S(f) : S(B) \longrightarrow S(A)$ is the continuous map defined by $S(f)(\mathcal{F}) = f^{-1}[\mathcal{F}]$.

The functor B: If X is a Boolean space, $B(X)$ is the BA of clopen subsets of X. If $Y \xrightarrow{g} X$ is a continuous map, $B(g) : B(X) \longrightarrow B(Y)$ is the BA-morphism defined by $B(g)(u) = g^{-1}[u]$.

Dual categorical notions correspond to each other in **BA** and **BS**. For instance, f is epic (resp., monic) in **BA** iff $B(f)$ is monic (resp., epic) in **BS**. Recall that an arrow is monic (epic) in a category if it is left (resp., right) cancellable. The following result is folklore, but we include it for lack of a convenient reference (in [HBA], epic is <u>defined</u> as onto, Definition 2.5, p. 29).

PROPOSITION 4.9. *Let* $A \xrightarrow{f} B$ *be a BA-morphism and* $X \xrightarrow{g} Y$ *be a continuous map of Boolean spaces. Then,*

a) f *and* g *are monics in* **BA** *and* **BS**, *respectively, iff they are injective.*

b) f *and* g *are epics in* **BA** *and* **BS**, *respectively, iff they are surjective.*

PROOF. a) is quite straightforward. It is clearly enough to prove that monic implies injective. For g, given $x \neq y$ in X consider the continuous maps ι_x, ι_y from the one point space to X, with values x and y, respectively. If g is monic, then $g \circ \iota_x \neq g \circ \iota_y$, that is, $g(x) \neq g(y)$, proving that g is injective.

For f, given $x \neq y$ in A, consider the BA-morphisms ι_x, ι_y defined in the four-element BA $\{1, a, -a, -1\}$ which send a to x and y, respectively. The same argument as above shows that f is injective.

b) The argument here is more sophisticated. Again, we need only verify that epic implies surjective.

(i) Assume that g is epic and that there is $y \in Y \setminus g[X]$. Let $Z = g[X]$. Since compact spaces are normal, Z is a closed subset of Y, and $y \notin Z$, Urysohn's Theorem (see [E]) applies to yield a continuous map h from Y into the real unit interval $[0, 1]$ such that $h_{|Z} = 0$, while $h(y) = 1$. Let $k : Y \longrightarrow [0,1]$ be the function constantly equal to 0. It is clear that $h \circ g = k \circ g$, even though $h \neq k$, contradicting the epic nature of g. Notice that this proof works for the category of compact Hausdorff spaces.

(ii) For the case of f we start with a BA version of the 'going up' theorem. Ultrafilters are always assumed to be proper.

Fact. Let $C \subseteq B$ be BA's. For every $b \in B \setminus C$ there is an ultrafilter $\mathcal{F}$ in C and ultrafilters $\mathcal{F}_1, \mathcal{F}_2$ in B such that

(i) $\mathcal{F}_i \cap C = \mathcal{F}$; (ii) $b \in \mathcal{F}_1$ and $-b \in \mathcal{F}_2$.

Proof. Recall that $b^{\leftarrow} = \{x \in B : x \leq b\}$, $b^{\rightarrow} = \{x \in B : b \leq x\}$ are the principal ideal and filter, respectively, generated by b in B. Since $b \notin C$, the ideal $I = b^{\leftarrow} \cap C$ and the filter $F = b^{\rightarrow} \cap C$ are disjoint in C. By Stone's Separation Theorem, there is a ultrafilter $\mathcal{F}$ in C such that $F \subseteq \mathcal{F}$ and $\mathcal{F} \cap I = \emptyset$.

We now verify that both $\mathcal{F} \cup \{b\}$ and $\mathcal{F} \cup \{-b\}$ have the finite intersection property (fip). If $\mathcal{F} \cup \{b\}$ does not have the fip, then for some $a \in \mathcal{F}$, we have $a \wedge b = \bot$. This means that $b \leq -a$, i.e. $-a \in F \subseteq \mathcal{F}$, which is impossible because $a \in \mathcal{F}$ and this is a proper filter. Similarly, if $\mathcal{F} \cup \{-b\}$ did not have the fip, then for some $a \in \mathcal{F}$, we would have $a \leq b$, untenable, because $a \in I$ and $\mathcal{F}$ is disjoint from this ideal, by construction.

To finish the proof, choose ultrafilters $\mathcal{F}_1$ containing $\mathcal{F} \cup \{b\}$, and $\mathcal{F}_2$ containing $\mathcal{F} \cup \{-b\}$. Since $\mathcal{F}$ is maximal in C, it is clear that $\mathcal{F}_i \cap C = \mathcal{F}$, ending the proof of the Fact. ◇

Now, suppose that f is epic and that there is $b \in B \setminus f[A]$. Let $\mathcal{F}, \mathcal{F}_i$, $i = 1, 2$ be the ultrafilters in $f[A]$ and B constructed in the Fact and let

$$B \xrightarrow[h_1]{h_2} \mathbb{Z}_2$$

be the quotient maps induced by $\mathcal{F}_1, \mathcal{F}_2$, respectively. For $a \in A$,

$h_1(f(a)) = -1$ iff $f(a) \in \mathcal{F}_1 \cap f[A]$ iff $f(a) \in \mathcal{F}_2 \cap f[A]$
        iff $h_2(f(a)) = -1$,

which shows that $h_1 \circ f = h_2 \circ f$. But this contradicts the assumption that f is epic, since $h_1(b) = -1$, while $h_2(b) = 1$, that is, $h_1 \neq h_2$. □

2. The Boolean Hull of a Reduced Special Group

With notation as in Chapter 3, let B_G be the BA of clopens in X_G. Define a map ε_G by

$$\varepsilon_G : G \longrightarrow B_G, \quad \text{where} \quad \varepsilon_G(a) = [a = -1],\, a \in G.$$

PROPOSITION 4.10. *Let $(G, \equiv_G, -1)$ be a rsg. Then*

a) ε_G is an injective group homomorphism from $(G, \cdot, 1, -1)$ into $(B_G, \triangle, \emptyset, X_G)$, where $\triangle$ denotes symmetric difference in B_G.

b) If u is an element in B_G, then there is a family $\{F_i : 1 \leq i \leq n\}$ of finite subsets of G, such that $u = \bigcup_{i \leq n} \bigcap_{a \in F_i} \varepsilon_G(a)$.

PROOF. a) It is clear that $\varepsilon_G(1) = \emptyset$ and that $\varepsilon_G(-1) = X_G$. Elementary computations will show that $[ab = -1] = [a = -1] \triangle [b = -1]$, and so ε_G is a group homomorphism. On the other hand, $[a = -1] = [b = -1]$ implies $[ab = -1] = \emptyset$, and so $[ab = 1] = X_G$. Since X_G separates points in G (Fact 3.4.(c)), we have $ab = 1$, that is, $a = b$. Thus, ε_G is injective.

b) Recall that the collection $\mathcal{B}$ given in $[\star]$ of Notation 1.1.(a) is a basis for the topology in X_G. Thus, every $u \in B_G$ can be written as a finite union of finite intersections of elements of type $[a = 1]$ and $[b = -1]$. Since $[a = 1] = [-a = -1]$, u can written as a finite union of finite intersections of sets of the form $\varepsilon_G(a)$, $a \in G$. $\square$

PROPOSITION 4.11. *Let G be a rsg, $\varphi = \bigotimes_{i \leq n} \langle 1, a_i \rangle$ be a Pfister form on G, and $a \in G$. Let $\Delta = D_G(\varphi)$. Then*

a) $\{\sigma \in X_G : \text{sgn}_\sigma(\varphi) = 2^n\} = \{\sigma \in X_G : \Delta \subseteq \ker \sigma\} = \bigcap_{i=1}^n [a_i = 1]$.

b) $a \in D_G(\varphi)$ iff $\varepsilon_G(a) \subseteq \bigcup_{i \leq n} \varepsilon_G(a_i)$.

PROOF. a) Because φ is a Pfister form, the values of $\text{sgn}_\sigma(\varphi) = \prod_{i=1}^n (1 + \sigma(a_i))$ are either 0 or 2^n, for any $\sigma \in X_G$. Clearly, $\text{sgn}_\sigma(\varphi) = 2^n$ iff $\sigma \in \bigcap_{i=1}^n [a_i = 1]$. This shows that the first and the last terms of (a) are equal.

The equality of the second and the third terms in (a) follows from Corollary 2.8 : if $\Delta \subseteq \ker \sigma$, then, since $D_G(1, a_i) \subseteq \Delta$, we have $\sigma \in \bigcap_{i=1}^n [a_i = 1]$.

Conversely, if $\sigma \in [a_i = 1]$ for all $i \leq n$, since $\ker \sigma$ is saturated (Corollary 2.29.(2)), Corollary 2.8 implies that $D_G(\varphi) = \Delta \subseteq \ker \sigma$.

b) By taking complements, the desired conclusion is equivalent to

$$a \in \Delta \quad \text{iff} \quad \bigcap_{i=1}^n [a_i = 1] \subseteq [a = 1].$$

If $a \in \Delta$, then $\{\sigma : \Delta \subseteq \ker \sigma\} \subseteq [a = 1]$, and the conclusion follows from (a). Conversely, assume that $\bigcap_{i=1}^n [a_i = 1] \subseteq [a = 1]$. If $a \notin \Delta$, then Theorem 2.11 and (a) yield

2. THE BOOLEAN HULL OF A REDUCED SPECIAL GROUP 69

$$\emptyset \neq \{\sigma : \Delta \subseteq ker\sigma\} \cap [a = -1] = \bigcap_{i=1}^{n} [a_i = 1] \cap [a = -1],$$

a contradiction. □

The following are some interesting consequences of Proposition 4.11.

COROLLARY 4.12. *The map ε_G is an injective SG-morphism from G into $Sg(B_G)$. In fact, for all $a, b \in G$,*

$$b \in D_G(1, a) \quad \text{iff} \quad \varepsilon_G(b) \subseteq \varepsilon_G(a).$$

PROOF. Corollary 4.4.(c) and Proposition 4.11 applied to the Pfister form $\langle 1, a \rangle$ yield,

$$b \in D_G(1, a) \ \text{iff} \ \varepsilon_G(b) \subseteq \varepsilon_G(a) \ \text{iff} \ \varepsilon_G(b) \in D_{B_G}(1, \varepsilon_G(a)),$$

and the conclusion follows from Lemma 1.12.(a). □

COROLLARY 4.13. *Let $(G, \equiv, -1)$ be a rsg, and T be a subgroup of G. $\overline{T}$ denotes the saturation of T (Definition 2.7). Then,*

a) $\overline{T} = \{y \in G : \exists \text{ finite subset } F \subseteq T \text{ such that } \varepsilon_G(y) \subseteq \bigcup_{x \in F} \varepsilon_G(x)\}$.

b) $\overline{T}$ is proper subgroup of G iff $\varepsilon_G(T) = \{\varepsilon_G(x) : x \in T\}$ generates a proper ideal in B_G.

PROOF. a) It is clear from Propositions 2.6 and 4.11.(b) that y is in the right-hand side of (a) iff it is represented by a Pfister form over T iff $y \in \overline{T}$.

b) It follows from (a) that $\overline{T} = G$ iff $-1 \in \overline{T}$ iff $\varepsilon_G(-1) = X_G$ is a finite union of elements of $\varepsilon_G(T)$. But $\varepsilon_G(T)$ generates a proper ideal in B iff it $\top$ is not the union of a finite subset of T. Consequently, $\overline{T}$ is proper iff $\varepsilon_G(T)$ generates a proper ideal in B. □

NOTATION 4.14. (a) If G is a rsg, $\Sigma(G)$ denotes the set, partially ordered by inclusion, of proper saturated subgroups of G. If B is a BA, $\mathcal{I}(B)$ denotes the set, partially ordered by inclusion, of proper ideals in B.

(b) Given a saturated subgroup Δ of a rsg G, let

$$\mathcal{I}(\Delta) = \{u \in B_G : \exists \text{ a finite subset } F \subseteq \Delta \text{ such that } u \leq \bigcup_{g \in F} \varepsilon_G(g)\},$$

denote the ideal generated by Δ in B_G. It is clear that $\Delta \subseteq \Gamma$ implies $\mathcal{I}(\Delta) \subseteq \mathcal{I}(\Gamma)$.

(c) If I is an ideal in B_G, let

$$\Sigma(I) = \{g \in G : \varepsilon_G(g) \in I\}.$$

It follows from Corollary 4.13 that $\Sigma(I)$ is a saturated subgroup of G. In fact, if we identify G with its image in B_G, $\Sigma(I)$ is simply $I \cap G$. Clearly, $I \subseteq J$ implies $\Sigma(I) \subseteq \Sigma(J)$.

Corollary 4.13.(b) implies that, in fact, we have increasing maps
$$\Sigma : \mathcal{I}(B_G) \longrightarrow \Sigma(G) \quad \text{and} \quad \mathcal{I} : \Sigma(G) \longrightarrow \mathcal{I}(B_G).$$
$\diamond$

The main properties of these maps are given by

PROPOSITION 4.15. *Let G be a reduced special group and B_G be its associated BA. With notation as above, we have*

a) $\Sigma \circ \mathcal{I} = id_{\Sigma(G)}$.

b) Σ *and* $\mathcal{I}$ *are inverse bijective correspondences between the maximal saturated subgroups of G and the maximal ideals in B_G.*

PROOF. a) Let $\Delta \in \Sigma(G)$; it is clear that $\Delta \subseteq \Sigma(\mathcal{I}(\Delta))$. Thus, it remains to be proved that if $g \in G$ is such that $\varepsilon_G(g) \in \mathcal{I}(\Delta)$, then g is already in Δ.

If $\varepsilon_G(g) \in \mathcal{I}(\Delta)$, then, by definition, there is a finite subset $F \subseteq \Delta$ such that $\varepsilon_G(g) \subseteq \bigcup_{a \in F} \varepsilon_G(a)$. By Proposition 4.11.(b), this means that $g \in D_G(\varphi)$, where φ is the Pfister form $\bigotimes_{a \in F} \langle 1, a \rangle$. Since Δ is saturated, Proposition 2.6 yields $g \in \Delta$, as needed.

b) Since both $\mathcal{I}$, Σ are increasing, by (a) it is sufficient to verify that $[\mathcal{I} \circ \Sigma](I) = I$, for every maximal ideal I in B_G.

Suppose that I is a proper maximal ideal in B_G and let g be an element in G. Note that, since I is maximal, we have $\varepsilon_G(g) \in I$ or $\varepsilon_G(-g) = -\varepsilon_G(g) \in I$. But this means that the proper saturated subgroup $\Sigma(I)$ satisfies $g \in \Sigma(I)$ or $-g \in \Sigma(I)$, for all $g \in G$. By Proposition 2.9, $\Sigma(I)$ is a maximal saturated subgroup of G.

Since $\mathcal{I}(\Sigma(I)) \subseteq I$, to prove equality it suffices to verify that $\mathcal{I}(\Sigma(I))$ is maximal. We show that for all $u \in B_G$, $u \in \mathcal{I}(\Sigma(I))$ or $-u \in \mathcal{I}(\Sigma(I))$.

Set $\Delta = \Sigma(I)$ and let u be an element of B_G. By Proposition 4.10.(b), there are finite sets $F_i \subseteq G$, $1 \leq i \leq n$, such that

$$(*) \qquad\qquad u = \bigcup_{i=1}^{n} \bigcap_{a \in F_i} \varepsilon_G(a).$$

Two cases must be discussed :

(i) For all $i \leq n$, there is $a_i \in F_i \cap \Delta$. In this case, we have $u \leq \bigcup_{i=1}^{n} \varepsilon_G(a_i)$, which shows that $u \in \mathcal{I}(\Delta)$.

(ii) There is $i \leq n$, such that $F_i \cap \Delta = \emptyset$. In this situation, since Δ is maximal saturated, we must have $-F_i = \{-a : a \in F_i\} \subseteq \Delta$. Taking complements in (*) yields $-u \leq \bigcup_{a \in F_i} \varepsilon_G(-a)$, that is, $-u \in \mathcal{I}(\Delta)$, ending the proof. $\square$

Remark. For ideals which are not maximal, it is not in general true that $\mathcal{I} \circ \Sigma = id_{\mathcal{I}(B_G)}$. An example are the fans of rank ≥ 3. It will be seen in the next section (Proposition 5.19.(c)), that the Boolean hull of a fan of rank

$\kappa \geq 2$ is the free BA on κ generators. Therefore, if p, q are distinct elements of a fan F, neither of which are equal to 1 or -1, the ideal I generated by $p \wedge q$ in B_F, when intersected with F, is equal to $\{1\}$. Consequently, $\Sigma(I) = \{1\}$ and so $\mathcal{I}(\Sigma(I)) = \{1\} \neq I$.

In Proposition 7.28 it is shown that $\mathcal{I}$ and Σ are inverse bijective correspondences between $\Sigma(G)$ and $\mathcal{I}(B_G)$ iff G is a Boolean algebra, i.e., $G = B_G$. ◇

We now establish the existence of a functor from the category of **RSG** to **BA**, in fact right adjoint to the functor $Sg : \mathbf{BA} \longrightarrow \mathbf{RSG}$ defined in Corollary 4.6.

DEFINITION 4.16. Let G and H be reduced special groups and let $G \xrightarrow{f} H$ be a SG-morphism. Let $f^* : X_H \longrightarrow X_G$ be the continuous map dual to f given by the Duality Theorem 3.19. We define $B(f)$ to be the Stone dual of f^*, that is the BA-morphism $B(f) : B_G \longrightarrow B_H$ given by

$$B(f)(u) = f^{*-1}[u], \quad u \in B_G.$$ ◇

THEOREM 4.17. *Let G and H be reduced special groups. With notation as above:*

1. *The correspondence B defined by*

$$G \mapsto B_G \qquad (G \xrightarrow{f} H) \mapsto (B_G \xrightarrow{B(f)} B_H),$$

is a functor from the category of reduced special groups (with SG-morphisms) to the category of Boolean algebras (with BA-morphisms).

2. *For all SG-morphisms $G \xrightarrow{f} H$, we have*

[BH] $$\varepsilon_H \circ f = B(f) \circ \varepsilon_G,$$

that is, the following diagram is commutative:

[BH]

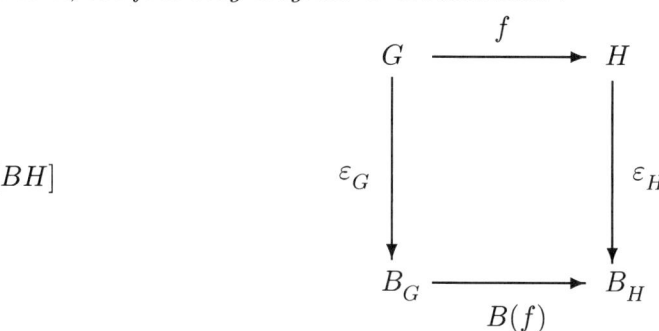

3. *(Uniqueness) Given a SG-morphism $G \xrightarrow{f} H$ and a BA-morphism $F : B_G \longrightarrow B_H$ such that the diagram*

$(**)$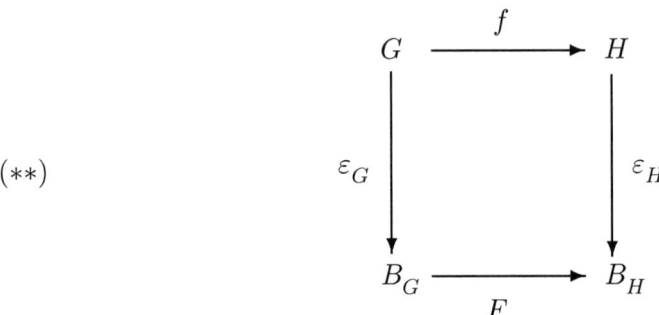

commutes, then $F = B(f)$.

*4. The pair (B_G, ε_G) is a **hull** for G in the category of BA's : given a BA, B, any SG-morphism $G \xrightarrow{f} Sg(B)$ factors through ε_G, i.e., the following diagram of special groups is commutative :*

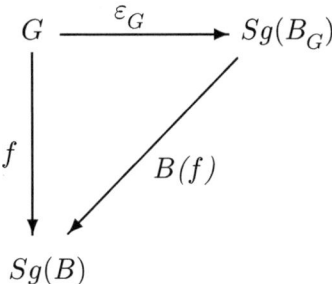

(modulo the identification of B with the BA of clopens in $S(B)$, via the canonical map).

*5. The functor in (1) is right adjoint to the (forgetful) functor Sg from **BA** to **RSG**.*

Remark. In view of item (4), the BA B_G will in the sequel be referred to as the **Boolean hull** of the rsg G. ◇

PROOF. 1) The definition of B on SG-morphisms (Definition 4.16) gives at once $B(id_G) = id_{B_G}$ and $B(f \circ g) = B(f) \circ B(g)$, which shows (1).

2) Let $g \in G$. We have :

$$B(f)(\varepsilon_G(g)) = f^{*-1}[\varepsilon_G(g)] = \{\sigma \in X_G : f^*(\sigma)(g) = (\sigma \circ f)(g) = -1\}$$
$$= \{\sigma \in X_H : \sigma(f(g)) = -1\} = \varepsilon_H(f(g)),$$

proving [BH].

3) Commutativity of the diagrams [BH] (in (2)) and $(**)$ (in (3)) of the statement shows that

$$F \circ \varepsilon_G = \varepsilon_H \circ f = B(f) \circ \varepsilon_G,$$

and hence that $F_{|Im(\varepsilon_G)} = B(f)_{|Im(\varepsilon_G)}$. Since $Im(\varepsilon_G)$ generates B_G (Proposition 4.10) and F and $B(f)$ are BA-homomorphisms, it follows that $F = B(f)$. Indeed, if $u = \bigcup_i \bigcap_j \varepsilon_G(g_{ij})$, then

$$F(u) = \bigcup_i \bigcap_j F(\varepsilon_G(g_{ij})) = \bigcup_i \bigcap_j B(f)(\varepsilon_G(g_{ij})) = B(f)(u).$$

4) The commutative diagram in (2) gives

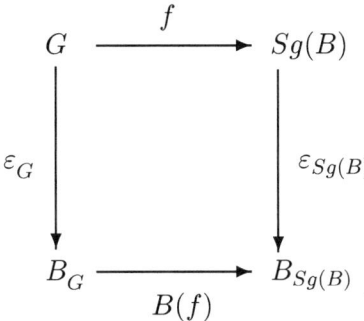

It suffices to show

FACT 4.18. If B is a BA, then $B_{Sg(B)} = B$ and $\varepsilon_{Sg(B)} = id_B$.

Proof of Fact 4.18. We use Proposition 4.8 and its proof. Since $X_{Sg(B)} = S(B)$, then $B_{Sg(B)} = B(X_{Sg(B)})$ is just the Boolean algebra of clopens of $S(B)$, and Stone duality gives a natural isomorphism between B and $B(S(B))$, namely $b \mapsto \{\sigma : \sigma(b) = \top\}$, $b \in B$. Thus, without loss of generality we may **identify** B and $B(S(B))$. Hence, $B = B_{Sg(B)}$.

For $b \in B$,

$$\varepsilon_{Sg(B)}(b) = [b =\!\!-1] = \{\sigma \in S(B) : \sigma(b) = \top\},$$

and so, after the identification above, $\varepsilon_{Sg(B)}(b) = b$. ◇

5) The commutativity of the diagram in (4) entails the adjointness of B and Sg, because for every pair of objects (G, B), where G is a reduced special group and B is a BA, it establishes a natural bijective correspondence between $Hom_{SG}(G, Sg(B))$ and $Hom_{BA}(B_G, B)$; see [McL], Chapter IV, p. 77 ff. □

CHAPTER 5

Embeddings

In this chapter we carry out a detailed study of embeddings of special groups, mostly in the reduced case. We consider four types of morphisms:

5.1. Complete embeddings.

5.2. Morphisms having a retract.

5.3. Pure embeddings.

5.4. Isotropy-reflecting morphisms.

The first, and largest, of these classes consists of those SG-homomorphisms which reflect isometry of forms of arbitrary dimension. Our main result, Theorem 5.2, shows that they are exactly the SG-morphisms $G \xrightarrow{f} H$ whose associated Boolean morphism $B_G \xrightarrow{B(f)} B_H$ (defined in Theorem 4.17) is injective. Most of the results in section 5.1 follow from this characterization. Amongst them stand out the facts that the canonical embedding ε_G of a rsg G into its Boolean hull B_G is complete, and that every SG-character of G extends uniquely to B_G. In 5.10 we give an example of a non-complete inclusion of rsg's, which we use to obtain information concerning the categories of rsg's and abstract spaces of orders.

The classes of morphisms examined in the remaining sections form an increasing sequence, all three included in that of complete embeddings.

Some of the standard constructions of special groups yield morphisms which have a retract.

Pure embeddings are morphisms which reflect positive-existential sentences of the natural first-order language L_{SG} for special groups (this language will be examined in Chapter 11). We have borrowed the name from established terminology in the theory of modules (see, e.g., [Ho], p. 56). Pure embeddings are rather natural objects and some examples arise in the context of fields.

In the last section we exhibit some natural examples of isotropy-reflecting SG-morphisms arising from field theory and from abstract order spaces.

1. Complete Embeddings

DEFINITION 5.1. Let G, H be special groups. A group homomorphism $f : G \longrightarrow H$ such that $f(-1) = -1$ is a **complete embedding** if for all forms φ and ψ over G

$$\varphi \equiv_G \psi \quad \text{iff} \quad f \star \varphi \equiv_H f \star \psi.$$

Clearly, the composition of complete embeddings is complete.

We say that $G \subseteq H$ is a **complete subgroup** of H if the canonical embedding of G into H is complete, i.e., for all n-forms $\langle a_1, \ldots, a_n \rangle$, $\langle b_1, \ldots, b_n \rangle$, over G,

$$\langle a_1, \ldots, a_n \rangle \equiv_G \langle b_1, \ldots, b_n \rangle \quad \text{iff} \quad \langle a_1, \ldots, a_n \rangle \equiv_H \langle b_1, \ldots, b_n \rangle. \quad \diamond$$

It is clear that any complete embedding is a SG-morphism. That, in fact, it is injective, comes from the observation that if $f(a) = f(b)$, then $f(ab) = 1$, and so

$$\langle f(1), f(ab) \rangle \equiv_H \langle f(1), f(1) \rangle \quad \text{implies} \quad \langle 1, ab \rangle \equiv_G \langle 1, 1 \rangle,$$

and the discriminant axiom [SG3] yields $ab = 1$, i.e. $a = b$.

In what follows we deal mostly with reduced special groups. Our first result gives several equivalent conditions for a map to be a complete embedding. The characterization in terms of the associated Boolean algebra will prove especially fruitful.

THEOREM 5.2. *Let $G \xrightarrow{f} H$ be a morphism of reduced special groups. Let f^* be the dual to f (Theorem 3.19), and $B(f)$ the associated Boolean morphism (Definition 4.16). The following are equivalent :*

1. f^* *is surjective.* 2. $Im(f^*)$ *is dense in* X_G.

3. $B(f)$ *is injective.*

4. $B(f)$ *is an isomorphism from* B_G *onto the subalgebra of* B_H *generated by* $Im(\varepsilon_H \circ f)$.

5. f *is a complete embedding.*

6. f *is injective and for every Pfister form φ over G,*

$$D_H(f \star \varphi) \cap Im(f) \subseteq f[D_G(\varphi)].$$

or, equivalently, $D_H(f \star \varphi) \cap Im(f) = f[D_G(\varphi)]$.

7. *For every pair of Pfister forms φ, ψ over G*

$$f \star \varphi \equiv_H f \star \psi \quad \text{implies} \quad \varphi \equiv_G \psi.$$

8. *For all Pfister forms φ over G,*

$$f \star \varphi \text{ hyperbolic} \quad \text{implies} \quad \varphi \text{ hyperbolic}.$$

PROOF. (1) $\Rightarrow$ (2) is obvious, while (2) $\Rightarrow$ (1) comes from the fact that any dense compact set in a Hausdorff space X must be equal to X.

(1) $\Leftrightarrow$ (3). Since $B(f)$ is just the Stone dual of f^*, it suffices to recall that 'monic' in **BA** iff 'epic' in the category **BS** of Boolean spaces, together with Proposition 4.9.

(3) $\Leftrightarrow$ (4). Since $\text{Im}(\varepsilon_G)$ generates B_G as a Boolean algebra (Proposition 4.10.(b)), commutativity of the diagram

[BH]

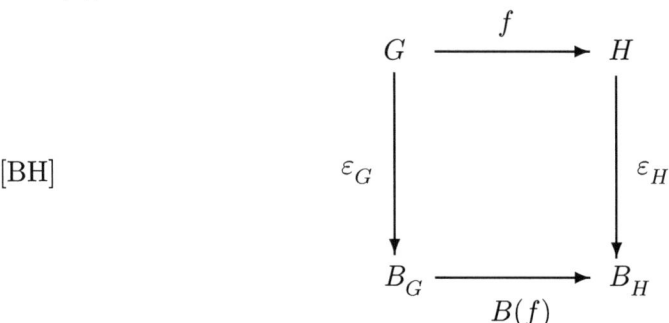

entails that $\text{Im}(B(f))$ is identical with the subalgebra B_f^* generated by $\text{Im}(\varepsilon_H \circ f)$. Clearly, injectivity of $B(f)$ is equivalent to $B(f)$ being an isomorphism between B_G and $\text{Im}(B(f)) = B_f^*$.

(1) $\Rightarrow$ (5). Commutativity of [BH] shows that f is injective, as $B(f)$ is injective by (3).

Let $\varphi = \langle a_1, \ldots, a_n \rangle$, $\psi = \langle b_1, \ldots, b_n \rangle$ be forms over G; assume that $f \star \varphi \equiv_H f \star \psi$. Then, by Proposition 3.7

(+) $\qquad \forall \gamma \in X_H, \sum_{i=1}^n \gamma(f(a_i)) = \sum_{i=1}^n \gamma(f(b_i))$.

We need to show $\varphi \equiv_G \psi$, i.e.

(++) $\qquad \forall \sigma \in X_G, \sum_{i=1}^n \sigma(a_i) = \sum_{i=1}^n \sigma(b_i)$.

Fix $\sigma \in X_G$. By (1), there is $\gamma \in X_H$ so that $\sigma = f^*(\gamma) = \gamma \circ f$; thus, (++) follows from (+).

(5) $\Rightarrow$ (6). We have already noted that a complete embedding is injective. Let φ be a Pfister form over G and $a \in \text{Im}(f) \cap D_H(f \star \varphi)$, say $a = f(b)$. Then, $f \star \varphi = a(f \star \varphi) = f \star (b\varphi)$. Since both forms φ, $b\varphi$ have coeficients in G, assumption (5) yields $\varphi \equiv_G b\varphi$. Thus, $b \in D_G(\varphi)$ and hence $a \in f[D_G(\varphi)]$. The inclusion $f[D_G(\varphi)] \subseteq D_H(f \star \varphi) \cap \text{Im}(f)$ holds because f is a SG-homomorphism.

(6) $\Rightarrow$ (7). We start with the following observation :

Fact. Let G be a reduced special group and φ, ψ Pfister forms of the same dimension over G. Then

$$\varphi \equiv_G \psi \quad \text{iff} \quad D_G(\varphi) = D_G(\psi).$$

Proof of Fact. By n-transitivity of $\equiv_G$, it suffices to show that $D_G(\varphi) = D_G(\psi)$ implies $\varphi \equiv_G \psi$. If $\varphi = \bigotimes_{i=1}^n \langle 1, a_i \rangle$, since $a_i \in D_G(\psi)$, Proposition 2.2.(f) yields

$$\varphi \otimes \psi = \langle 1, a_1 \rangle \otimes \ldots \otimes \langle 1, a_n \rangle \otimes \psi = \bigotimes_{i=1}^{n-1} \langle 1, a_i \rangle \otimes (\psi \oplus a_n \psi)$$
$$\equiv_G \bigotimes_{i=1}^{n-1} \langle 1, a_i \rangle \otimes (\psi \oplus \psi).$$

By induction on n we get $\varphi \otimes \psi \equiv_G 2^n \cdot \psi$. A similar computation with the coefficients of ψ in place of those of φ, yields

$$2^n \cdot \varphi \equiv_G \varphi \otimes \psi \equiv_G 2^n \cdot \psi,$$

and the conclusion follows from Proposition 1.6.(e.4), as required. $\diamondsuit$

Let φ, ψ be Pfister forms over G, such that $f \star \varphi \equiv_H f \star \psi$. It follows from the equality version of (6) and Lemma 1.12.(c), that $f[D_G(\varphi)] = f[D_G(\psi)]$. Since f is injective, we conclude $D_G(\varphi) = D_G(\psi)$, and the Fact yields $\varphi \equiv_G \psi$, as desired.

(7) $\Rightarrow$ (8) is straightforward.

(8) $\Rightarrow$ (2). Assume that $\text{Im}(f^*)$ is not dense in X_G. Since $\mathcal{B}$ in $[\star]$ of Notation 1.1.(a) is a basis for the topology in X_G, there are $a_1, \ldots, a_n \in G$ such that $u = \bigcap_{i=1}^n [a_i = 1]$ is non-empty and disjoint from $\text{Im}(f^*)$. Note that $f^{*-1}[u] = \emptyset$; thus, we have $f^{*-1}[u] = B(f)(u) = \bot$. Recalling that $\varepsilon_G(a_i)$ is the complement of $[a_i = 1]$ and that $B(f)$ is a Boolean morphism we arrive at

(H) $\quad \top = B(f)(\bigcup_{i=1}^n \varepsilon_G(a_i)) = \bigcup_{i=1}^n B(f)(\varepsilon_G(a_i)) = \bigcup_{i=1}^n \varepsilon_H(f(a_i)),$

where we have used the commutativity of diagram [BH]. Let φ be the Pfister form $\bigotimes_{i=1}^n \langle 1, a_i \rangle$; by Proposition 4.11, (H) implies that -1 is represented by $f \star \varphi$. Thus, the Pfister form $f \star \varphi$ over H is hyperbolic (Proposition 2.2.(ℓ.2)). But then, φ should be hyperbolic, which is impossible, because for all $\sigma \in u = \bigcap_{i=1}^n [a_i = 1]$, $sgn_\sigma(\varphi) = 2^n$. This ends the proof. $\square$

It is useful to restate the particular case of Theorem 5.2 for inclusions.

PROPOSITION 5.3. *Let $G \subseteq H$ be rgs's, and let $\iota : G \longrightarrow H$ be the inclusion map. The following are equivalent :*

1. *Every character $\sigma \in X_G$ extends to a character in X_H (i.e., X_G is a quotient of X_H).*

2. *B_G is isomorphic (via $B(\iota)$) to a subalgebra of B_H.*

3. *G is a complete subgroup of H.*

4. *For any Pfister form φ over G,*

$$D_H(\varphi) \cap G = D_G(\varphi).$$

5. *For every pair of Pfister forms φ, ψ over G,*

$$\varphi \equiv_G \psi \quad \text{iff} \quad \varphi \equiv_H \psi.$$

6. *A Pfister form φ over G is hyperbolic in G iff it is hyperbolic in H.*

PROOF. (1) is equivalent to ι^* being surjective. With this observation, the equivalences follow directly from Theorem 5.2. □

Remark. If $G \overset{\iota}{\hookrightarrow} H$, the AOS X_G is said to be a **quotient** of X_H if the restriction map $\iota^* : X_H \longrightarrow X_G$ is surjective; cf. [Ma3]; Def. 3.1. ◇

COROLLARY 5.4. *Let G be a rsg and let $\varepsilon_G : G \longrightarrow B_G$ be the canonical embedding of G into B_G.*

a) ε_G is a complete embedding.

*b) Every $\sigma \in X_G$ extends **uniquely** to B_G.*

c) The AOS $(X_G, G, -1)$ is a quotient of the AOS $(S(B_G), B_G, -1)$.

PROOF. a) By Fact 4.18, we have $B_{B_G} = B_G$. By Theorem 4.17.(5), $B_G \xrightarrow{B(\varepsilon_G)} B_G$ is id_{B_G}. In particular, $B(\varepsilon_G)$ is injective, and so the conclusion follows from (3) ⇔ (5) in Theorem 5.2.

b) Identifying G with its image in B_G, Proposition 5.3 gives the existence of an extension for any $\sigma \in X_G$. Uniqueness follows from the fact that any SG-morphism from B_G to $\mathbb{Z}_2$ is a BA-morphism (Proposition 4.5). Since $\varepsilon_G[G]$ generates B_G (Proposition 4.10), any two characters in $S(B_G)$ that coincide on $\varepsilon_G[G]$ must be the same.

c) Follows at once from (b) and the remark above. □

An interesting consequence of Corollary 5.4 is the following result of Kula, Marshall and Sladek (see [KMS]; Cor. 4.8).

PROPOSITION 5.5. *Every AOS is a quotient of the space of orders of a Pythagorean, SAP field.*

PROOF. The result is an immediate consequence of the following two remarks (cf. [P1] or [Cr] for undefined notions):

1. For every BA, B, there is a field K (necessarily Pythagorean and SAP) such that $Sg(B)$ and $G(K)$ are isomorphic as special groups.

In [Cr], Thms. 5, 8, Craven constructs a Pythagorean, SAP field K and a map ρ with the following properties:

(i) $\rho : \mathcal{X}(K) \longrightarrow S(B)$ is a homeomorphism between the space of orders $\mathcal{X}(K)$ and the Stone space $S(B)$.

(ii) ρ carries every subbasic clopen $[a = -1]$ of $\mathcal{X}(K)$, $a \in \dot{K}$, onto the subbasic clopen of $S(B)$ determined by some element of B.

1. COMPLETE EMBEDDINGS

(iii) ρ induces a <u>group</u> homomorphism of $\langle B, \triangle, 1 \rangle$ onto $\langle \dot{K}/\dot{K}^2, \cdot, 1 \rangle$.

[A simpler construction of a (not necessarily Pythagorean) field K having property (i) is given in [P1], Thms. 6.7-6.9 and 9.4.]

Only (i) will be used in the sequel. Stone duality gives $B = C(S(B), \pm 1)$, where $C(\cdot, \pm 1)$ denotes the set of continuous functions with values ± 1 under pointwise (Boolean) operations. By means of ρ we identify the (special group of the) given BA, B, with the special group $C(X(K), \mathbb{Z}_2)$ under coordinate-wise defined group operation and binary isometry. A group homomorphism is explicitly defined by:

$$f(a/\dot{K}^2)(P) = sgn_P(a) \qquad \text{for } a \in \dot{K} \text{ and } P \in X(K).$$

The reader may check by hand that property (iii) holds, using the SAP property of K to prove surjectivity.

Clearly, $f(-1/\dot{K}^2)$ is the function with constant value -1. To establish (1) it only remains to be shown that f preserves, in both directions, the representation relations of domain and codomain. For $a, b \in \dot{K}$ and $P \in X(K)$ we have:

$$a/\dot{K}^2 \in D_{G(K)}(1, b/\dot{K}^2) \quad \text{and} \quad sgn_P(b) = 1 \quad \text{imply} \quad sgn_P(a) = 1,$$

(see Definition 1.28(a)), which shows that f preserves representation. The converse follows from the Separation Theorem 2.11 which, whenever $a/\dot{K}^2 \notin D_{G(K)}(1, b/\dot{K}^2)$, yields an order $P \in X(K)$ such that $sgn_P(b) = 1$ and $sgn_P(a) = -1$.

2. Let $(X, G, -1)$ be an AOS. Let K be a field so that $(\chi(K), G(K), -1)$ is isomorphic to $(S(B_G), B_G, -1)$. The result now follows from Corollary 5.4(c). $\square$

The following remark shows that the canonical embedding $G \xrightarrow{\varepsilon_G} B_G$ arises in a natural way *also in the context of fields*.

Remark. Assume $G = G_{red}(K)$ is the reduced special group of a formally real field K, defined in section 3 of Chapter 1. By item (1) of 5.5, $B_G = G(F) = G_{red}(F)$ for some (non-unique) Pythagorean, SAP field F. We write ε_K for the embedding $\varepsilon_{G_{red}(K)}$ and ask whether:

(i) The field F can be an extension of K.

(ii) If so, is the embedding ε_K induced in some natural sense by the inclusion $i : K \longrightarrow F$?

Both questions are answered affirmatively by the following results:

Proposition A ([HJ]; Prop. 8.7, p. 478). *For any formally real field K, there is a pseudo real closed extension F such that:*

(1) *K is relatively algebraically closed in F.*

(2) *The restriction map $\rho_{F|K} : X(F) \longrightarrow X(K)$ is a homeomorphism (i.e., every order of K extends <u>uniquely</u> to F).* ◇

Proposition B ([Cr]; Thm. 7, p. 233). *For any formally real field K, there is a maximal algebraic extension to which every order of K extends uniquely; moreover, every such extension is Pythagorean.* ◇

These results together imply:

(†) *Every formally real field has a Pythagorean, SAP extension to which every order extends uniquely.*

(Such an extension may not be algebraic or have property (1) of Proposition A.)

Indeed, using successively Propositions A and B one gets fields $K \subseteq L \subseteq F$ such that:

(i) *L is pseudo real-closed.*

(ii) *F is Pythagorean and algebraic over L.*

(iii) *The restriction maps $\rho_{F|L}$ and $\rho_{L|K}$ are homeomorphisms, and hence so is $\rho_{F|K}$.*

It follows from (i) and (ii) that F is also pseudo real-closed, and hence SAP; cf. [P2], Thm. 3.1, pp. 148-150, and Prop. 1.3, p. 136. Therefore F satisfies the conclusion of (†)

Fixing an extension, F, as in (†), we define the maps:

(a) $\hat{i} : G_{red}(K) \longrightarrow G(F)$, induced by the inclusion $i : K \hookrightarrow F$:

$$\hat{i}(a/\Sigma \dot{K}^2) = a/\dot{F}^2, \quad \text{for } a \in \dot{K}.$$

(b) $j = B(\hat{i}) : B_{G_{red}(K)} \longrightarrow G(F)$, induced by $\rho_{F|K}$:

$$j(U) = \rho_{F|K}^{-1}[U], \quad \text{for } U \in B_{G_{red}(K)}.$$

Since $\rho_{F|K}$ is a homeomorphism, j is an isomorphism of BA's. Furthermore, $\varepsilon_K = j^{-1} \circ \hat{i}$, i.e., modulo the identification of $G(F)$ with $B_{G_{red}(K)}$ given by j^{-1}, the embedding ε_K is identical to $\hat{i}$. Indeed, for $a \in \dot{K}$,

$$\varepsilon_K(a/\Sigma \dot{K}^2) = \{P \in X(K) \mid a <_P 0\} = [a = -1]_K,$$

and

$$j([a = -1]_K) = \rho_{F|K}^{-1}[[a = -1]_K] = \{Q \in X(F) \mid a <_Q 0\} = [a = -1]_F.$$

On the other hand, $[a = -1]_F = a/\dot{F}^2 = \hat{i}(a/\Sigma \dot{K}^2)$, which proves our assertion. ◇

1. COMPLETE EMBEDDINGS

The following are further consequences of Theorem 5.2.

COROLLARY 5.6. *Any injective SG-morphism from a BA to a rsg is a complete embedding. Alternatively, any BA is a complete subgroup of any rsg in which it is contained.*

PROOF. We give two proofs; comparing them is revealing.

First Proof. Let $B \xrightarrow{f} H$ be an injective SG-morphism from the the BA B into the rsg H. In the commutative diagram

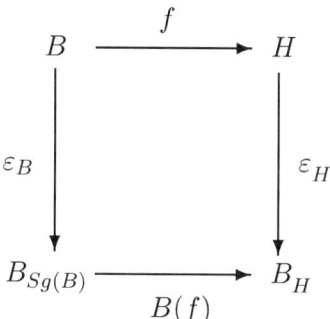

we have $B_{Sg(B)} = B$ and $\varepsilon_B = id_B$ (Fact 4.18); then $\varepsilon_H \circ f = B(f)$ is injective, and we conclude by Theorem 5.2.

Second Proof. To ease notation we do the proof for the case $B \subseteq H$. Let $a_1, \ldots, a_n, b_1, \ldots, b_n$ be elements of B and assume that $\langle a_1, \ldots, a_n \rangle \equiv_H \langle b_1, \ldots, b_n \rangle$. Since $H \subseteq B_H$, this isometry also holds in B_H. By the Horn-Tarski conditions (Theorem 7.1), for any $k \leq n$ we have

$$(*) \qquad \bigcup_{p \in S^{n+1,k}} \bigcap_{i<k} a_{p_i} = \bigcup_{p \in S^{n+1,k}} \bigcap_{i<k} b_{p_i} \quad (\text{in } B_H).$$

Since B is a subalgebra of B_H and the coefficients a_i, b_i are in B, $(*)$ holds also <u>in B</u> which, by Theorem 7.1 again, entails $\langle a_1, \ldots, a_n \rangle \equiv_B \langle b_1, \ldots, b_n \rangle$. □

Remark. The two proofs of Corollary 5.6 combined yield a stronger result; see Example 5.38 below. ◇

We remark the following significant consequence of the preceding Corollary.

COROLLARY 5.7. *Let B be a BA and φ, ψ be forms over B of the same dimension. Then, $\varphi \equiv_B \psi$ iff $\varphi \equiv_{B_0} \psi$, where B_0 is the subalgebra of B generated by the coefficients of φ and ψ.*

PROOF. Let B_0 be as in the statement. By Corollary 5.6, B_0 is a complete subgroup of B and the conclusion follows immediately. □

COROLLARY 5.8. *Let G be a special group with $1 \neq -1$ and let $\mathbb{Z}_2 \xrightarrow{\delta_G} G$ be defined by $\delta_G(\pm 1) = \pm 1$. Then, δ_G is a complete embedding iff G is formally real.*

PROOF. Clearly, δ_G is a SG-morphism. If δ_G is not a complete embedding, then there are forms s, t over $\mathbb{Z}_2$ of dimension $n \geq 1$, such that $s \equiv_G t$, but s is not isometric to t over $\mathbb{Z}_2$. Let p_s = number of 1's in s and $n_s = n - p_s$. Similarly, one defines p_t and n_t. Since s is not isometric to t over $\mathbb{Z}_2$, $n_s \neq n_t$, say $n_t - n_s = k > 0$. Since $s \equiv_G t$, Witt cancelation yields

$$p_s\langle 1 \rangle \oplus n_s \langle -1 \rangle \equiv_G p_t \langle 1 \rangle \oplus n_t \langle -1 \rangle \quad \text{implies} \quad p_s \langle 1 \rangle \equiv_G p_t \langle 1 \rangle \oplus k \langle -1 \rangle. \tag{I}$$

If $2^m \geq p_s$, we may add $(2^m - p_s)\langle 1 \rangle$ to both sides of the last isometry in (I) to get

$$2^m \langle 1 \rangle \equiv_G k \langle -1 \rangle \oplus (2^m - k) \langle 1 \rangle,$$

that is, $-1 \in Sat(G)$ and G is not formally real.

For the converse, we first observe that Corollary 5.6 shows that δ_G is a complete embedding whenever G is reduced. Now suppose G is formally real and let $\pi : G \longrightarrow G_{red}$ be the canonical projection. Note that $\pi \circ \delta_G = \delta_{G_{red}}$ and so is a complete embedding. Thus, if s, t are forms over $\mathbb{Z}_2$ such that $\delta_G \star s \equiv_G \delta_G \star t$, then $\delta_{G_{red}} \star s \equiv_{G_{red}} \delta_{G_{red}} \star t$, wherefrom we conclude $s \equiv_{\mathbb{Z}_2} t$, as needed. □

The following result will be useful in model-theoretic considerations concerning reduced special groups in Chapter 11.

PROPOSITION 5.9. *The class of reduced special groups has the amalgamation property for **complete** embeddings.*

PROOF. Given rsg's G, H_1, H_2 and complete embeddings $g_i : G \longrightarrow H_i$ (i = 1, 2), we look for another rsg H and complete embeddings $h_i : H_i \longrightarrow H$ such that $h_2 \circ g_2 = h_1 \circ g_1$.

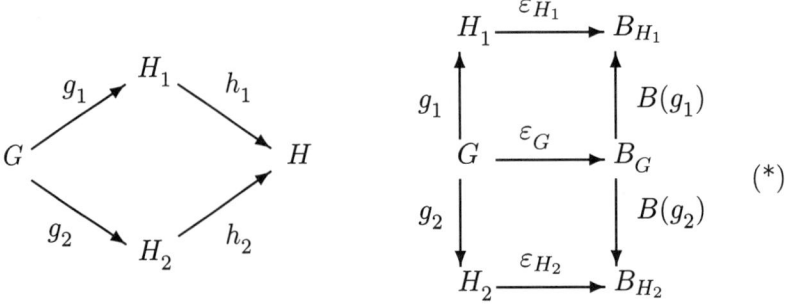

By Theorem 4.17, (*) is a commutative diagram. By Theorem 5.2 above, the maps $B(g_1)$, $B(g_2)$ are, in addition, injective. Now, BA's have the amalgamation property (it suffices to consider the free product of B_{H_1} and B_{H_2} along the maps $B(g_1)$, $B(g_2)$, respectively; for details see [HBA], vol. 1, section 11.3, pp. 168-172). In other words, there is a BA, B, and injective BA-morphisms $F_i : B_{H_i} \longrightarrow B$ ($i = 1, 2$) so that $F_1 \circ B(g_1) = F_2 \circ B(g_2)$. The commutativity of (*) entails $F_1 \circ \varepsilon_{H_1} \circ g_1 = F_2 \circ \varepsilon_{H_2} \circ g_2$. Since each F_i is a complete embedding (Corollary 5.6), the maps $h_i = F_i \circ \varepsilon_{H_i}$ ($i = 1,2$), being compositions of complete embeddings, are also complete. □

We now present an example of a non-complete reduced special subgroup of a rsg. It will turn out that one can find **fans** inside a whole class of BA's, which are not complete subgroups of the BA in which they are contained. This example serves to answer a number of natural questions about (non-complete) embeddings of reduced special groups.

EXAMPLE 5.10. **A non-complete reduced special subgroup.**

Let B be any BA containing 10 pairwise disjoint elements distinct from $\bot$; for example we may take B for the power algebra of any set with at least 10 elements. We fix $x_1, \ldots, z_3$ to be any nine of these elements and so

$$(*) \qquad \bigvee_{i=1}^{3} (x_i \vee y_i \vee z_i) \neq \top.$$

Next, define the elements p, q, u, v, w of B as follows:

$$(**) \qquad p = \bigvee_{i=1}^{3} x_i \vee y_i, \qquad q = \bigvee_{i=1}^{3} y_i \vee z_i$$

$$(***) \qquad u = x_1 \vee y_1 \vee y_2 \vee z_1, \qquad v = x_2 \vee y_1 \vee y_3 \vee z_2,$$
$$w = x_3 \vee y_2 \vee y_3 \vee z_3.$$

Recalling that the x_i's, y_i's, z_i's are pairwise disjoint, it is readily verified that the following equations hold:

$$u \wedge v \wedge w = \bot,$$
$$p \wedge q = (u \wedge v) \vee (u \wedge w) \vee (v \vee w)$$
$$p \vee q = u \vee v \vee w.$$

Theorem 7.1 yields $\langle u, v, w \rangle \equiv_B \langle 1, p, q \rangle$.

We now show that in the subgroup of B generated by $\{p, q, u, v, w\}$, these forms **are no longer isometric**.

Let F be the subgroup of B generated $\{p, q, u, v, w, \top\}$ endowed with the isometry $\equiv_F$ induced by B, and with $-1 = \top$ (just as in B).

PROPOSITION 5.11. *$(F, \equiv_F, -1)$ is a fan, and thus a reduced special group.*

PROOF. We start with the following

Fact. *Let Δ be a subgroup of a rsg $(G, \equiv_G, -1)$ such that $-1 \notin \Delta$. Suppose that Δ satisfies the following conditions: For all $a, b \in \Delta$,*

(i) $\varepsilon_G(a) \vee \varepsilon_G(b) \neq \top$; (ii) $\varepsilon_G(a) \wedge \varepsilon_G(b) = \bot$ implies $a = 1$ or $b = 1$,

where ε_G is the canonical embedding of G in B_G. Let K be the pre-special subgroup of G generated by Δ and -1. Then, K is a reduced special group, in fact, the fan on the group K.

Proof of Fact. Note that $K = \Delta \cup -\Delta$. Let $a \in \Delta$ and suppose $b \in K$ is such that $b \in D_G(1,a)$. Then, $\langle b, ba \rangle \equiv_{B_G} \langle 1, a \rangle$, and so from the definition of isometry in B_G, we have

(1) $\quad \varepsilon_G(b) \wedge \varepsilon_G(ba) = \varepsilon_G(1) \wedge \varepsilon_G(a) = \bot = 1,$

and

(2) $\quad \varepsilon_G(b) \vee \varepsilon_G(ba) = \varepsilon_G(1) \vee \varepsilon_G(a) = \varepsilon_G(a) \neq \top.$

We have two cases to discuss :

(i) $b \in \Delta$: In this case $ab \in \Delta$, and (1), (2), together with the assumptions (i) and (ii) yield $b = 1$ or $ab = 1$, i.e. $a = b$.

(ii) $-b \in \Delta$: Then $-b, -ba \in \Delta$, and equation (1) yields (taking complements) the equation $\varepsilon_G(-b) \vee \varepsilon_G(-ba) = \top$, contrary to assumption (ii).

The reasoning above shows that $D_K(1,a) = \{1,a\}$. A similar reasoning with $a \in -\Delta$ shows that, in fact, for all $x \in K$, $D_K(1,x) = \{1,x\}$. Thus, K is a fan.

To prove Proposition 5.11, apply the Fact to the subgroup Δ generated by $\{p, q, u, v, w\}$, with $G = B$ and $\varepsilon_G = id_B$; no join of two elements in this subgroup is $\top$, since all are contained in $\bigvee_{i=1}^{3} (x_i \vee y_i \vee z_i) \neq \top$. Now a tedious, but straightforward, computation will show that equations (**) and (***) imply that the meet of any two (of the 16) elements of Δ is $\bot$ iff one of them is $\bot$. □

To prove the assertion of 5.10, note that $\langle 1, p, q \rangle \not\equiv_F \langle u, v, w \rangle$: these forms are not isotropic (clear for the first, since e.g. $p \wedge q \neq \bot$; for the second, use 7.1 and the fact that $u \vee v \vee w \neq \top$). Hence, they are isometric in a fan iff $\{1, p, q\} = \{u, v, w\}$, which clearly is not the case. ◇

REMARK 5.12. a) The point of the construction above is that the group F **is not freely embedded** in B : the elements q, u, v, w are free generators of F as a special group (Proposition 5.19.(b)), but are not free in B in the Boolean algebraic sense, since $u \wedge v \wedge w = \bot$.

b) From Proposition 5.3.(1) we know that there ought to be a character of F (as a rsg) which does not extend to B. An example is the character $\sigma : F \longrightarrow \mathbb{Z}_2$ given by

$$\sigma(q) = \sigma(u) = \sigma(v) = \sigma(w) = -1 \, (= \top)$$

which does not extend to a (necessarily Boolean) character $B \longrightarrow \mathbb{Z}_2$, since any such extension, λ, would have to verify

$$\lambda(u \wedge v \wedge w) = 1 \quad \text{and} \quad \lambda(u \wedge v \wedge w) = \sigma(u) \wedge \sigma(v) \wedge \sigma(w) = -1. \quad \diamond$$

Example 5.10 answers a number of natural questions.

COROLLARY 5.13. *The category of reduced special groups (with SG-morphisms) does not have the amalgamation property.*

PROOF. Consider the (non-complete) inclusion $\iota : F \longrightarrow B$ and the (complete) natural embedding ε_F of F into its Boolean hull, B_F. B and B_F cannot be amalgamated over F along these maps because, on the one hand, $\iota(u \wedge v \wedge w) = \bot$ (in B), and on the other, $\varepsilon_F(u \wedge v \wedge w) \neq \bot$ (in B_F), since they are part of a set of free generators of the BA B_F (Proposition 5.19.(c)). $\square$

PROPOSITION 5.14. *The category of reduced special groups has no injectives.*

PROOF. We observe first that an injective rsg is necessarily a BA. Let I be an injective reduced special group. By Corollary 5.20.(c), there is a fan F_1 and a surjective SG-morphism $f : F_1 \longrightarrow I$. By injectivity there is a SG-morphism g making diagram (*) commutative.

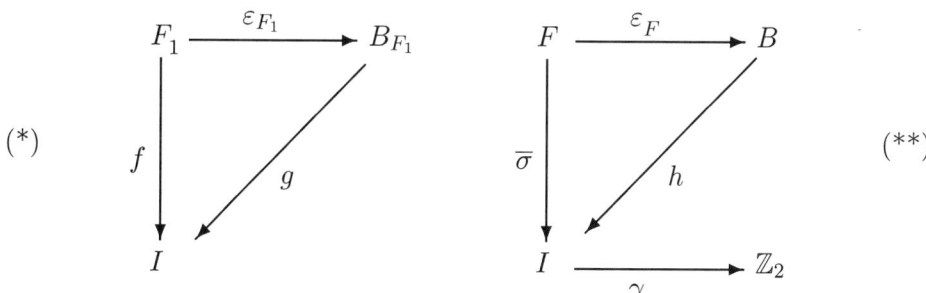

Obviously g is surjective, and it follows from Proposition 4.7.(c) that I is (the special group of) a BA.

Next, we note that I is clearly an injective object in the category of (non-trivial) BA's.

Now let σ be the character of the fan $F \subseteq B$ of Example 5.10, constructed in Remark 5.12.(b), which does not extend to B. We may conceive σ as a SG-morphism $\overline{\sigma}$ from F into I, taking only values ± 1. Thus, if γ is any character of I (i.e., a BA-morphism from I into $\mathbb{Z}_2$), we may write $\sigma = \gamma \circ \overline{\sigma}$. By injectivity, there is a SG-morphism $h : B \longrightarrow I$ such that diagram (**) is commutative. By Proposition 4.5, h is also a BA-morphism. But then $\gamma \circ h$ is a BA-morphism from B to $\mathbb{Z}_2$ extending σ, a contradiction. $\square$

COROLLARY 5.15. *The category* **AOS** *has no projectives.*

PROOF. This follows directly from Proposition 5.14 and the Duality Theorem 3.19. □

We end this section showing that the operation of Boolean hull commutes with quotients. Although we need Theorem 7.12 in section 7, we think it is better to present it here, as a nice application of the preceding discussion.

THEOREM 5.16. *Let G be a rsg, $G \xrightarrow{\varepsilon_G} B_G$ its Boolean hull. Let Δ be a proper saturated subgroup of G, and $\mathcal{I}(\Delta)$ be the ideal generated by Δ in B_G,*
$$\mathcal{I}(\Delta) = \{\, x \in B_G : \text{there is a finite } F \subseteq \Delta \text{ such that } x \subseteq \bigcup_{a \in F} \varepsilon_G(a)\}.$$
Then the Boolean hull of the quotient G/Δ is naturally isomorphic to $B_G/\mathcal{I}(\Delta)$.

PROOF. To ease notation, we identify G with $\varepsilon_G(G)$ in B_G, set $I = \mathcal{I}(\Delta)$, and $B = B_G/I$. From Corollary 4.13.(b), we know that I is a proper ideal in B_G. Define
$$f : G/\Delta \longrightarrow B_G/I \quad \text{setting} \quad f(g/\Delta) = g/I.$$
Since $\Delta \subseteq I$, f is well defined and $f(-1/\Delta) = -1/I$. Since the equivalence relation determined by I in B_G is a congruence with respect to all Boolean operations, in particular with respect to symmetric difference $\triangle$, f is a group homomorphism. As a first step in the proof we show that

Fact A. f is a complete embedding.

Proof of Fact A. We wish to apply item (7) of Theorem 5.2. Thus, if φ, ψ, are Pfister forms over G, we must prove
$$[f \circ \pi] \star \varphi \equiv_B [f \circ \pi] \star \psi \quad \Rightarrow \quad \pi \star \varphi \equiv_{G/\Delta} \pi \star \psi,$$
where π is the canonical projection $G \longrightarrow G/\Delta$.

(*)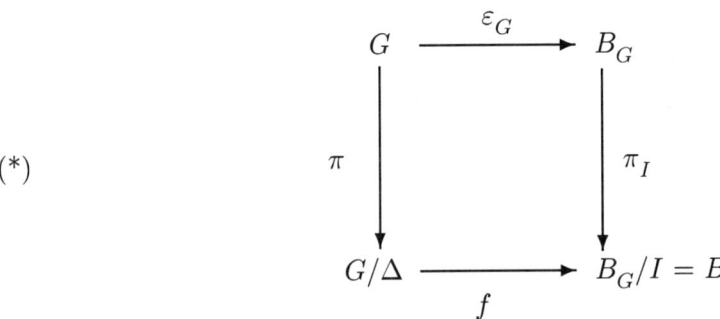

Notice that, since we have identified G with $\varepsilon_G(G)$ in B_G, $f \circ \pi$ is simply π_I, the natural quotient map from B_G to B (see diagram (*), above). Thus, by Propositions 4.7 and 2.21.(1), there is a Pfister form $\theta = \bigotimes_{j=1}^m \langle 1, c_j \rangle$, with coefficients in I, such that $\varphi \otimes \theta \equiv_{B_G} \psi \otimes \theta$. By the definition of $I = \mathcal{I}(\Delta)$, for each c_j there is a finite set $F_j \subseteq \Delta$ such that $c_j \subseteq \bigcup F_j$. Now, Theorem 7.12 tells us that $c_j \in D_{B_G}(\theta_j)$, where $\theta_j = \bigotimes_{a \in F_j} \langle 1, a \rangle$, $1 \leq j \leq m$. Notice that all θ_j are Pfister forms over Δ. Now we prove

Fact B. Set $\Lambda = \bigotimes_{j=1}^m \theta_j$. Then, $\varphi \otimes \Lambda \equiv_G \psi \otimes \Lambda$.

<u>Proof of Fact B</u>. Since G is a complete subgroup of B_G, it is sufficient to verify the desired isometry in B_G. We have

$$\varphi \otimes \theta \otimes \Lambda = \varphi \otimes \bigotimes_{j=1}^m \langle 1, c_j \rangle \otimes \bigotimes_{j=1}^m \theta_j \equiv_{B_G} \varphi \otimes \bigotimes_{j=1}^m (\langle 1, c_j \rangle \otimes \theta_j)$$
$$\equiv_{B_G} \varphi \otimes \bigotimes_{j=1}^m (\theta_j \oplus \theta_j) \equiv_{B_G} 2^m \cdot (\varphi \otimes \Lambda).$$

Consequently, $\varphi \otimes \theta \equiv_{B_G} \psi \otimes \theta$ yields, tensored on both sides with Λ,

$$2^m \cdot (\varphi \otimes \Lambda) \equiv_{B_G} 2^m \cdot (\psi \otimes \Lambda),$$

and we conclude $\varphi \otimes \Lambda \equiv_{B_G} \psi \otimes \Lambda$ from Proposition 1.6.(e.5), ending the proof of Fact B.

Since Λ has coefficients in Δ, Fact B and Proposition 2.21 yield $\pi \star \varphi \equiv_{G/\Delta} \pi \star \psi$, as needed to finish the proof of Fact A.

Since π_I, ε_G are epics, and diagram (*) is commutative, f is itself epic. Thus, f is an epic complete embedding from G/Δ into $B = B_G/I$. By Theorem 4.17.(3) we have a commutative diagram

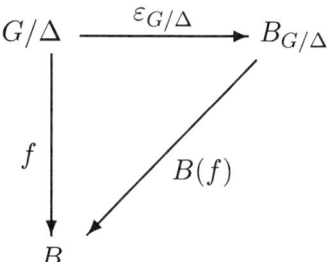

By Theorem 5.2, $B(f)$ is injective (f is a complete embedding). On the other hand, the fact that f is epic, implies that $B(f)$ is also epic. Now Proposition 4.9.(b) applies to show that $B(f)$ is a BA-isomorphism between $B_{G/\Delta}$ and $B_G/\mathcal{I}(\Delta)$, ending the proof of the theorem. □

2. SG-morphisms with a retract

LEMMA 5.17. *Let G, H be special groups and let $G \xrightarrow{f} H$, $H \xrightarrow{g} G$, be SG-morphisms. Assume that g is a retract of f, i.e., $g \circ f = id_G$. Then:*

i) f reflects positive-existential L_{SG}-formulas.

ii) $\ker g \cap \operatorname{Im} f = \{1\}$.

iii) $a \cdot f(g(a)) \in \ker g$ *for all* $a \in H$.

iv) The group homomorphism $(H/\ker g) \longrightarrow G$ *induced by g is an isomorphism of* L_{SG}-*structures of* $(H/\ker g, \equiv^*_{H/\ker g}, -1)$ *onto* $(G, \equiv_G, -1)$.

In particular,

v) $(H/\ker g, \equiv^*_{H/\ker g}, -1)$ *is a special group.*

Remark. We do not assume that G or H are reduced; in connection with item (v), see Proposition 2.14.(2). $\diamond$

PROOF. (i) is clear since any homomorphism preserves positive-existential formulas (in any language). Explicitly, if φ is a positive-existential L_{SG}-formula on n free variables, $a_1, \ldots, a_n \in G$, and $H \models \varphi[f(a_1), \ldots, f(a_n)]$, then $G \models \varphi[g(f(a_1)), \ldots, g(f(a_n))]$, i.e., $G \models \varphi[a_1, \ldots, a_n]$, as $g \circ f = id_G$.

(ii) Let $x \in \ker g \cap \text{Im } f$. Then, $x = f(y)$ for some $y \in G$, and $g(x) = 1$. Hence, $1 = g(f(y)) = y$, which implies $x = 1$.

(iii) $g(a \cdot f(g(a))) = g(a) \cdot g f(g(a)) = g(a)^2 = 1$.

(iv) Let $\widehat{g} : H/\ker g \longrightarrow G$ denote the group homomorphism induced by g, i.e., $\widehat{g}(\overline{a}) = g(a)$, where $\overline{a} = a/\ker g$, for $a \in H$. Clearly, g and $\widehat{g}$ are surjective, $\widehat{g}$ is injective, and $\widehat{g}(\overline{-1}) = -1$. Next we show, for $a_i, b_i \in H$ ($i = 1, 2$):

(*) $\langle \overline{a_1}, \overline{a_2} \rangle \equiv^*_{H/\ker g} \langle \overline{a_1}, \overline{a_2} \rangle \Leftrightarrow \langle g(a_1), g(a_2) \rangle \equiv_G \langle g(b_1), g(b_2) \rangle$,

which proves that $\widehat{g}$ is a L_{SG}-isomorphism.

($\Rightarrow$) By 2.13 there are elements $a'_i, b'_i \in H$ so that $a_i a'_i, b_i b'_i \in \ker g$ ($i = 1, 2$), and $\langle a'_1, a'_2 \rangle \equiv_H \langle b'_1, b'_2 \rangle$. Taking images by g on both sides of this isometry yields the conclusion.

($\Leftarrow$) The isometry on the right-hand side of (*) gives $\langle f(g(a_1)), f(g(a_2)) \rangle \equiv_H \langle f(g(b_1)), f(g(b_2)) \rangle$. Using (iii) and Definition 2.13 yields the isometry on the left-hand side. $\square$

Some of the standard constructions of special groups give raise to morphisms having a retract.

EXAMPLE 5.18. *The canonical embedding of a special group into any of its extensions has a retract. In particular, the inclusion of the special subgroup of basic elements,* $Ba(G)$ *(Definition 1.15), into G, has a retract.*

PROOF. Let G be a rsg, Δ a group of exponent 2, and $G[\Delta]$ the extension of G by Δ. Notation as in 1.10, we have a natural injective SG-morphism $\iota : G \longrightarrow G[\Delta]$, given by $\iota(g) = (g, 1) = g \cdot 1$. This map has a retract: the SG-morphism $pr : G[\Delta] \longrightarrow G$, given by $pr(g \cdot \delta) = g$. The assertion about $Ba(G)$ follows from Proposition 1.19. $\square$

Remark. This example shows, in particular, that any special group is both a complete subgroup and a quotient of any of its extensions. ◇

Before our next application, we prove some simple facts about fans. Notation and definitions are in Example 1.7.

PROPOSITION 5.19. *Let F be a fan and let $|F|$ denote the underlying group. Let $\mathcal{B}$ be a basis of $|F|$ as a vector space over $\mathbb{F}_2$, containing -1. Set $\mathcal{B}' = \mathcal{B} \setminus \{-1\}$. Then :*

(a) *A group homomorphism from F into any rsg is an SG-morphism iff it sends -1 to -1.*

(b) *F is a free object of the category **RSG**, having $\mathcal{B}'$ as a set of free generators.*

(c) *The Boolean algebra B_F associated to F by Theorem 4.17 is the free BA having $\mathcal{B}'$ as a set of free generators.*

PROOF. Item (a) was already observed in Remark 1.13.

b) Let G be a reduced special group and let $f : \mathcal{B}' \longrightarrow G$ be an arbitrary (set-theoretic) map. Set $f(-1) = -1$ and extend f by linearity to a group homomorphism (still called f) of F into G. By (a), f is a SG-morphism.

c) Given a BA, A, and a map $\mathcal{B}' \longrightarrow A$, item (b) provides an extension to a SG-morphism $f : F \longrightarrow A$. Then, $B(f) : B_F \longrightarrow A$ is a BA-morphism extending the given map. □

COROLLARY 5.20. (a) *There is, up to isomorphism, exactly one fan of each cardinality $\kappa \geq 2$.*

In particular,

(b) *The fan structure defined on a given group of exponent 2 is, up to isomorphism, independent of the choice of the distinguished element -1, provided that $-1 \neq 1$.*

(c) *Every rsg G is a homomorphic image of any fan of rank $\kappa \geq dim_{\mathbb{F}_2}(G)$.*

Remark. The **rank** of a fan is $\kappa - 1$, where κ is the cardinality of any of its basis as $\mathbb{F}_2$-vector space, containing -1. ◇

PROOF. (a) is clear from Proposition 5.19, since a free object in a set-based category is unique up to the cardinality of any set of free generators.

(c) Given G and a cardinal $\kappa \geq dim_{\mathbb{F}_2}(|G|)$, construct the fan F_κ on the unique group of exponent 2 of dimension κ over $\mathbb{F}_2$. Let $\mathcal{B}$ and $\mathcal{B}_0$ be basis of F_κ and G, respectively, over $\mathbb{F}_2$, containing -1. Let $f : \mathcal{B}' \longrightarrow \mathcal{B}'_0$ be a surjection . By Proposition 5.19.(b), f extends to a SG-morphism $\overline{f}$ from F_κ to G. Clearly, $\overline{f}$ is surjective. □

EXAMPLE 5.21. *Any injective SG-morphism of a rsg, G, into a fan, has a retract. Hence, G is also a fan.*

PROOF. Let $G \xrightarrow{f} F$ be an injective SG-morphism, where F is a fan. Let $\mathcal{B}_0$ be a basis for $|G|$ as an $\mathbb{F}_2$-vector space, containing -1. Since f is an injective group homomorphism and $f(-1) = -1$, $f[\mathcal{B}_0]$ is linearly independent in $|F|$; enlarge to a basis $\mathcal{B}_1$ of $|F|$ over $\mathbb{F}_2$. Let $f' : \mathcal{B}_1 \longrightarrow \mathcal{B}_0$ be a surjection extending $(f^{-1})|_{[f[\mathcal{B}_0]]}$. By Proposition 5.19, f' extends to a SG-morphism g from F to G. Since $(f' \circ f)|_{\mathcal{B}_0} = id_{\mathcal{B}_0}$, it follows by linearity that $g \circ f = id_G$. Straightforward checking shows that G is a fan. □

If G_i, $i \in I$, is a family of special groups, it is readily verified that $G = \prod_{i \in I} G_i$ with coordinatewise defined operations and special relation, together with the distinguished element -1 = constant sequence equal to -1, is a special group. Further, the natural projections $\pi_i : G \longrightarrow G_i$ are SG-morphisms. Moreover, G is reduced iff each coordinate group is reduced. This construction gives the **product** of the family G_i in **SG** or **RSG**.

EXAMPLE 5.22. *The diagonal embedding of a rsg G into G^I has a retract.* ◊

There is no canonical way of embedding a rsg G in a product of which it is a component. But there are many non-canonical embeddings. Below we construct a family of such embeddings, each of which has a retract.

Let G, H be rsg's, and Δ be a maximal saturated subgroup of G not containing -1 (i.e., Δ is the kernel of a $\sigma \in X_G$). Then, $G = \Delta \cup -\Delta$. We define $\iota_\Delta : G \longrightarrow G \times H$ by

$$\iota_\Delta(g) = \begin{cases} (g, 1) & \text{if } g \in \Delta \\ (g, -1) & \text{if } g \notin \Delta \end{cases}$$

We leave it to the reader to verify that ι_Δ is an injective group homomorphism. To see that it is a SG-morphism, note first that $\iota_\Delta(-1) = (-1, -1) = -1$ (in $G \times H$). Now we must check that for all $g_1, g_2 \in G$,

$$g_1 \in D_G(1, g_2) \Rightarrow \iota_\Delta(g_1) \in D_{G \times H}(1, \iota_\Delta(g_2)).$$

Assume $g_1 \in D_G(1, g_2)$. Since Δ is saturated, if $g_2 \in \Delta$, then $g_1 \in \Delta$ and we get $\iota_\Delta(g_1) = (g_1, 1) \in D_{G \times H}((1, 1), (g_2, 1))$, because $\iota_\Delta(g_2) = (g_2, 1)$. Now suppose $g_2 \in -\Delta$. If $g_1 \in \Delta$, then $\iota_\Delta(g_1) = (g_1, 1) \in D_{G \times H}((1, 1), (g_2, -1))$, because representation is coordinatewise. If $g_1 \in -\Delta$, then $\iota_\Delta(g_1) = (g_1, -1)$, and the same argument applies.

EXAMPLE 5.23. *Each of the maps $\iota_\Delta : G \longrightarrow G \times H$ has a retract.*

PROOF. The projection $\pi_G : G \times H \longrightarrow G$ is a SG-morphism, and a retract for ι_Δ. □

The following example is a slight variant of Example 5.22. Given a Boolean space, X, and a special group, G, the set $C(X, G)$ of all continuous functions of X into G (with the discrete topology) is a special group, under pointwise defined operations, relations and constants. (The special groups $C(X, G)$ and, more generally, filtered Boolean powers of special groups, will be studied in Chapter 6.) Clearly, G is embedded into $C(X, G)$ as a sg by the diagonal map which assigns to each $g \in G$ the function with constant value g. The evaluation at any point of X is a SG-morphism of $C(X, G)$ into G, and a retract to the diagonal embedding. Thus, we have:

EXAMPLE 5.24. *For every Boolean space X and every special group G, the diagonal embedding of G into $C(X, G)$ has a retract.* ◇

Remark. As a consequence of arguments of a different type, Corollary 10.6 below shows that any embedding of a **complete** Boolean algebra into a reduced special group has a retract. ◇

3. Pure Embeddings

DEFINITION 5.25. A SG-morphism $G \xrightarrow{f} H$ is called a **pure embedding** if it reflects positive-existential L$_{\text{SG}}$-formulas, that is, for every such formula $\varphi(v_1, \ldots, v_n)$ on n free variables and all $a_1, \ldots, a_n \in G$,

$$(*) \qquad H \models \varphi[f(a_1), \ldots, f(a_n)] \Rightarrow G \models \varphi[a_1, \ldots, a_n]. \qquad \diamond$$

Clearly, any such map is injective (consider the formula $\varphi(v_1, v_2) : v_1 = v_2$), and f is a pure embedding if and only if the implication (*) holds for all positive-primitive (abbreviated pp) L$_{\text{SG}}$-formulas φ, i.e., all formulas of form

$$\varphi(v_1, \ldots v_n) : \qquad \exists x_1 \ldots x_m \bigwedge_{j=1}^{k} \theta_j(x_1, \ldots, x_m, v_1, \ldots, v_n)$$

where the θ_j are atomic, $m \geq 0, k \geq 1$. Incorporating the parameters $a_1, \ldots, a_n \in G$ into the formula φ, it suffices to check (*) for L$_{\text{SG}}$-sentences with parameters in G. Since any atomic L$_{\text{SG}}$-formula is equivalent (modulo the axioms [SG0]-[SG5]) to either one of form $v \in D(1, w)$ or one of form $v = 1$, the pp-sentences of L$_{\text{SG}}$ with parameters in G are

$$(**) \qquad \exists x_1 \ldots x_m [\bigwedge_{j=1}^{\ell} a_j t_j(\overline{x}) \in D(\langle 1, b_j s_j(\overline{x}) \rangle) \wedge \bigwedge_{r=1}^{k} c_k u_k(\overline{x}) = 1],$$

where $a_j, b_j, c_k \in G$ and t_j, s_j, u_k are L$_{\text{SG}}$-terms on (some of) the variables $\overline{x} = (x_1, \ldots, x_m)$, i.e., just products of some of these variables.

The assertion that it suffices to check purity just on pp-formulas follows from the easily verified fact that any positive-existential first-order formula (in any language) is equivalent to a finite disjunction of conjunctions of pp-formulas. Furthermore, for reduced special groups one may even omit all conjuncts of type $cu(\bar{x}) = 1$ in (**), since $v = 1$ is equivalent to $v \in D(1,1)$.

Item (i) in Lemma 5.17 shows that SG-morphisms having a retract are pure embeddings. Below we shall see that this inclusion is proper, even for reduced special groups. However, a general model-theoretic argument shows that these classes are not too far apart.

Proposition. Let $\mathcal{A}$, $\mathcal{B}$, be structures with (arbitrary) language L, and let $f : \mathcal{A} \longrightarrow \mathcal{B}$ be an L-homomorphism. The following are equivalent:

(1) f is a pure L-embedding (i.e., the implication (*) in 5.25 holds for every positive-existential L-formula).

(2) There is an elementary extension $\mathcal{A}'$ of $\mathcal{A}$ and an L-homomorphism $g : \mathcal{B} \longrightarrow \mathcal{A}'$ such that the following diagram commutes:

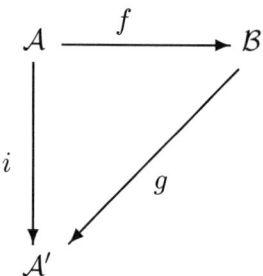

where i denotes the inclusion of $\mathcal{A}$ into $\mathcal{A}'$.

(3) Same as (2), where $\mathcal{A}'$ is only required to be an $\exists^+$-extension of $\mathcal{A}$, i.e., the inclusion $i : \mathcal{A} \longrightarrow \mathcal{A}'$ is (only) a pure L-embedding. ◇

Remark. The Proposition is proven by a slight variant of a standard model-theoretic argument which we omit; see, for example, Hodges [Ho], §6.5, pp. 294 ff. (especially, Thms. 6.5.7 and 6.5.8), or the proof of Lemma 5.2.1 in Chang-Keisler [CK] (and also their Exercise 5.2.6). ◇

Next we give two natural examples of pure SG-embeddings.

EXAMPLE 5.26. Let F be a formally real field and let $F(X)$ be the field of rational functions in one variable over F. The SG-morphisms $G(F) \longrightarrow G(F(X))$ and $G_{red}(F) \longrightarrow G_{red}(F(X))$ induced by the inclusion $F \hookrightarrow F(X)$, are pure embeddings.

PROOF. We do it for the case of G_{red}, the other case being simpler. We write $\bar{a}$ for the class modulo sums of squares of an element $a \in \dot{F}$, and similarly for $F(X)$.

By the considerations following Definition 5.25 we need only prove:

3. PURE EMBEDDINGS

(+) $\quad G_{red}(F(X)) \models \exists v_1 \ldots v_n \varphi(v_1, \ldots, v_n) \Rightarrow$

$$\Rightarrow G_{red}(F) \models \exists v_1 \ldots v_n \varphi(v_1, \ldots, v_n),$$

where

$$\varphi(v_1, \ldots v_n) : \quad \bigwedge_{j=1}^{k} a_j t_j(\overline{v}) \in D(1, b_j s_j(\overline{v})),$$

with $a_j, b_j \in \dot{F}$ (we can harmlessly replace elements of $G_{red}(F)$ by representatives from $\dot{F}$), and $t_j(\overline{v}), s_j(\overline{v})$, are L_{SG}-terms on the variables $v_1, \ldots, v_n$; thus, $t_j(\overline{v}) = \prod_{\ell=1}^{n} v_\ell^{\varepsilon_j(\ell)}$, $s_j(\overline{v}) = \prod_{\ell=1}^{n} v_\ell^{\eta_j(\ell)}$, where $\varepsilon_j(\ell), \eta_j(\ell) \in \{0, 1\}$, and $v_\ell^0 = 1, v_\ell^1 = v_\ell$ ($\ell = 1, \ldots, n$). The antecedent of (+) means that there are non-zero rational functions $R_1, \ldots, R_n \in F(X)$ so that, for $j = 1, \ldots, n$:

$$\overline{a_j} \cdot \prod_{\ell=1}^{n} \overline{R_\ell}^{\varepsilon_j(\ell)} \in D_{G_{red}(F(X))}(\langle 1, \overline{b_j} \cdot \prod_{\ell=1}^{n} \overline{R_\ell}^{\eta_j(\ell)} \rangle).$$

According to the definition of representation in G_{red} (1.28.(a)), there are $S_j, T_j \in \sum F(X)^2$ such that the following identities hold for $j = 1, \ldots, n$:

(++) $\quad a_j \prod_{\ell=1}^{n} R_\ell(X)^{\varepsilon_j(\ell)} = S_j(X) + (b_j \prod_{\ell=1}^{n} R_\ell(X)^{\eta_j(\ell)}) \cdot T_j(X).$

Since $char(F) = 0$ there is $y \in F$ which is not a zero of either the numerator or the denominator of any of the R_ℓ's ($\ell = 1, \ldots, n$), nor of any of the rational functions occurring in the sums of squares $S_j, T_j \neq 0$ ($j = 1, \ldots, k$). Thus, $R_\ell(y), S_j(y), T_j(y)$ are defined and $\neq 0$ for $\ell = 1, \ldots, n$, and also for $j = 1, \ldots, k$ whenever $S_j, T_j \neq 0$. Evaluating the identity (++) at y, and using 1.28.(a) again, shows that

$$\overline{a_j} \cdot \prod_{\ell=1}^{n} \overline{R_\ell(y)}^{\varepsilon_j(\ell)} \in D_{G_{red}(F)}(\langle 1, \overline{b_j} \cdot \prod_{\ell=1}^{n} \overline{R_\ell(y)}^{\eta_j(\ell)} \rangle),$$

for $j = 1, \ldots, k$. Thus, the consequent of (+) holds for the values $\overline{R_\ell(y)}$ of v_ℓ ($\ell = 1, \ldots, n$). $\square$

Remark. Under mild assumptions, the preceding example generalizes to the function field of irreducible affine varieties over F. $\diamond$

For our next example we observe that the fact that a special group has two distinguished elements makes it impossible to mimic the definition of direct sum, usual in group theory (in this connection, see Proposition 10.11). The closest we can get is

DEFINITION 5.27. Let $G_i, i \in I$, be a family of rsg's. We denote by $\bigoplus_{i \in I}^{*} G_i$ the following pre-special subgroup of $\prod_{i \in I} G_i$:

$$\bigoplus_{i\in I}^* G_i = \{x \in \prod_{i\in I} G_i : \text{ There is a finite subset } J \text{ of } I \text{ such that either}$$
$$x_i = 1, \forall\, i \in I \setminus J \text{ or } x_i = -1, \forall\, i \in I \setminus J\}$$

together with the isometry induced by that of $\prod_{i\in I} G_i$, and having $-\mathbb{1} = (-1_{G_i} : i \in I)$ as distinguished element. $\diamond$

EXAMPLE 5.28. *For reduced special groups G_i, $i \in I$, the inclusion of $\bigoplus_{i\in I}^* G_i$ into $\prod_{i\in I} G_i$ is a pure embedding. In particular, $\bigoplus_{i\in I}^* G_i$ is a (reduced) special group.*

PROOF. We may assume I infinite, since both groups are equal for finite I. We set $H = \bigoplus_{i\in I}^* G_i$, $G = \prod_{i\in I} G_i$, and assume that G verifies a formula of type

$$(*) \qquad \exists v_1 \ldots v_n \bigwedge_{j=1}^k a^j t^j(\overline{v}) \in D(1, b^j s^j(\overline{v})),$$

where $a^j, b^j \in H$, and t^j, s^j are products of the variables $v_1, \ldots, v_n$. Then, for each coordinate $i \in I$, we have:

$$G_i \models \exists v_1 \ldots v_n \bigwedge_{j=1}^k a_i^j t^j(\overline{v}) \in D(1, b_i^j s^j(\overline{v})).$$

For $a \in H$ let J_a denote a finite subset of I outside which a is constant, either 1 or -1. Let $J = \bigcup_{j=1}^k (J_{a^j} \cup J_{b^j})$; then, for $j = 1, \ldots, k$, a_i^j, b_i^j take on a constant value, either 1 or -1, for all $i \in I \setminus J$. Let $x^1, \ldots, x^n \in G$ be elements satisfying the formula $(*)$ in G. For $i \in I \setminus J$ we have:

$$(**) \qquad G_i \models \bigwedge_{j=1}^k \pm t^j(x_i^1, \ldots, x_i^n) \in D(1, \pm s^j(x_i^1, \ldots, x_i^n)),$$

as $a_i^j, b_i^j \in \{\pm 1\}$.

Now, fix an index $i_0 \in I \setminus J$ and pick a SG-character $\sigma \in X_{G_{i_0}}$ (exists because G_{i_0} is reduced). Then $(**)$ yields:

$$(***) \; \mathbb{Z}_2 \models \bigwedge_{j=1}^k \pm t^j(\sigma(x_{i_0}^1), \ldots, \sigma(x_{i_0}^n)) \in D(\langle 1, \pm s^j(\sigma(x_{i_0}^1), \ldots, \sigma(x_{i_0}^n))\rangle).$$

Next, we define elements y^ℓ, $\ell = 1, \ldots, n$, as follows:

$$y_i^\ell = \begin{cases} x_i^\ell & \text{if } i \in J \\ \sigma(x_{i_0}^\ell) & \text{if } i \in I \setminus J. \end{cases}$$

Clearly, $y^\ell \in H$; from $(***)$ we get, for $i \in I \setminus J$:

$$\mathbb{Z}_2 \models \bigwedge_{j=1}^k \pm t^j(y_i^1, \ldots, y_i^n) \in D(1, \pm s^j(y_i^1, \ldots, y_i^n)),$$

which implies the validity of this formula in G_i, i.e.,

$$(****) \qquad G_i \models \bigwedge_{j=1}^k a_i^j t^j(y_i^1, \ldots, y_i^n) \in D(1, b_i^j s^j(y_i^1, \ldots, y_i^n)).$$

On the other hand, the choice of the x^ℓ's and the definition of the y^ℓ's guarantees that (****) holds for $i \in J$ as well. Hence (****) holds for all $i \in I$ which shows:
$$G \models \bigwedge_{j=1}^k a^j t^j(y^1, \ldots, y^n) \in D(1, b^j s^j(y^1, \ldots, y^n)).$$

Since $y^1, \ldots, y^n \in H$, we have proved that formula (*) holds in H. □

Next we prove:

PROPOSITION 5.29. *The inclusion of $\bigoplus_{i \in I}^* G_i$ into $\prod_{i \in I} G_i$ does not have a retract, whenever I is infinite.*

PROOF. We claim:

(+) $\quad \Delta \cap \bigoplus_{i \in I}^* G_i \neq \{1\}$ for every saturated subgroup Δ of $\prod_{i \in I} G_i$ such that $\Delta \neq \{1\}$.

This claim, together with item (iii) of Lemma 5.17, proves the Proposition. Indeed, it suffices to prove that $\ker h \neq \{1\}$ for any SG-homomorphism $h: \prod_{i \in I} G_i \longrightarrow \bigoplus_{i \in I}^* G_i$ which is the identity on $\bigoplus_{i \in I}^* G_i$. Since I is infinite, $\prod_{i \in I} G_i \setminus \bigoplus_{i \in I}^* G_i \neq \emptyset$; let a be in this set. Since $h(a) \in \bigoplus_{i \in I}^* G_i$, $a \neq h(a)$, whence $a \cdot h(a) \neq 1$, and $a \cdot h(a) \in \ker h$ by 5.17(iii).

In order to prove (+), let $a \in \Delta \setminus \{1\}$; then, $a = (a_i)_{i \in I}$ and $a_{i_0} \neq 1$ for some $i_0 \in I$. Let $b = (b_i)_{i \in I}$, where $b_i = 1$ if $i \neq i_0$, and $b_{i_0} = a_{i_0}$. Clearly, $b_i \in D_{G_i}(1, a_i)$ for all $i \in I$, and then $b \in D_{\prod_{i \in I} G_i}(1, a)$. Since Δ is saturated, $b \in \Delta$. Since $b \in \bigoplus_{i \in I}^* G_i$ and $b \neq 1$, the claim is proved. □

Proposition 5.29, together with Examples 5.18, 5.22, 5.21 and Corollary 10.6 yields:

COROLLARY 5.30. *Let I be an infinite set, and for $i \in I$ let G_i be a rsg. Then:*

(i) $\prod_{i \in I} G_i$ *is neither an extension nor a power of $\bigoplus_{i \in I}^* G_i$.*

(ii) $\prod_{i \in I} G_i$ *is never a fan.*

(iii) $\bigoplus_{i \in I}^* G_i$ *is never a complete Boolean algebra.* ◇

4. Isotropy-Reflecting Morphisms

The last class of SG-morphisms which we shall consider is that of **isotropy-reflecting** (equivalently, **anisotropy-preserving**) morphisms, that is, SG-homomorphisms $f: G \longrightarrow H$ such that for every form φ over G,

$$f_\star \varphi \text{ isotropic in } H \quad \Rightarrow \quad \varphi \text{ isotropic in } G.$$

Since isotropy of a form φ is expressed by a positive-existential L_{SG}-sentence having as parameters the coefficients of φ, we have:

FACT 5.31. *Any pure SG-embedding is isotropy-reflecting.* ◇

The next result was pointed out by A. Petrovich.

PROPOSITION 5.32. *Every isotropy-reflecting SG-morphism is a complete embedding.*

PROOF. Let $f : G \longrightarrow H$ be an isotropy-reflecting SG-morphism, and assume $f * \varphi \equiv_H f * \psi$, where $\varphi = \langle a_1, \ldots, a_n \rangle$, $\psi = \langle b_1, \ldots, b_n \rangle$ are forms over G. By induction on n we prove $\varphi \equiv_G \psi$. The case $n = 1$ follows from:

(i) f is injective.

If $f(a) = f(b)$, then $\langle f(a), -f(b) \rangle \equiv_H \langle 1, -1 \rangle$. By assumption $\langle a, -b \rangle$ is isotropic in G, and then [SG 3] gives $a = b$.

(ii) Induction step. Assume the statement holds for arbitrary forms of dimension $n - 1$ ($n \geq 2$). The assumption $f * \varphi \equiv_H f * \psi$ implies that $\langle -f(b_1), f(a_1), \ldots, f(a_n) \rangle$ is isotropic in H. Hence $\langle -b_1, a_1, \ldots, a_n \rangle$ is isotropic in G, i.e., $\langle -b_1, a_1, \ldots, a_n \rangle \equiv_G \langle 1, -1, c_2, \ldots, c_n \rangle$, for some $c_2, \ldots, c_n \in G$. Since $\langle 1, -1 \rangle \equiv_G \langle -b_1, b_1 \rangle$, Witt's cancellation theorem (1.6.(b)) yields:

(*) $\qquad\qquad \langle a_1, \ldots, a_n \rangle \equiv_G \langle b_1, c_2, \ldots, c_n \rangle.$

Taking images under f gives $\langle f(b_1), \ldots, f(b_n) \rangle \equiv_H \langle f(a_1), \ldots, f(a_n) \rangle \equiv_H \langle f(b_1), f(c_2), \ldots, f(c_n) \rangle$, whence $\langle f(b_2), \ldots, f(b_n) \rangle \equiv_H \langle f(c_2), \ldots, f(c_n) \rangle$. By the induction hypothesis, $\langle b_2, \ldots, b_n \rangle \equiv_G \langle c_2, \ldots, c_n \rangle$. Substituting this isometry in (*) yields $\varphi \equiv_G \psi$, as required. □

REMARK 5.33. The canonical embedding $\varepsilon_G : G \longrightarrow B_G$ of a rsg into its Boolean hull, whenever G itself is not a Boolean algebra, gives an example of a complete embedding (Corollary 5.4(a)) which is not isotropy-reflecting, as follows from Proposition 7.17 below. ◇

In the remainder of this section we shall exhibit some natural examples of isotropy-reflecting SG-morphisms.

EXAMPLE 5.34. **Odd-degree field extensions**.

Let K be any field and let L be an extension of K of odd degree. Springer's theorem [L1; Thm. VII.2.3, pp. 198-199] shows that the SG-homomorphism $G(K) \longrightarrow G(L)$ induced by the inclusion $K \hookrightarrow L$ is isotropy-reflecting. ◇

EXAMPLE 5.35. **Totally quadratic reduced field extensions**.

Let F be a formally real field and $\beta \in (\sum F^2) \setminus F^2$. The SG-homomorphism $\iota : G_{red}(F) \longrightarrow G_{red}(F(\sqrt{\beta}))$ induced by the inclusion $F \hookrightarrow F(\sqrt{\beta})$ is isotropy-reflecting.

PROOF. Given a form $\varphi = \langle a_1/F^2, \ldots, a_n/F^2 \rangle$ over $G(F)$ ($a_1, \ldots, a_n \in \dot{F}$), we denote by $\overline{\varphi} = \langle a_1/\sum F^2, \ldots, a_n/\sum F^2 \rangle$ its image in $G_{red}(F)$. To ease notation we write $K = F(\sqrt{\beta})$.

Let φ be a form of dimension n over $G(F)$ such that $\iota * \overline{\varphi}$ is isotropic in $G_{red}(K)$. We may also assume $2^k \varphi$ anisotropic in $G(F)$ for all $k \in \mathbb{N}$; otherwise, its image $2^k \overline{\varphi}$ in $G_{red}(F)$ would also be isotropic, which implies that $\overline{\varphi}$ is isotropic in $G_{red}(F)$, by Proposition 1.6.(e.3). Thus, we have $\iota * \overline{\varphi} \equiv_{G_{red}(K)} \langle 1, -1 \rangle \oplus \theta$, where θ is a form of dimension $n-2$ over $G_{red}(K)$. We may write $\theta = \langle \overline{b_1}, \ldots, \overline{b_{n-2}} \rangle$, where $b_i \in \dot{K}$ and $\overline{b_i} = b_i / \sum \dot{K}^2$; we set $\psi = \langle b_1, \ldots, b_{n-2} \rangle$.

Since $G_{red}(K) = G(K)/Sat(G(K))$, (cf. 2.24), there is $k \in \mathbb{N}$ such that
$$2^k \varphi \equiv_{G(K)} 2^k(\langle 1, -1 \rangle \oplus \psi),$$
which shows that $2^k \varphi$ is isotropic over $G(K)$. Since we assumed that this form is anisotropic over $G(F)$, [L1; Lemma VII.3.1, p. 200] applies, yielding an element $a \in \dot{F}$ and a form τ over F such that
$$2^k \varphi \equiv_{G(F)} a \langle 1, -\beta \rangle \oplus \tau.$$

Reducing this isometry modulo $Sat(G(F))$, we get:
$$2^k \overline{\varphi} \equiv_{G_{red}(F)} \overline{a} \langle 1, \overline{-\beta} \rangle \oplus \overline{\tau} = a \langle 1, -1 \rangle \oplus \overline{\tau} \equiv_{G_{red}(F)} \langle 1, -1 \rangle \oplus \overline{\tau},$$
i.e., $2^k \overline{\varphi}$ is isotropic in $G_{red}(F)$. Proposition 1.6.(e.3) implies that $\overline{\varphi}$ is isotropic in $G_{red}(F)$, as required. □

EXAMPLE 5.36. **Relative Pythagorean closures**.

The preceding Example 5.35 is the initial step of an inductive process which yields further examples of isotropy-reflecting SG-morphisms. Here is a description of the results, without proofs.

Let Ω be a prime-closed extension of a formally real field F; see [Be; Chapter II] for details. We denote by F^Ω the largest Galois extension of F in Ω to which every order of F can be extended; F^Ω exists and is the intersection of all the Ω-real closures of F in Ω; cf. [Be; Thm. II.3.6, p. 81]). The field F^Ω is called the Ω-**Pythagorean closure of** F; F^Ω is infinite-dimensional over F, unless F itself is Pythagorean. When Ω is the algebraic closure of F, F^Ω is the intersection of all the real closures of F. When Ω is the maximal 2-extension of F, F^Ω is the Pythagorean closure of F. In [DM5] we prove:

Theorem. *The SG-morphism* $G_{red}(F) \longrightarrow G_{red}(F^\Omega)$ *induced by the inclusion* $F \hookrightarrow F^\Omega$ *is isotropy-reflecting.* ◇

EXAMPLE 5.37. **Marshall's isotropy theorem.**

In [Ma 4; Thm. 1.4, p. 604] Marshall proved:

Theorem. *Let $(X, G, -1)$ be an abstract order space (cf. Definition 3.9) and let φ be a form over G which is anisotropic in X. Then, there is a <u>finite</u> subspace $(Y, G/Y^\perp, -1)$ of $(X, G, -1)$ such that (the image in $G/Y^\perp$ of) the form φ is anisotropic in Y.* ◇

The translation of this result in the dual terms of reduced special groups (see Theorem 3.19) is as follows:

Theorem. *Let G be a rsg. Let $\mathcal{F}$ denote the family of all saturated subgroups of G of finite index. For $\Delta \in \mathcal{F}$, let $\pi_\Delta : G \longrightarrow G/\Delta$ be the canonical quotient map. Then, the SG-morphism $\theta : G \longrightarrow \prod_{\Delta \in \mathcal{F}} G/\Delta$ given by $\theta(g) = (\pi_\Delta(g))_{\Delta \in \mathcal{F}}$, for $g \in G$, is isotropy-reflecting.* ◇

Another version of Marshall's isotropy theorem runs as follows: the family $\mathcal{F}$ ordered under <u>reverse</u> inclusion is a right-directed poset. The family of rsg's $\{G/\Delta | \Delta \in \mathcal{F}\}$ with the transition functions $f_{\Delta\Gamma} : G/\Delta \longrightarrow G/\Gamma$ defined by $f_{\Delta\Gamma}(g/\Delta) = g/\Gamma$, for $g \in F$ and $\Delta, \Gamma \in \mathcal{F}$, $\Delta \subseteq \Gamma$, is a projective system of finite rsg's. It is obvious that the values of θ lie in the projective limit, H, of this system.

Theorem. *With notation as above, the following holds:*

(a) *The group H, endowed with the isometry relation induced by the product $\prod_{\Delta \in \mathcal{F}} G/\Delta$, is a reduced special group.*

(b) *The map $\theta : G \longrightarrow H$ is an isotropy-reflecting SG-morphism.* ◇

The proof will appear in [DM3; §8].

EXAMPLE 5.38. **Boolean algebras** (again).

Any injective SG-morphism of a Boolean algebra to a reduced special group is isotropy-reflecting.

PROOF. Let $B \xrightarrow{f} H$ be such a morphism. The first proof of Corollary 5.6 shows that $B(f) = \varepsilon_H \circ f : B \longrightarrow B_H$ and (hence) is injective.

Let φ be a form over B such that $f * \varphi$ is isotropic in H. Then, $\varepsilon_H * (f * \varphi) = B(f) * \varphi$ is isotropic in B_H. Let $n = dim(\varphi)$. Propositions 7.17 and 7.16 below show that (in B_H) $\mathcal{HT}_1(B(f) * \varphi) = \top$ and $\mathcal{HT}_n(B(f) * \varphi) = \bot$, where $\mathcal{HT}_k(\psi)$ denotes the k-th Horn-Tarski invariant of ψ (defined in Chapter 7). Since $B(f)$ is a Boolean homomorphism and the $\mathcal{HT}_k$'s involve only Boolean operations, $\mathcal{HT}_k(B(f) * \varphi) = B(f)(\mathcal{HT}_k(\varphi))$. Injectivity of $B(f)$ yields, then, $\mathcal{HT}_1(\varphi) = \top$ and $\mathcal{HT}_n(\varphi) = \bot$ (in B), which, by 7.17 and 7.16 again, imply isotropy of φ in B. □

We do not know, at present, whether any of the examples of isotropy-reflecting SG-morphisms presented above is also a pure embedding. A. Petrovich has given an example of a morphism of the type introduced in 5.35 which does not have a retract.

CHAPTER 6

Special Groups of Continuous Functions

In this Chapter we discuss how to construct a wide class of special groups, by considering certain subgroups of continuous functions defined in a topological space, with values in a special group, known as filtered powers. In the first section we establish certain elementary properties of locally constant functions and define the concept of filtered power, showing that if the codomain is a pre-special group, the same is true of the filtered power. In section 2, we give sufficient conditions for a filtered power to be special group, introducing the notion of SG-filtered power. The main results in this section are Theorems 6.10 and 6.13. Preparatory to the characterization of the space of orders and the Boolean hull of a reduced SG-filtered power, we discuss, in section 3, germs of sections and the stalk construction. The results in section 3 are used in section 4 to give a concrete characterization of the space of orders of a reduced SG-filtered power (Theorem 6.27). Section 5 is dedicated to obtaining a workable description of the Boolean hull of a reduced SG-filtered power T. In particular, Theorem 6.34 guarantees that if the groups in the filtration determining T are complete subgroups of its codomain, then the Boolean hull of T can be naturally identified with the filtered power of the Boolean hulls of the component groups, with values in the Boolean hull of the codomain.

1. Filtered Powers

Recall that a **Boolean space** is a compact, Hausdorff topological space, with a basis of clopen (both open and closed) sets.

DEFINITION 6.1. Let X be a Boolean space. Write $B(X)$ for the Boolean algebra of clopens in X. If $x \in X$,
$$\nu_x = \{U \in B(X) : x \in U\},$$
is the filter of clopen neighborhoods of x in X. For $U \in B(X)$, $-U$ is its complement in X, consistent with our notation in Chapter 4. ◇

NOTATION 6.2. Let X be a Boolean space and let A be a set, considered as a space with the discrete topology (all points are open).

a) Write $C(X, A)$ for the set of all continuous functions, defined in X, with values in A.

b) For each $a \in A$, $\widehat{a}$ is the constant function on X, with value a. Clearly, $\widehat{\ }$ is a bijection from A to the constant functions in $C(X, A)$.

c) For f in $C(X, A)$ and $a \in A$, define
$$[f = \widehat{a}] = \{x \in X : f(x) = a\};$$
$[f = \widehat{a}]$ is clopen in X and $\bigcup_{a \in A} [f = \widehat{a}] = X$. ◇

Since X is compact, the image of every element in $C(X, A)$ is finite. $C(X, A)$ will inherit any structure that is present on A, by defining it pointwise. For instance, if A is a group, the operation
$$[f \cdot g](x) = f(x) \cdot g(x),$$
for $f, g \in C(X, A)$, defines the structure of group in $C(X, A)$, whose identity is $\widehat{1}$, indicated by $\mathbb{1}$. Moreover, the map $a \mapsto \widehat{a}$ is an isomorphism of A onto the set of constant functions in $C(X, A)$. If A is a group of exponent 2, the same will be true of $C(X, A)$.

If A is a group with distinguished elements, 1 and -1, let $\widehat{1} = \mathbb{1}$ and $\widehat{-1} = -\mathbb{1}$ be the corresponding distinguished elements in $C(X, A)$.

Recall that a **partition** of a clopen set U is a finite collection of non-empty clopens, $P = \{p_1, \ldots, p_n\}$, that are pairwise disjoint and whose union is U. For instance, each f in $C(X, G)$ gives rise to a partition of X, namely, $\{[f = \widehat{a}] : a \in \text{Im } f\}$.

If P, Q are partitions, Q is **finer than** P (written $P \prec Q$) if there is a map $\alpha : Q \longrightarrow P$ such that $q \subseteq \alpha(q)$, $\forall q \in Q$.

If P, Q are partitions, $P \vee Q = \{p \cap q : p \in P, q \in Q \text{ and } p \cap q \neq \emptyset\}$ is a common refinement, with refinement maps given by $p \cap q \mapsto p$ (or q) from $P \vee Q$ to P (or Q).

If P is a partition of $U \in B(X)$ and we assign to each member of P an element of A, this map determines a unique element of $C(U, A)$, namely, the continuous function whose restriction to each $p \in P$ is constant, with value the element of A assigned to p. The following simple result is very useful.

LEMMA 6.3. *Let $U \in B(X)$ and let $\{f_1, \ldots, f_n\}$ a finite set in $C(X, A)$. Then, there is a partition P of U such that for all $1 \leq i \leq n$ and $p \in P$, f_i is constant in the clopen set p. The set of partitions that satisfy this property has a least upper bound in the order $\prec$.*

PROOF. For each $1 \leq i \leq n$, let
$$P_i = \{U \cap [f_i = \widehat{a}] : a \in \text{Im } f_i \text{ and } U \cap [f_i = \widehat{a}] \neq \emptyset\}.$$
Each P_i is a partition of U and we may take for P any common refinement of the P_i. In particular, P may be $P_1 \vee \ldots \vee P_n$, the coarsest satisfying our requirements. □

Let X be a Boolean space and G a special group. An element of $C(X,G)$ is said to be a **characteristic function** if its only values are ± 1. The set of characteristic functions is a subgroup of $C(X,A)$, containing $\mathbb{1}$ and $-\mathbb{1}$. If u is a clopen in X, write η_u for the characteristic function of u in $C(X,G)$, that is, the map defined by

$$\eta_u(x) = \begin{cases} -1 & x \in u \\ 1 & x \notin u. \end{cases}$$

It is clear that $-\eta_u = -\mathbb{1} \cdot \eta_u$ is the characteristic function of $-u$ in $C(X,G)$. The map $u \mapsto \eta_u$ has the following properties, where $\triangle$ denotes symmetric difference :

[ch 1] : $\eta_{(u \triangle v)} = \eta_u \cdot \eta_v$ [ch 2] : $\eta_u = -\mathbb{1}$ iff $u = X$.

Hence, $u \mapsto \eta_u$ is a group isomorphism from $B(X)$ (with $\triangle$) to the group of characteristic functions in $C(X,G)$.

DEFINITION 6.4. If G is a special group and X is a Boolean space, we define binary isometry in $C(X,G)$ by the clause

$$\langle f, g \rangle \equiv \langle h, k \rangle \quad \text{iff} \quad \forall\, x \in X,\ \langle f(x), g(x) \rangle \equiv_G \langle h(x), k(x) \rangle,$$

where $\equiv_G$ is the isometry in G.

If $\varphi = \langle f_1, \ldots, f_n \rangle$ is a form in $C(X,G)$ and $x \in X$, write $\varphi(x)$ for the form $\langle f_1(x), \ldots, f_n(x) \rangle$ in G. $\diamond$

DEFINITION 6.5. Let X be a Boolean space, G a special group and I a non-empty set. Let $\Sigma = \{(F_i, G_i) : i \in I\}$ be a collection of pairs, such that F_i is a closed set in X and G_i is a pre-special subgroup of G. Σ is called an **I-filtration** of (X,G) if it satisfies

$$\forall\, i, j \in I, \quad F_i \subseteq F_j \text{ implies } G_i \subseteq G_j.$$

In particular, for all I, $\{(X,G)\}$ is an I-filtration, called the **trivial filtration**.

If Σ is an I-filtration of (X,G), define

$$C(X,G;\Sigma) = \{f \in C(X,G) : \forall\, i \in I,\ \forall\, x \in X\ (x \in F_i \Rightarrow f(x) \in G_i)\},$$

called the **filtered Boolean power** of G by X over Σ. Note that if Σ is the trivial filtration on (X,G) then $C(X,G;\Sigma)$ is simply $C(X,G)$. $\diamond$

LEMMA 6.6. *Let Σ be an I-filtration of (X,G). With isometry as in Definition 6.4,*

a) $T = C(X,G;\Sigma)$ is a pre-special group, which is reduced if G is reduced. Moreover, if $\bigcup_{i \in I} F_i$ is not dense in X, $C(X,G;\Sigma)$ is reduced iff G is reduced.

b) $C(X,G)$ is a pre-special group, which is reduced iff G is reduced.

PROOF. Quite clearly, (a) implies (b). The axioms for pre-special group are readily verified, since isometry is defined pointwise and G is a special group; if G is reduced and $f \in T$, then $\langle f, f \rangle \equiv \langle \mathbb{1}, \mathbb{1} \rangle$ implies $\langle f(x), f(x) \rangle \equiv_G \langle 1, 1 \rangle$, for all x in X. Thus, f must be constantly equal to 1, that is, $f = \mathbb{1}$, showing that T is reduced.

Now assume that T is reduced and that $\bigcup_{i \in I} F_i$ is not dense in X. Select a non-empty clopen u in X such that $u \cap F_i = \emptyset$, for all $i \in I$. For each $a \in G$, consider u_a defined by

$$u_a(x) = \begin{cases} a & x \in u \\ 1 & x \notin u \end{cases}$$

Since $u \cap F_i = \emptyset$, for all i, $u_a \in T$. Moreover, for $a, b, c, d \in G$,

$$\langle a, b \rangle \equiv_G \langle c, d \rangle \quad \text{iff} \quad \langle u_a, u_b \rangle \equiv \langle u_c, u_d \rangle.$$

Thus, $\langle a, a \rangle \equiv_G \langle 1, 1 \rangle$ implies $u_a = \mathbb{1}$, and so $a = 1$ in G. □

2. SG-filtered Powers

We are interested in conditions that make a filtered Boolean power into a special group. Our most general results in this direction are Theorems 6.10 and 6.13, below. The considerations that follow are preparatory to their proof. The main difficulty to be overcome is the fact that the intersection of special subgroups (even if complete), might not be a special subgroup. In other situations, filtered powers will preserve the structure of its codomain. As an example, we state

LEMMA 6.7. *Let $\Sigma = \{(F_i, B_i) : i \in I\}$ be an I-filtration on (X, B), where X is a Boolean space, B is a Boolean algebra and B_i is a subalgebra of B. Then $C(X, B; \Sigma)$ is a Boolean algebra.*

PROOF. Write T for $C(X, G; \Sigma)$; clearly $\widehat{\top}$ and $\widehat{\bot}$ are in T. Since the operations of $\wedge$ (meet) and $\vee$ (join) are defined pointwise, it is straightforward to check that if $f, g \in T$ then

(i) $f \wedge g$ and $f \vee g$ are in T;

(ii) The operations $\wedge$ and $\vee$, defined in T, together with the constants $\widehat{\top}$ (as top) and $\widehat{\bot}$ (as bottom), satisfy the axioms for a Boolean algebra.

In particular, the complement of $f \in T$, $-f$, is given by $[-f](x) = -f(x)$. □

If L is a lattice, write $x \wedge y$ for the inf of x, y in L. If $S \subseteq L$, $\bigwedge S$ denotes the infimum of S in L (if it exists). We shall always assume that our lattices have a least and a largest element, indicated $\bot$ and $\top$, respectively.

DEFINITION 6.8. Let X be Boolean space, G a special group and L a complete lattice. Let $\Sigma = \{(F_l, G_l) : l \in L\}$ be a L-filtration of (X, G).

a) For each $x \in X$, define a map $\lambda_\Sigma : X \longrightarrow L$ by
$$\lambda_\Sigma(x) = \bigwedge \{l \in L : x \in F_l\},$$
where $\bigwedge \emptyset = \top$. When Σ is clear from context, write λ for λ_Σ.

For $x \in X$, set
$$F_x = F_{\lambda(x)} \quad \text{and} \quad G_x = G_{\lambda(x)}.$$
Note that for the trivial filtration, $F_x = X$ and $G_x = G$, for all $x \in X$.

b) Σ is said to be **continuous** if

1. $\forall\, S \subseteq L,\ F_{\bigwedge S} = \bigcap_{s \in S} F_s$ 2. $F_\top = X$ and $G_\top = G$.

c) Σ is a **SG-filtration** if it is continuous, if for every $l \in L$, G_l is a special subgroup of G, and it satisfies

[SG] Every non-empty clopen in X has a partition P such that, for each $p \in P$, there is $x \in p$ satisfying
$$\lambda(x) = \bigwedge \{\lambda(y) : y \in p\}. \qquad \diamond$$

Note that the trivial filtration is **SG**, and that a filtration over any complete linear order is continuous. Continuity furnishes a neat criterion for an element to be in a filtered power.

LEMMA 6.9. *Suppose Σ is a continuous L-filtration of (X, G). Then,*

a) $\forall\, l, l' \in L,\ l \leq l' \Rightarrow G_l \subseteq G_{l'}$.

b) For all $x \in X$, $x \in F_x$.

c) For $f \in C(X, G)$, the following are equivalent

(1) $f \in C(X, G; \Sigma)$ (2) $\forall\, x \in X,\ f(x) \in G_x$.

PROOF. Item (a) is straightforward, while (b) comes directly from continuity. To verify (c), write T for $C(X, G; \Sigma)$.

(1) $\Rightarrow$ (2). Let x be a point in X; by the continuity of Σ,
$$x \in F_x = \bigcap \{F_l : x \in F_l\},$$
and so by the very definition of T, $f(x) \in G_x$.

(2) $\Rightarrow$ (1). If $x \in F_l$, then $\lambda(x) \leq l$ and so $F_x \subseteq F_l$. By (a), $G_x \subseteq G_l$. Thus, if f satisfies (2), then $f(x) \in G_x \subseteq G_l$, and $f \in T$. $\square$

With the concepts introduced in Definition 6.8, we prove

THEOREM 6.10. *Let $\Sigma = \langle F_l, G_l \rangle_{l \in L}$ be a **SG**-filtration on (X, G) and let $T = C(X, G; \Sigma)$. Then, T is a special group such that, for n-forms $\varphi = \langle f_1, \ldots, f_n \rangle$, $\psi = \langle g_1, \ldots, g_n \rangle$ over T and $f \in T$, we have :*

a) $\varphi \equiv \psi$ *iff* $\forall\, x \in X,\ \varphi(x) \equiv_{G_x} \psi(x)$.

b) $f \in D_T(\varphi)$ iff $\forall\, x \in X$, $f(x) \in D_{G_x}(\varphi(x))$.

In particular, (a) and (b) hold for $C(X,G)$, with $G_x = G$, for all $x \in X$.

Moreover,

c) T is a complete subgroup of $C(X,G)$ iff $\forall\, x \in X$, G_x is a complete subgroup of G.

d) T is reduced   iff $\forall\, x \in X$, G_x is reduced. If $\bigcup_{l \in L} F_l$ is not dense in X, then T is reduced iff G is reduced.

PROOF. By Lemma 6.6, to show that T is a special group, it suffices to prove 3-transitivity. But this will follow directly from (a), since G_x is a special group, for all $x \in X$. In fact, except for (b) and the last clause in (d) (a restatement of Lemma 6.6.(a)), all assertions follow easily from (a). Thus, we give a proof only of (a) and (b).

Proof of (a). Since each G_l is a special subgroup of G and isometry in T is defined pointwise, it is clear that (a) holds for 2-forms over T. We proceed by induction on dimension. Assume, then, that (a) holds for $(n-1)$-forms.

(i) If $\varphi \equiv \psi$, then there are h, k, α_3, ..., α_{n-1} in T such that

$\langle f_1, h \rangle \equiv \langle g_1, k \rangle$; $\langle f_2, \ldots, f_n \rangle \equiv \langle h, \alpha_3, \ldots, \alpha_{n-1} \rangle$;   and

$\langle g_2, \ldots, g_n \rangle \equiv \langle k, \alpha_3, \ldots, \alpha_{n-1} \rangle$.

By induction, for each $x \in X$, we get

* $\langle f_1(x), h(x) \rangle \equiv_{G_x} \langle g_1(x), k(x) \rangle$,
* $\langle f_2(x), \ldots, f_n(x) \rangle \equiv_{G_x} \langle h(x), \alpha_3(x), \ldots, \alpha_{n-1}(x) \rangle$,
* $\langle g_2(x), \ldots, g_n(x) \rangle \equiv_{G_x} \langle k(x), \alpha_3(x), \ldots, \alpha_{n-1}(x) \rangle$,

conditions which are equivalent to $\varphi(x) \equiv_{G_x} \psi(x)$.

(ii) Now suppose that $\varphi(x) \equiv_{G_x} \psi(x)$, for all $x \in X$. By Lemma 6.3, there is a partition P of X such that the f_i's and g_i's are constant on any element of P. For each $p \in P$, the property that Σ is **SG** yields a partition Q_p of p, together with a point t_q in each $q \in Q_p$, such that t_q is a point of minimum of λ on q. In particular, for all $y \in q$, $G_{t_q} \subseteq G_y$. Let $Q = \bigcup_{p \in P} Q_p$; then Q is a partition of X such that for each $q \in Q$,

   (i) For all $1 \leq j \leq n$, f_j and g_j are constant on q;

   (ii) There is $t_q \in q$ such that for all $y \in q$, $G_{t_q} \subseteq G_y$.

Fix $q \in Q$; to ease notation put $G_q = G_{t_q}$. Let a_j, b_j be the constant value of f_j and g_j in q, respectively. Observe that $f_j(t_q) = a_j$ and $g_j(t_q) = b_j$; thus, $a_j, b_j \in G_q$, for all j (Lemma 6.9).

Since $\varphi(x) \equiv_{G_x} \psi(x)$, for all $x \in X$, this relation in t_q will imply $\langle a_1, \ldots, a_n \rangle \equiv_{G_q} \langle b_1, \ldots, b_n \rangle$. G_q being a special group, there are c_q, d_q, $z_{3q}, \ldots, z_{nq}$ in G_q such that

i) $\langle a_1, c_q \rangle \equiv_{G_q} \langle b_1, d_q \rangle$;

ii) $\langle a_2, \ldots, a_n \rangle \equiv_{G_q} \langle c_q, z_{3q}, \ldots, z_{nq} \rangle$;

(iii) $\langle b_2, \ldots, b_n \rangle \equiv_{G_q} \langle d_q, z_{3q}, \ldots, z_{nq} \rangle$.

Because $G_q \subseteq G_y$, for all $y \in q$, we have

(*)   Items (i), (ii) and (iii) hold for all $y \in q$, with G_y in place of G_q.

By the method described above, we construct, for each $q \in Q$, a family $\{c_q, d_q, z_{3q}, \ldots, z_{nq}\}$, such that (i), (ii) and (iii) hold for all $y \in q$, with G_y in place of G_q. Define an element γ of $C(X, G)$ by the following clause :

For each $q \in Q$, $\gamma_{|q}$ is the constant function whose value is c_q.

Note that if $x \in q$, $\gamma(x) = c_q \in G_q \subseteq G_x$, for all $x \in q$. Hence, for all $x \in X$, $\gamma(x) \in G_x$, and Lemma 6.9.(b) yields $\gamma \in T$.

Similarly, we can construct

∗ δ, as the element of T that restricted to each q, is constantly equal to d_q;

∗ For $3 \leq k \leq n$, ζ_k, as the element of T that restricted to q is equal to z_{kq}.

The definitions of γ, δ, ζ_k and (*) guarantee that for all $x \in X$,

∗ $\langle f_1(x), \gamma(x) \rangle \equiv_{G_x} \langle g_1(x), \delta(x) \rangle$;

∗ $\langle f_2(x), \ldots, f_n(x) \rangle \equiv_{G_x} \langle \gamma(x), \zeta_3(x), \ldots, \zeta_n(x) \rangle$;

∗ $\langle g_2(x), \ldots, g_n(x) \rangle \equiv_{G_x} \langle \delta(x), \zeta_3(x), \ldots, \zeta_n(x) \rangle$.

By induction, we get

$$\langle f_1, \gamma \rangle \equiv \langle g_1, \delta \rangle, \quad \langle f_2, \ldots, f_n \rangle \equiv \langle \gamma, \zeta_3, \ldots, \zeta_n \rangle, \text{ and}$$
$$\langle g_2, \ldots, g_n \rangle \equiv \langle \delta, \zeta_3, \ldots, \zeta_n \rangle,$$

precisely what is needed to conclude that $\varphi \equiv \psi$ in T.

Proof of (b). It follows from (a) that if f is represented by φ, then $f(x)$ is represented by $\varphi(x)$, for all $x \in X$. For the converse, we proceed as in the proof of part (a). Thus, select a partition P of X such that all the components of φ and f are constant on any element of P (Lemma 6.3). Since Σ is **SG**, each $p \in P$ has a partition Q_p where, for each $q \in Q_p$, $\lambda_{|q}$ attains its minimum at a point $t_q \in q$. Set $Q = \bigcup_{p \in P} Q_p$ and $G_{t_q} = G_q$.

For each $q \in Q$, let

$$f(q) = f(t_q) \quad \text{and} \quad \varphi(q) = \varphi(t_q),$$

be the constant value of f and the constant form corresponding to φ on q, respectively. We know that $f(q)$ is an element of G_q, while $\varphi(q)$ is a n-form in G_q. Moreover, $f(q) \in D_{G_q}(\varphi(q))$. For $q \in Q$, select $c_{2q}, \ldots, c_{nq} \in G_q$, such that $\langle f(q), c_{2q}, \ldots, c_{nq} \rangle \equiv_{G_q} \varphi(q)$. The same argument as in the proof of (a) shows that, for each $2 \leq k \leq n$, the c_{kq} give rise to a map γ_k in T, whose value in each q is exactly c_{kq}. Furthermore, for all $x \in X$,

$$\langle f(x), \gamma_2(x), \ldots, \gamma_n(x) \rangle \equiv_{G_x} \varphi(x),$$

and so, by (a), $\langle f, \gamma_2, \ldots, \gamma_n \rangle \equiv \varphi$, that is, $f \in D_T(\varphi)$. $\square$

COROLLARY 6.11. *Let $T = C(X, G; \Sigma)$ be an **SG** filtered power, with G a reduced special group. Let U, V be clopens in X and f an element of T. Then,*

a) $U \subseteq V$ iff $\eta_U \in D_T(\mathbb{1}, \eta_V)$.

b) $f \in D_T(\mathbb{1}, \eta_U)$ iff $-U \subseteq [f = \mathbb{1}]$.

PROOF. Both statements follow from item (b) in Theorem 6.10 and the fact that for all $x \in X$, $D_{G_x}(1, -1) = G_x$ and $D_{G_x}(1, 1) = \{1\}$. $\square$

REMARK 6.12. Theorem 6.10 applies of course to $T = C(X, 2)$ and so this structure is a reduced special group. We have already observed that the map $U \mapsto \eta_U$ is a group isomorphism between T and the BA of clopens in X. Moreover, Corollary 6.11 tells us that under this isomorphism, the representation partial order in T corresponds to the natural partial order in the BA of clopens of X. This remark is the origin of our Chapter 4. $\diamond$

We now show that there is a reasonable supply of **SG**-filtrations.

THEOREM 6.13. *Let L be a complete lattice. All L-filtrations are assumed to be such that G_l are special groups.*

a) *If L is finite, any continuous L-filtration is **SG**. Moreover,*

(∗) $\forall\, x \in X$, *there is $U \in \nu_x$, such that $\lambda(x) \leq \lambda(y)$, for all $y \in U$.*

b) *If L is a complete linear order, any continuous L-filtration is **SG**. Moreover,*

(usc)   *If K is a compact set in X, $\lambda_{|K}$ attains its minimum on K.*

PROOF. Our proof will be divided into two parts : in the first part we deal with the case of finite lattices; the linear order case shall be dealt with in the second part.

Part I. L is finite

Our first step is to check that statement (∗) in (a) holds true. Given $x \in X$, consider the set $A = \{l \in L : x \notin F_l\}$; since A is finite and x is in the open set $\bigcap_{l \in A} (X - F_l)$, there is $U \in \nu_x$, such that $U \cap F_l = \emptyset$, for all l in A. Thus, for all $y \in U$ we have

$$\{l \in L : y \in F_l\} \subseteq \{l \in L : x \in F_l\},$$

which shows that $\lambda(x) \leq \lambda(y)$, as desired. Now, we establish

Fact. *Suppose Σ is an L-filtration satisfying (*) in (a). If K is a non-empty compact in X and V is a clopen neighborhood of K, then there is a disjoint covering of K by clopens $\{U_1, \ldots, U_m\}$, together with $\{t_1, \ldots, t_m\} \subseteq K$, such that for all $1 \leq j \leq m$, $t_j \in U_j \subseteq V$ and for all $y \in U_j$, $\lambda(t_j) \leq \lambda(y)$.*

PROOF. By (*), for each $x \in K$, there is $V_x \in \nu_x$, which may be chosen inside V, such that $\lambda(x) \leq \lambda(y)$, for all $y \in V_x$. Since K is compact, there is a finite collection V_{x_i}, $1 \leq i \leq n$, such that $K \subseteq \bigcup_{i \leq n} V_{x_i}$. We shall define, by simultaneous induction, (W_k, t_k), such that for all k

(i) W_k is a clopen contained in V;   (ii) For all $y \in W_k$, $\lambda(t_k) \leq \lambda(y)$;

(iii) $t_k \in W_k \setminus \bigcup \{W_j : j < k\}$.   (iv) $K \subseteq \bigcup W_k \subseteq V$.

The definition of the desired sequence is as follows :

∗ $W_1 = \bigcup \{V_j : j \leq n \text{ and } \lambda(x_j) \geq \lambda(x_1)\}$ and $t_1 = x_1$.

It is clear that (W_1, t_1) satisfies the first two conditions above.

∗ Having defined (W_j, t_j), for $j \leq k$, let

$$\alpha = \{l \leq n : \text{For all } j \leq k, \text{ it is not the case that } \lambda(x_l) \geq \lambda(t_j)\}.$$

If $\alpha = \emptyset$, then the sequence of W's and t's stops at k ($= m$). If $\alpha \neq \emptyset$, let $i = \min \alpha$ and define $U_{k+1} = \bigcup \{V_j : j \leq n \text{ and } \lambda(x_j) \geq \lambda(x_i)\}$ and $t_{k+1} = x_i \in V_i \subseteq W_{k+1}$.

Clearly, W_{k+1} is a clopen contained in V, and $t_{k+1} \in (W_{k+1} - W_j)$, for $j \leq k$. Moreover, from the construction and the fact that x_i is a point of minimum value of λ in each V_i, it follows that t_k is a point of minimum for λ in W_k. Finally, since each of V_i will be contained in some W_k, K is contained in the union of the W_k. □

To finish the proof of Part I, set, for $k \leq m$,

$$U_k = W_k \setminus \bigcup_{j < k} W_j;$$

then the U_k are a pairwise disjoint cover of K contained in V and t_k is a point of minimum of λ in U_k, for all $k \leq m$.

Part II. L is a complete linear order

As in Part I, we first verify (usc) in (b). Let K be a non-empty compact in X. Consider the set $\alpha = \{\lambda(y) : y \in K\}$ and $\gamma = \bigwedge \alpha$; it is enough to prove that there is $x \in K$ such that $\lambda(x) = \gamma$. Moreover, since Σ is continuous, $F_\gamma = \bigcap_{x \in K} F_x$ and $y \in F_y \cap K$, for all $y \in K$.

Since L is a linear order, the family of closed sets $\{F_x \cap K : x \in K\}$ has the finite intersection property. Thus, K being compact, $F_\gamma \cap K$ is not empty and any of its elements will be points of minimum of λ in K. Consequently, if U is a clopen in X, the partition of U needed to verify **SG** may consist of U itself. $\square$

REMARK 6.14. Condition (*) in Theorem 6.13 is a strong form of upper semicontinuity (ucs) of the map $x \mapsto \lambda(x)$, with respect to the order of L. Although one might be used to the slogan *ucs maps assume a minimum on compacts*, it should be noted that this holds when the codomain of such functions are complete linear orders. Condition **SG** (or the statement of the Fact in the proof of 6.13) may be considered a working substitute for the existence of a minimum. $\diamond$

REMARK 6.15. There are other constructions, akin to filtered Boolean powers, in which X is not necessarily compact. Important examples are bounded filtered powers and filtered powers of compact support. Analogues of 6.10 and 6.13 are still true, with no essential modifications in the proofs presented above. We briefly recall the aforementioned constructions.

Let X be a **quasi-Boolean** space, that is, a Hausdorff topological space with a basis of **compact** clopens. If A is any set with the discrete topology, $C(X, A)$ denotes the set of all continuous functions from X to A. Now define

$$C_b(X, A) = \{f \in C(X, A) : \text{the image of } f \text{ in } A \text{ is finite}\},$$

called the **bounded quasi-Boolean power** of A by X.

If A is a special group with distinguished elements ± 1, define

$$C_c(X, A) = \left\{f \in C(X, A) : \begin{array}{l} \text{Outside of a compact set in } X, \\ f \text{ is constant and equal to 1 or } -1, \end{array}\right\}$$

called the **quasi-Boolean power of compact support** of A by X.

One can consider the pertinent definitions of filtered power in each of the contexts described herein, to obtain results which are analogous to the ones proven above when X is a Boolean space.

Observe that when X is compact, all these structures are none other than $C(X, A)$. We have chosen to present only the compact case, essentially because the others are isomorphic to filtered powers on compactifications of X. For bounded powers, the compactification γX is that determined by the following properties

(i) $X \hookrightarrow \gamma X$ is a dense embedding of topological spaces;

(ii) For all sets A and $f \in C_b(X, A)$, f has a unique extension to $C(\gamma X, A)$.

For structures of compact support, the needed compactification is the one-point compactification of X. A typical example follows. ◊

EXAMPLE 6.16. The filtered power construction extends, for reduced special groups, the notion of sum in Definition 5.27. To see this, let I be a set and G_i, $i \in I$, a family of special subgroups of a group G.

Let $\alpha I = \{\infty\} \cup I$ be the one-point compactification of the discrete topology on I. The compact sets in I are the finite sets; thus, the neighborhoods of ∞ in αI consists of itself joined with a cofinite set in I; αI is a Boolean space, where each singleton, as well as the join of $\{\infty\}$ with a cofinite set in I, is clopen.

Let L be the complete lattice consisting of $I \cup \{\bot, \top\}$, where the join or the meet of any two distinct elements is $\top$ or $\bot$, respectively. Consider the L-filtration Σ on $(\alpha I, G)$, determined by the following conditions:

i) For $i \in I$, $F_i = \{i, \infty\}$ and G_i is the group originally associated to i.

ii) $F_\bot = \{\infty\}$ and $G_\bot = \mathbb{Z}_2$.

Clearly, Σ is a continuous L-filtration. Moreover, given a clopen set u in αI, there are two possibilities:

- $\infty \in u$, in which case the map λ has ∞ as a point of minimum on u;

- $\infty \notin u$, in which case u is finite and so can be written as the union of a partition consisting of clopen points.

It follows immediately from the above that Σ is a **SG**-filtration on $(\alpha I, G)$. Moreover, $C(\alpha I, G; \Sigma)$ is exactly $\bigoplus_{i \in I}^* G_i$.

If the G_i's are reduced, there is no loss of generality in assuming that they are all special subgroups of a special group G; in fact, we can find a special group G in which they are all completely embedded. To verify this, proceed as follows:

Let X_i be the space of orders of G_i and $B_i = B(X_i)$. Let $X = \prod_{i \in I} X_i$ and let p_i be the canonical projection onto X_i; write B for the algebra of clopens of the Boolean space X. Then,

1. For $i \in I$, the map $U \in B_i \mapsto p_i^{-1}(U) \in B$, is an injective BA-morphism from B_i to B and so a complete embedding (Corollary 5.6).

2. Let $G_i \xrightarrow{\varepsilon_{G_i}} B_i$ be the complete embedding associated to the Boolean hull of G_i (Corollary 5.4). Then $f_i \circ \varepsilon_{G_i}$ is a complete embedding of G_i into B.

Continuity at ∞ and the fact that $G_\bot = \mathbb{Z}_2$ easily yield that $C(\alpha I, B; \Sigma)$ coincides with $\bigoplus_{i \in I}^* G_i$. ◊

3. Stalks and Germs

To be able to identify the space of orders of a **SG** filtered power T it will be important to characterize the germs of elements of T. It turns out that the restriction of such a filtered power to a clopen set in X is itself a special group. The inductive limit of such groups over the clopen neighborhoods of a point x, *the germs of T at x*, will also be a special group. It will be shown that the isometry of n-forms in any restriction of T to a clopen set U is equivalent to the isometry of associated forms in the stalks at the points of U. In a certain sense, although a corollary of it, this result is a refinement of Theorem 6.10.

COROLLARY 6.17. *Let X be a Boolean space and G a special group. Let $\Sigma = \{(F_l, G_l) : l \in L\}$ be a **SG**-filtration of (X, G) and let T be the associated filtered power. For U clopen in X, let*

$$\Sigma_{|U} = \{(U \cap F_l, G_l) : l \in L\},$$

be the restriction of Σ to U. Then,

a) *For all $x \in U$, $\lambda_\Sigma(x) = \lambda_{\Sigma_{|U}}(x)$. Hence, G_x and F_x are the same, whether computed with respect to λ_Σ or to $\lambda_{\Sigma_{|U}}$.*

b) *For all clopen U in X, $\Sigma_{|U}$ is a **SG**-filtration of (U, G) and $T_U =_{def} C(U, G; \Sigma_{|U})$ is a special group with the isometry relation induced by $C(U, G)$.*

c) *If $U \subseteq V$ are clopens in X, there is a surjective SG-morphism*

$$\gamma_{VU} : T_V \longrightarrow T_U, \text{ given by } \quad f \mapsto f_{|U},$$

such that for all $U \subseteq V \subseteq W$

$$\gamma_{UU} = Id_{T_U} \quad \text{and} \quad \gamma_{WU} = \gamma_{VU} \circ \gamma_{WV}.$$

d) *For $x \in X$ and $U \in \nu_x$, there is a SG-morphism*

$$ev_x : T_U \longrightarrow G_x, \text{ given by } f \mapsto f(x),$$

such that if $U \subseteq V$ are in ν_x, then $\gamma_{VU} \circ ev_x = ev_x$.

PROOF. Item (a) is immediate from the definitions of λ_Σ, F_x and G_x. It is clear that $\Sigma_{|U}$ is continuous and so it follows from (a) that it is also **SG**. Thus, T_U is a special group, for $U \in B(X)$. Moreover, if φ, ψ are n-forms in T_U, then (a) and Theorem 6.10 yield

$$\varphi \equiv_{T_U} \psi \quad \text{iff} \quad \forall\, x \in U, \ \varphi(x) \equiv_{G_x} \psi(x). \tag{I}$$

If $f \in T_V$ and $U \subseteq V$, it is immediate that $f_{|U}$ is in T_U, as well as that these maps have the properties stated in (c). Observe that (I) implies that

112   6. SPECIAL GROUPS OF CONTINUOUS FUNCTIONS

γ_{VU} is a SG-morphism. To see that it is in fact surjective, just note that if $f \in T_U$, then the map defined in V by

$$g(y) = \begin{cases} f(y) & y \in U \\ 1 & y \in V \setminus U, \end{cases}$$

is an extension of f that belongs to T_V. It remains to comment on (d). Lemma 6.9.(c) tells us that if $f \in T_U$, then $f(x) \in G_x$, for all $x \in X$. Thus, ev_x is well defined, and obviously has the commutative property in (d). It follows from (I) (or Theorem 6.10) that ev_x is a SG-morphism. □

REMARK 6.18. Item (c) in Corollary 6.17 says that a **SG** filtered power gives rise to a presheaf basis of special groups on X. Although we shall refrain from adopting a systematic sheaf theoretic point of view, it is clear to those familiar with this language that many of our results and definitions have that flavor. ◇

DEFINITION 6.19. Let Σ be a **SG** filtration on (X, G) and let $T = C(X, G; \Sigma)$ be the associated filtered power. Notation is as in Corollary 6.17. For each $x \in X$, let T_x be the inductive limit of the inductive system of special groups

$$\mathcal{T}(x) = (T_U, \{\gamma_{VU} : U \subseteq V \text{ clopen neighborhoods of } x\}).$$

T_x is called the **stalk** of T at x and its elements, the **germs** of T at x.

For each $f \in T_U$, let f_x be the germ of f at x, that is, the equivalence class of all $g \in T_U$ such that for some $W \in \nu_x$, $W \subseteq U$, we have $f_{|W} = g_{|W}$.

If $\varphi = \langle f_1, \ldots, f_n \rangle$ is a form over T_U, we indicate by φ_x the form $\langle f_{1x}, \ldots, f_{nx} \rangle$ over T_x.

Observe that the canonical SG-morphisms that come with the inductive limit,

$$\gamma_{Ux} : T_U \longrightarrow T_x,$$

where $U \in \nu_x$, satisfy, for $U \subseteq V$,

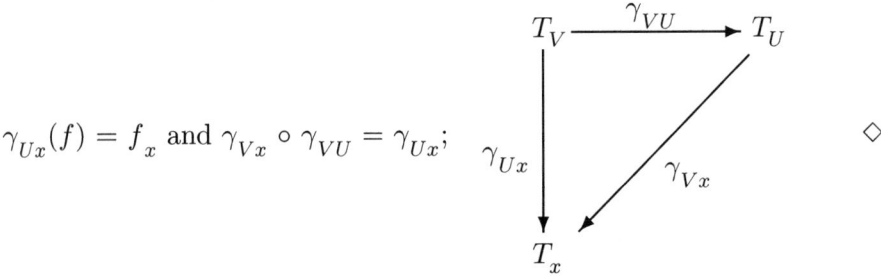

$\gamma_{Ux}(f) = f_x$ and $\gamma_{Vx} \circ \gamma_{VU} = \gamma_{Ux}$;   ◇

THEOREM 6.20. For a **SG**-filtration Σ on (X, G), let $T = C(X, G; \Sigma)$ be the associated filtered power. Let $U \in B(X)$ and let φ, ψ be n-forms over T_U.

a) For $x \in X$, there is an injective SG-morphism $\gamma_x : T_x \longrightarrow G_x$, such that

For all clopen neighborhoods U of x,
$$\gamma_x \circ \gamma_{Ux} = ev_x,$$
where ev_x is the evaluation map of 6.17.

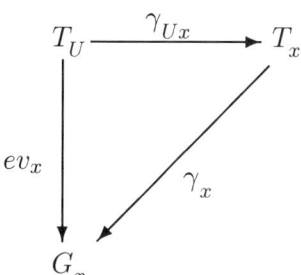

Moreover, if Σ satisfies

(*) $\forall\, x \in X$, there is $U \in \nu_x$, such that $\lambda(x) \leq \lambda(y)$, for all $y \in U$,

then, γ_x is an isomorphism of special groups and we identify T_x with G_x and γ_{Ux} with ev_x.

b) The following conditions are equivalent:

(1) $\varphi \equiv_{T_U} \psi$.   (2) $\forall\, x \in U,\ \varphi_x \equiv_{T_x} \psi_x$.

PROOF. a) For $x \in X$, if $U \in \nu_x$, Corollary 6.17.(d) guarantees that $ev_x : T_U \longrightarrow G_x$ is a morphism of the inductive system $\mathcal{T}(x)$ to the special group G_x. Since T_x is the inductive limit of $\mathcal{T}(x)$, its defining universal property yields a SG-morphism $\gamma_x : T_x \longrightarrow G_x$, such that for all $U \in \nu_x$, $\gamma_x \circ \gamma_{Ux} = ev_x$, as desired.

To see that γ_x is injective, let $a \neq b$ be germs in T_x and let $f, g \in T_U$ be such that $f_x = a$ and $g_x = b$. Because $a \neq b$, f and g are distinct in every clopen neighborhood of x. Taking into account that these maps are locally constant, it follows that $f(x) \neq g(x)$.

Now assume that Σ satisfies (*) and fix x in X and a clopen $U \in \nu_x$, for which the conclusion of (*) holds. We show that γ_x is isomorphism in two steps :

i) $\underline{\gamma_x \text{ is onto}}$. For $a \in G_x$, let u_a be the map

$$u_a(y) = \begin{cases} a & y \in U \\ 1 & y \notin U \end{cases} \tag{I}$$

defined in the proof of 6.6. Since G_x is contained in G_y, for all $y \in U$, Lemma 6.9 entails $u_a \in T$. By Theorem 6.20.(a),

$$\gamma_x(\gamma_{Tx}(u_a)) = ev_x(u_a) = a,$$

showing that a is a value of γ_x.

ii) $\underline{\gamma_x \text{ is an isomorphism}}$. We verify that for all n-forms φ, ψ over T_x,

$$\varphi \equiv_{T_x} \psi \quad \text{iff} \quad \gamma_x \star \varphi \equiv_{G_x} \gamma_x \star \psi \qquad (**)$$

It is clearly enough to show the "only if" direction in (**). For $1 \leq j \leq n$, let a_j be the jth entry of $\gamma_x \star \varphi$ and b_j be corresponding entry of $\gamma_x \star \psi$. By hypothesis

$$\langle a_1, \ldots, a_n \rangle \equiv_{G_x} \langle b_1, \ldots, b_n \rangle. \qquad (II)$$

Consider the maps u_{a_j}, u_{b_j} determined, as in (I), by U, a_j and b_j, respectively. Just as before, we have $u_{a_j}, u_{b_j} \in T$. By Theorem 6.10, the isometry in (II) implies that

$$\langle u_{a_1}, \ldots, u_{a_n} \rangle \equiv_T \langle u_{b_1}, \ldots, u_{b_n} \rangle. \qquad (III)$$

Notice that for each $1 \leq j \leq n$,

$$\gamma_x(\gamma_{Tx}(u_{a_j})) = ev_x(u_{a_j}) = a_j \quad \text{and} \quad \gamma_x(\gamma_{Tx}(u_{b_j})) = ev_x(u_{b_j}) = b_j,$$

and so, γ_x being injective, we conclude

$$\gamma_{Tx} \star \langle u_{a_1}, \ldots, u_{a_n} \rangle = \varphi \quad \text{and} \quad \gamma_{Tx} \star \langle u_{b_1}, \ldots, u_{b_n} \rangle = \psi. \qquad (IV)$$

Since γ_{Tx} is a SG-morphism, (III) and (IV) yield $\varphi \equiv_{T_x} \psi$, as needed.

b) We need only prove (2) implies (1) because γ_{Ux} is a SG-morphism for each $x \in U$. Assume that $\varphi_x \equiv_{T_x} \psi_x$, for all $x \in U$. By (a) we have

$$\varphi(x) = \gamma_x \star (\gamma_{Ux} \star \varphi) = \gamma_x \star \varphi_x;$$

similarly, $\psi(x) = \gamma_x \star \psi_x$. From the fact that γ_x is a SG-morphism, we conclude $\varphi(x) \equiv_{G_x} \psi(x)$. Since x is arbitrary in U, Theorem 6.10 (or Corollary 6.17) yields $\varphi \equiv_{T_U} \psi$, ending the proof. □

EXAMPLE 6.21. Let $\Sigma = \{(F_l, G_l) : l \in L\}$ be an L-filtration of (X, G) and T the corresponding filtered power.

(I) If L is a finite lattice or T is the filtered power of 6.16, then $T_x = G_x$ and $\gamma_{Ux} = ev_x$, for all $x \in X$. To see this, just note that the Fact in the proof of Theorem 6.13 shows that (*) in Theorem 6.20 is verified.

It is left to the reader the verification of

(II) If L is a complete linear order, then

$$T_x = \begin{cases} G_x & \text{if } x \notin \overline{\bigcup\{F_l : l < \lambda(x)\}} \\ \bigcup\{G_l : l < \lambda(x)\} & \text{if } x \in \overline{\bigcup\{F_l : l < \lambda(x)\}}. \end{cases} \quad \diamond$$

Before our next result, we generalize Remark 1.9.13 in [LP3].

LEMMA 6.22. *Let* $\mathcal{G} = (G_i, \{f_{ij} : i \leq j\})$ *be an inductive system of special groups over the right directed set* I *(i.e.,* $\forall\ i, j\ \exists\ k$ *such that* $i \leq k$ *and* $j \leq k$*). Let* G *be the inductive limit of* $\mathcal{G}$ *and let* $G_i \xrightarrow{f_i} G$ *the canonical SG-morphisms associated to this construction.*

3. STALKS AND GERMS

a) For n-forms φ, ψ over G, the following are equivalent :

(1) $\varphi \equiv_G \psi$.

(2) There is $i \in I$ and forms φ_i, ψ_i over G_i, such that

(i) $\varphi_i \equiv_{G_i} \psi_i$; (ii) $f_i \star \varphi_i = \varphi$ and $f_i \star \psi_i = \psi$.

b) Let φ, ψ be forms over G_i. Then, $f_i \star \varphi \equiv_G f_i \star \psi$ iff $\exists\, k \geq i$ such that $f_{ik} \star \varphi \equiv_{G_k} f_{ik} \star \psi$.

c) Let $\mathcal{H} = (H_i, \{g_{ij} : i \leq j\})$ be an inductive system of special groups over the same set I, with limit $(H, \{g_i : i \in I\})$. Let α_i be SG-morphisms from G_i to H_i, such that, $\{\alpha_i : i \in I\}$ is a morphism of inductive systems (i.e. $g_{ij} \circ \alpha_i = \alpha_j \circ f_{ij}$, $\forall\, i \leq j$). Let α be the map from G to H, induced by the α_i's. Then, α is a SG-morphism. If all α_i's are complete embeddings, the same will be true of α.

PROOF. a) It is observed in [LP3] that G is a special group and that every f_i is a SG-morphism. Recall that if $x \in G_i$ and $y \in G_j$ are such that $f_i(x) = f_j(y)$, then there is $k \geq i, j$ such that $f_{ik}(x) = f_{jk}(y)$.

The equivalence of (1) and (2) is the definition of $\equiv_G$ for 2-forms. Lemma 1.13.(c) yields (2) implies (1) for forms of any dimension. It remains to check that (1) implies (2) for forms φ, ψ, of dimension ≥ 2, which we proceed to do by induction.

Assume the conclusion valid for forms of dimension $(n-1) \geq 2$ and let φ, ψ be n-forms. Write $\varphi = \langle a \rangle \oplus \varphi'$ and $\psi = \langle b \rangle \oplus \psi'$; thus, there are $x, y \in G$ and a $(n-2)$-form θ over G such that

* $\langle a, x \rangle \equiv_G \langle b, y \rangle$;   * $\varphi' \equiv_G \langle x \rangle \oplus \theta$;   * $\psi' \equiv_G \langle y \rangle \oplus \theta$.

Choose $i_1 \in I$ and $a_1, x_1, b_1, y_1 \in G_{i_1}$, such that

* $f_{i_1}(a_1) = a$, $f_{i_1}(x_1) = x$, $f_{i_1}(b_1) = b$, $f_{i_1}(y_1) = y$;
* $\langle a_1, x_1 \rangle \equiv_{G_{i_1}} \langle b_1, y_1 \rangle$.

By induction, select $i_2, i_3 \in I$, and $x_2, \varphi'_2, \theta_2, y_3, \psi'_3, \theta_3$ such that

* $\varphi'_2 \equiv_{G_{i_2}} \langle x_2 \rangle \oplus \theta_2$;   * $\psi'_3 \equiv_{G_{i_3}} \langle y_3 \rangle \oplus \theta_3$;   * $f_{i_2}(x_2) = x$, $f_{i_3}(y_3) = y$;
* $f_{i_2} \star \varphi'_2 = \varphi'$, $f_{i_3} \star \psi'_3 = \psi'$ and $f_{i_2} \star \theta_2 = f_{i_3} \star \theta_3 = \theta$.

Hence, there is a k in I, greater than i_1, i_2 and i_3, such that if we set

* $x_k = f_{i_1 k}(x_1) = f_{i_2 k}(x_2)$; $y_k = f_{i_1 k}(y_1) = f_{i_3 k}(y_3)$;
* $a_k = f_{i_1 k}(a_1)$ and $b_k = f_{i_1 k}(b_1)$;
* $\theta_k = f_{i_2 k} \star \theta_1 = f_{i_3 k} \star \theta_3$; $\varphi'_k = f_{i_2 k} \star \varphi'_2$ and $\psi'_k = f_{i_3 k} \star \psi'_3$;
* $\varphi_k = \langle a_k \rangle \oplus \varphi'_k$ and $\psi_k = \langle b_k \rangle \oplus \psi'_k$

then, $\varphi = f_k \star \varphi_k$ and $\psi = f_k \star \psi_k$; since the f_{ij} are SG-morphisms, we also get

$$\langle a_k, x_k \rangle \equiv_{G_k} \langle b_k, y_k \rangle, \quad \varphi'_k \equiv_{G_k} \langle x_k \rangle \oplus \theta_k \quad \text{and} \quad \psi'_k \equiv_{G_k} \langle y_k \rangle \oplus \theta_k,$$

isometries which imply $\varphi_k \equiv_{G_k} \psi_k$, ending the proof of (a). Item (b) is a straightforward consequence of (a).

Proof of (c). The fact that $\{\alpha_i : i \in I\}$ is a morphism of special groups, i.e., that all squares of type (I) below are commutative, makes it possible to define α as follows : For $a \in G$, select $i \in I$ and $x \in G_i$, such that $f_i(x) = a$. Set $\alpha(a) = g_i(\alpha_i(x))$.

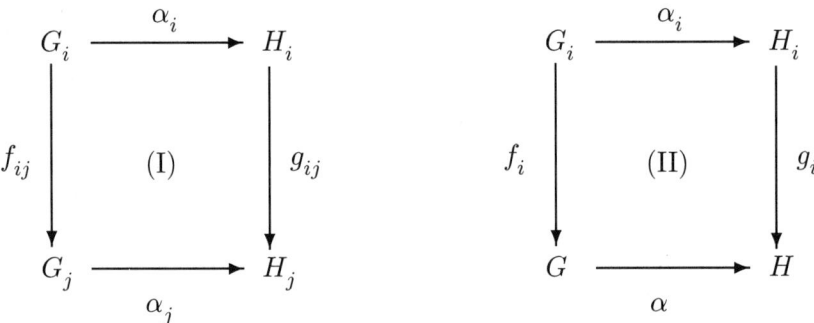

With this definition, α will also make all squares of type (II) commutative. It follows easily from (a) that α is a SG-morphism. Now suppose that each α_i is a complete embedding. To prove that α is injective, take $a, b \in G$ such that $\alpha(a) = \alpha(b)$. Select $i \in I$ and $x, y \in G_i$ such that $f_i(x) = a$ and $f_i(y) = b$. By the commutativity of squares of type (II), we get

$$\alpha(a) = \alpha(f_i(x)) = g_i(\alpha_i(x)) \quad \text{and} \quad \alpha(b) = \alpha(f_i(y)) = g_i(\alpha_i(y)).$$

Thus, $g_i(\alpha_i(x)) = g_i(\alpha_i(y))$, and so, there is $j \geq i$ such that $g_{ij}(\alpha_i(x)) = g_{ij}(\alpha_i(x))$. Applying the commutativity of squares of type (I) yields $\alpha_j(f_{ij}(x)) = \alpha_j(f_{ij}(y))$. Since α_j is injective, we get $f_{ij}(x) = f_{ij}(y)$, exactly the condition for a to be equal to b. The completeness of α follows from (a) and that of the α_i's. $\square$

Notational Convention. If T is a **SG** filtered power, $U \subseteq V$ are clopens and $f \in T_V$, to ease notation, we write $f_{|U}$ in place of $\gamma_{VU}(f)$. Similarly, f_x will denote the germ of f at x, that is, $\gamma_{Vx}(f)$. If f is indexed, say f_n, we write f_{nx} in place of $(f_n)_x$. $\diamond$

As an application of Lemma 6.22, we obtain a description of the Boolean hull of the stalks of a reduced **SG** filtered power.

NOTATION 6.23. Let $T = C(X, G; \Sigma)$ be a **SG** filtered power, with G reduced. We know from Lemma 6.6.(a) and Corollary 6.17 that T_U is a reduced special group for all clopen U in X; let

$$\varepsilon_U : T_U \longrightarrow B_U,$$

be the Boolean hull of T_U, as in Chapter 4. Since the Boolean hull is a functor (Theorem 4.17), for $U \subseteq V$, let $\beta_{VU} : B_V \longrightarrow B_U$ be the Boolean homomorphism associated to the restriction SG-morphism γ_{VU}, indicated in Chapter 5 by $B(\gamma_{VU})$. The functoriality of this construction guarantees that $\beta_{UU} = Id_{B_U}$; further, if $W \subseteq U \subseteq V$ then $\beta_{VW} = \beta_{UW} \circ \beta_{VU}$. Moreover, item (2) in Theorem 4.17, implies that all squares of type (A) are commutative.

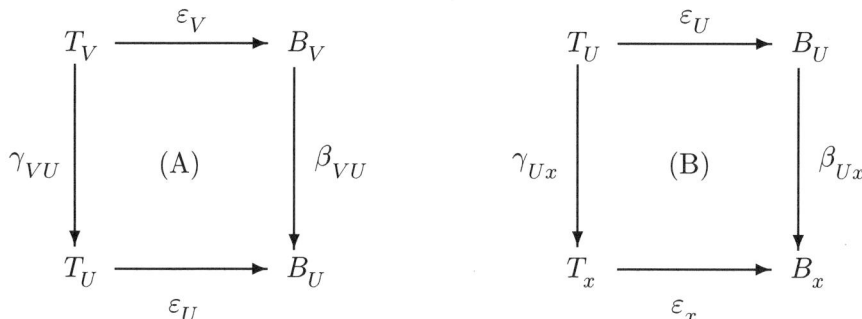

For $x \in X$, let $\mathcal{T}(x) = (T_U, \{\gamma_{VU} : U, V \in \nu_x, U \subseteq V\})$ be the inductive system associated to x by T, as in Definition 6.19. The observations above imply that
$$\mathcal{B}(x) = (B_U, \{\beta_{VU} : U, V \in \nu_x, U \subseteq V\},$$
is an inductive system of BA's and BA-homomorphisms, as well as that the family of complete embeddings ε_U determines a morphism of inductive systems, $\varepsilon(x) : \mathcal{T}(x) \longrightarrow \mathcal{B}(x)$.

Recall that a **Boolean polynomial** in the variables $Z_1, \ldots, Z_n$ is a term constructed from $\top, \bot, Z_1, \ldots, Z_n$ by successive application of join, meet and complementation. $\diamond$

PROPOSITION 6.24. *With notation as in 6.23, let $\mathcal{T}(x) \xrightarrow{\varepsilon(x)} \mathcal{B}(x)$ be the morphism of inductive systems associated to the Boolean hull $\varepsilon_U : T_U \longrightarrow B_U$. Let $(B_x, \{\beta_{Ux} : U \in \nu_x\})$ be the inductive limit of $\mathcal{B}(x)$ and $\varepsilon_x : T_x \longrightarrow B_x$ be the induced map of limits. Then,*

a) B_x is a Boolean algebra, each β_{Ux} is a BA-homomorphism and ε_x is a complete embedding. Moreover, all squares of type (B) in 6.23 are commutative.

b) If P, Q are Boolean polynomials in n variables, $a_1, \ldots, a_n \in B_V$ and $V \in \nu_x$, the following are equivalent

(1) $P(\beta_{Vx}(a_1), \ldots, \beta_{Vx}(a_n)) = Q(\beta_{Vx}(a_1), \ldots, \beta_{Vx}(a_n))$;

(2) $\exists\, U \in \nu_x,\, U \subseteq V$ such that,
$$P(\beta_{VU}(a_1), \ldots, \beta_{VU}(a_n)) = Q(\beta_{VU}(a_1), \ldots, \beta_{VU}(a_n)).$$

If $a_i = \varepsilon_V(f_i)$, where $\{f_1, \ldots, f_n\} \subseteq T_V$, then (2) holds with $\varepsilon_U(f_{i|U})$ in place of the $\beta_{VU}(a_i)$.

c) Let $\varepsilon_{T_x} : T_x \longrightarrow B_{T_x}$ be the Boolean hull of T_x. Then, there is an isomorphism of Boolean algebras, $f : B_{T_x} \longrightarrow B_x$, such that the following diagram is commutative :

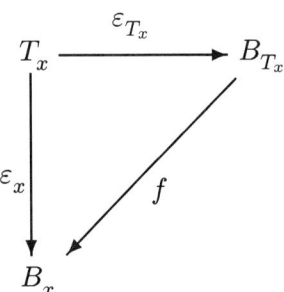

PROOF. a) It is well known that the inductive limit of BA's is a BA, that the β_{Ux} are BA-homomorphisms, and that the property stated in (b) holds. Its modification for elements of T_U is an easy consequence of the commutativity of squares of type (A).

It follows from the definition of ε_x that squares of type (B) are commutative, while Lemma 6.22.(c) yields that $\varepsilon_x : T_x \longrightarrow B_x$ is a complete embedding.

c) By Theorem 5.2, ε_x can be extended to an **injective** BA homomorphism $f : B_{T_x} \longrightarrow B_x$, satisfying $\varepsilon_x = f \circ \varepsilon_{T_x}$. To show that f is onto, it suffices to verify, by Proposition 4.10, that every $u \in B_x$ can be written as

(I) $$u = \bigvee_{i \leq n} \bigwedge_{a \in F_i} \varepsilon_x(a),$$

where F_i is a finite set in T_x, $1 \leq i \leq n$. Because once this is shown, since $\varepsilon_x = f \circ \varepsilon_{T_x}$ and f is a BA morphism, the element $\bigvee_{i \leq n} \bigwedge_{a \in F_i} \varepsilon_{T_x}(a)$ of B_{T_x} will be taken by f to u.

To prove (I), select $U \in \nu_x$ and $W \in B_U$ such that $\beta_{Ux}(W) = u$. Since B_U is the Boolean hull of T_U, there is a finite collection of finite sets $F_i \subseteq T_U$, $1 \leq i \leq n$, such that

$$W = \bigvee_{i \leq n} \bigwedge_{f \in F_i} \varepsilon_U(f).$$

Because β_{Ux} is a BA-homomorphism and squares of type (B) are commutative, this last equality yields

$$u = \beta_{Ux}(W) = \bigvee_{i \leq n} \bigwedge_{f \in F_i} \beta_{Ux}(\varepsilon_U(f)) = \bigvee_{i \leq n} \bigwedge_{f \in F_i} \varepsilon_x(\gamma_{Ux}(f)),$$

ending the proof of (I) and of the Proposition. □

Because of Proposition 6.24, the morphism $T_x \xrightarrow{\varepsilon_{T_x}} B_{T_x}$ will be identified with $T_x \xrightarrow{\varepsilon_x} B_x$ and we shall speak of this last map as the Boolean hull of T_x. In particular, for $U \in \nu_x$ and $f \in T_U$ we have

[BH] $$\varepsilon_x(f_x) = [f_x = -1],$$

where $f_x \in T_x$ is the germ of f at x.

4. The Space of Orders of a SG-filtered Power

To determine the space of orders of a reduced **SG**-filtered power, we need

NOTATION 6.25. Let $T = C(X, G; \Sigma)$ be a **SG**-filtered power and let U, V disjoint clopens in X. For $f \in T_U$ and $g \in T_V$, define $f \vee g$ to be the map whose domain is $U \cup V$, given by

$$f \vee g = \begin{cases} f(x) & x \in U \\ g(x) & x \in V. \end{cases}$$

Lemma 6.9.(c) guarantees that $f \vee g$ is in $T_{U \cup V}$. Let

$$\mathcal{D}_T = \{h \in T : \text{outside of } [h = \mathbb{1}], h \text{ is constant}\}.$$

Note that all constant functions in T, as well as all characteristic functions, are in $\mathcal{D}_T$; moreover if $f \in T$ and U is a clopen where f is constant, then $f_{|U} \vee \mathbb{1}_{|-U}$ is in $\mathcal{D}_T$. Consequently, we have

[$\mathcal{D}$] $\quad \forall f \in T, \forall x \in X, \exists U \in \nu_x$ and $\exists h \in \mathcal{D}_T$, such that $f_{|U} = h_{|U}$.

For $h \in \mathcal{D}_T$, let $spt(h)$ be the clopen complement of $[h = \mathbb{1}]$, called the **support** of h; if we write $[h = \mathbb{1}]$ as U, then

$$*h =_{def} -h_{|spt(h)} \vee \mathbb{1}_{|U} = h \cdot \eta_{spt(h)},$$

is an element of $\mathcal{D}_T$, corresponding to changing the sign of h only over its support.

Let $\mathcal{X}_T = \bigcup_{x \in X} \{x\} \times X_{T_x}$; there is a natural projection $\mathcal{X}_T \xrightarrow{p} X$, such that $p^{-1}(x)$ is $\{x\} \times X_{T_x}$, for all $x \in X$. For $h \in \mathcal{D}_T$, let

$$V(h) = \{(x, \sigma) \in \mathcal{X}_T : \sigma \in [h_x = -1]\},$$

where h_x is the germ of h at x. Observe that if (x, σ) is in $V(h)$, then x must be in the support of h. For otherwise, the corresponding σ would take $h_x = 1$ to -1, which is impossible. Thus, if h and k have disjoint supports, then $V(h)$ and $V(k)$ are also disjoint. Furthermore, if $U = [h = \mathbb{1}]$, then

$$[V(h)]^c = V(*h) \cup V(\eta_U),$$

a disjoint union of sets of type $V(k)$. Thus, we may take finite intersections of the family

$$\mathcal{B} = \{V(h) : h \in \mathcal{D}_T\},$$

as a basis of clopens for a topology on $\mathcal{X}_T$. For $\{h_1, \ldots, h_n\} \subseteq \mathcal{D}_T$, set

$$V(h_1, \ldots, h_n) = \bigcap_{i \leq n} V(h_i). \qquad \diamond$$

PROPOSITION 6.26. *With the conventions in Notation 6.25,*

a) $\mathcal{X}_T$ *is a Hausdorff space and p is continuous.*

b) With the induced topology of $\mathcal{X}_T$, $\{x\} \times X_{T_x}$ is homeomorphic to X_{T_x} and thus compact in $\mathcal{X}_T$.

c) If W is a clopen in $\mathcal{X}_T$ such that $p^{-1}(x) \subseteq W$, then $p(W)$ contains a clopen neighborhood of x.

PROOF. (a) To show that $\mathcal{X}_T$ is Hausdorff, let (x, σ), (y, μ) be distinct points in $\mathcal{X}_T$. We discuss two cases

(i) $x \neq y$: Select U clopen in X such that $x \in U$ and $y \notin U$. Then, $(x, \sigma) \in V(\eta_U)$, while (y, μ) cannot be in this clopen, because y is outside the support of η_U.

(ii) $x = y$: There is $a \in T_x$ such that $\sigma(a) = -1$ and $\mu(a) = 1$. By the definition of stalk and Corollary 6.17.(c), there is $f \in T$ such that $f_x = a$. Choose $U \in \nu_x$ on which f is constant and let $h = f_{|U} \vee \mathbb{1}_{|-U}$. Then, σ is in $[f_x = -1]$, while μ is not, showing that the clopen $V(h)$ separates (x, σ) from (x, μ).

To show that p is continuous, we verify that for U clopen in X,

$$p^{-1}(U) = V(\eta_U).$$

First, note that $spt(\eta_U) = U$; thus, $p(V(\eta_u)) \subseteq U$, that is, $V(\eta_U) \subseteq p^{-1}(U)$. For the reverse inclusion, it is enough to show that if $x \in U$ and $\sigma \in X_{T_X}$, then $(x, \sigma) \in V(\eta_U)$. But this follows directly from the fact that the germ of η_U at all points of U is -1. A similar argument will show that, in fact, $p(V(h)) = spt(h)$, for all $h \in \mathcal{D}_T$.

b) Fix x in X and consider the map $\alpha : \{x\} \times X_{T_x} \longrightarrow X_{T_x}$, given by $\alpha(x, \sigma) = \sigma$; clearly, α is bijective and so it suffices to verify that it is continuous, or equivalently, that for all $a \in T_x$, $\alpha^{-1}([a = -1])$ is open in $\{x\} \times X_{T_x}$. For $a \in T_x$ choose, just as in (ii) above, $f \in T$ such that $f_x = a$ and set $h = f_{|U} \vee \mathbb{1}_{|U^c}$, where U is a clopen containing x where f is constant. But then,

$$(x, \sigma) \in V(h) \quad \text{iff} \quad \sigma(f_x) = \sigma(a) = -1,$$

that is, $\alpha^{-1}([a = -1]) = V(h) \cap (\{x\} \times X_{T_x})$, as desired.

c) Since $p^{-1}(x)$ is compact and the finite intersections of sets of the form $V(h)$ is a basis for the topology on $\mathcal{X}_T$, there are finite sets $\{h_{ij} : 1 \leq j \leq m_i\} \subseteq \mathcal{D}_T$, $1 \leq i \leq n$, such that

(iii) For each $1 \leq i \leq n$, $V(h_{i1}, \ldots, h_{im_i}) \subseteq W$

(iv) $p^{-1}(x) \subseteq \bigcup_{i \leq n} \bigcap_{j \leq m_i} V(h_{ij})$.

To prove (c), it is sufficient to find $U \in \nu_x$, such that for all $y \in U$,
$$p^{-1}(y) \subseteq \bigcup_{i \leq n} \bigcap_{j \leq m_i} V(h_{ij}).$$

By the definition of $V(h_{i1}, \ldots, h_{im_i})$, (iv) is equivalent to the following equation in the Boolean hull B_x of T_x :

$$\bigvee_{i \leq n} \bigwedge_{j \leq m_i} \varepsilon_x(h_{ijx}) = \top, \tag{I}$$

where h_{ijx} is the germ of h_{ij} at x. To see this, notice that (iv) says that in B_x,
$$X_{T_x} = \bigcup_{i \leq n} \bigcap_{j \leq m_i} [h_{ijx} = -1].$$
But we have $\varepsilon_x(h_{ijx}) = [h_{ijx} = -1]$ (condition [BH], above), which immediately entails (I).

Since (I) is an equality of polynomials in B_x, by Proposition 6.24.(b) there is $U \in \nu_x$, such that same equation holds, in B_U, for $h_{ij}|_U$, that is

$$\bigvee_{i \leq n} \bigwedge_{j \leq m_i} \varepsilon_U(h_{ij}|_U) = \top. \tag{II}$$

But then, since β_{Uy} is a BA-homomorphism and squares of type (B) in Notation 6.25 are commutative, we have, for all $y \in U$,

$$\top = \beta_{Uy}(\top) = \beta_{Uy}(\bigvee_{i \leq n} \bigwedge_{j \leq m_i} \varepsilon_U(h_{ij}|_U)) = \bigvee_{i \leq n} \bigwedge_{j \leq m_i} \beta_{Uy}(\varepsilon_U(h_{ij}|_U))$$
$$= \bigvee_{i \leq n} \bigwedge_{j \leq m_i} \varepsilon_y(h_{ijy}),$$

which is equivalent to $p^{-1}(y) \subseteq \bigcup_{i \leq n} \bigcap_{j \leq m_i} V(h_{ij})$, ending the proof. $\square$

The next result describes the space of orders of a **SG**-filtered power.

THEOREM 6.27. *Let $T = C(X, G; \Sigma)$ be a **SG**-filtered power, with G a reduced special group. Let X_T be the space of orders of T. Then there are a continuous surjection $X_T \xrightarrow{\pi} X$ and a homeomorphism $\otimes : \mathcal{X}_T \xrightarrow{\otimes} X_T$, such that $p = \otimes \circ \pi$.*

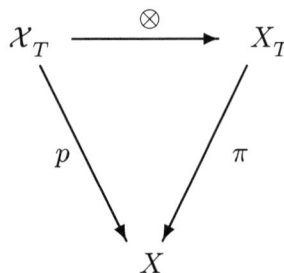

Moreover, if $x \in X$ and W is clopen set in X_T such that $\pi^{-1}(x) \subseteq W$, then there is $U \in \nu_x$ satisfying $U \subseteq \pi(W)$.

PROOF. Since G is reduced we know that T_U is reduced for all clopen U in X (Lemma 6.6). From Lemma 6.22, it follows that all stalks of T are reduced special groups.

Recall that if $f \in T$, then the symbol $[f = 1\!\!1]$ stands for the subset of X where f has value 1. On the other hand, $[f = 1]$ is the set of the elements τ in X_T, such that $\tau(f) = 1$.

Our first step is to construct π. To that end, let $\tau \in X_T$; recall that τ is a group homomorphism from T to $\mathbb{Z}_2$, such that $\tau(-1\!\!1) = -1$ and for all $f \in T$,

[ker] $\qquad\qquad f \in ker\ \tau \quad$ implies $\quad D_T(1\!\!1, f) \subseteq ker\ \tau.$

Consider $\mathcal{F}_\tau = \{U \in B(X) : \tau(\eta_U) = -1\}$. To complete the construction of π we need

Fact 1. $\mathcal{F}_\tau$ is an ultrafilter in $B(X)$. If x_τ is the limit of $\mathcal{F}_\tau$ in X, then

a) For all clopen U in X, $x_\tau \in U$ iff $\tau \in [\eta_u = -1] (\subseteq X_T)$.

b) For all $f, g \in T$, $f_{x_\tau} = g_{x_\tau}$ implies $\tau(f) = \tau(g)$.

PROOF. To verify that $\mathcal{F}_\tau$ is an ultrafilter in $B(X)$, we show

i) $\mathcal{F}_\tau$ is not empty and distinct from $B(X)$: Note that $X \in \mathcal{F}_\tau$, while $\emptyset \notin \mathcal{F}_\tau$, because $\eta_\emptyset = 1\!\!1$.

ii) $U \subseteq V$ and $U \in \mathcal{F}_\tau$, implies $V \in \mathcal{F}_\tau$: This follows easily from [ker] and Corollary 6.11.(b).

iii) $U, V \in \mathcal{F}_\tau$, implies $U \cap V \in \mathcal{F}_\tau$: Note that

 * $U \triangle V$ is not in $\mathcal{F}_\tau$, because $\tau(\eta_{(U \triangle V)}) = \tau(\eta_U)\tau(\eta_V) = 1$;

 * $U \cup V$ is in $\mathcal{F}_\tau$, because it contains U (and V);

Since $U \cup V = (U \triangle V) \triangle (U \cap V)$, the preceding observations yield

$$\tau(\eta_{(U\cap V)}) = \tau(\eta_{[U\cup V]})\tau(\eta_{[U\triangle V]}) = -1 \cdot 1 = -1,$$

verifying that $U \cap V \in \mathcal{F}_\tau$. Thus, $\mathcal{F}_\tau$ is a proper filter of clopens in X. Its maximality comes from the observation that, since $\eta_U \cdot \eta_{-U} = -\mathbb{1}$, either U or $-U$ must be in $\mathcal{F}_\tau$.

The (Hausdorff) compactness of X yields a **unique** limit x_τ for $\mathcal{F}_\tau$, satisfying

$$x_\tau \text{ is the unique element of } \bigcap \{U : U \in \mathcal{F}_\tau\}.$$

Furthermore, because $\mathcal{F}_\tau$ is an ultrafilter, we have $\nu_{x_\tau} \subseteq \mathcal{F}_\tau$. Item (a) follows directly from the these observations and the definition of $\mathcal{F}_\tau$.

Before proving (b), we establish

$$\text{For } f \in T, \quad x_\tau \in [f = \mathbb{1}] \quad \text{implies} \quad \tau(f) = 1. \tag{I}$$

To see this, note that if $u = [f = \mathbb{1}] \in \nu_{x_\tau} \subseteq \mathcal{F}_\tau$, then $f \in D_T(\mathbb{1}, \eta_{-u})$ (Corollary 6.11), with $\tau(\eta_{-u}) = 1$. Thus, [ker] implies $\tau(f) = 1$.

Now suppose that $f_{x_\tau} = g_{x_\tau}$; by the definition of stalk, there is $U \in \nu_{x_\tau}$, such that $f_{|U} = g_{|U}$. But then, $U \subseteq [f \cdot g = \mathbb{1}]$; by (I), the values of τ on f and g must be the same, ending the proof of Fact 1. □

We define $\pi(\tau)$ to be x_τ; by item (a) in Fact 1, π is continuous.

For $x \in X$, write c_x for the germ SG-morphism $\gamma_{Tx} : T \longrightarrow T_x$ of Definition 6.19, i.e., $c_x(f) = f_x$. We now prove

Fact 2. *With notation as above, the map*

$$\otimes : \bigcup_{x \in X} \{x\} \times X_{T_x} \longrightarrow X_T, \quad \text{given by} \quad \sigma \otimes c_x(f) = \sigma(c_x(f)),$$

is a bijection, such that for all $x \in X$,

a) $\forall \sigma \in X_{T_x}$, $\pi(\sigma \otimes c_x) = x$. *In particular, π is surjective and $p = \otimes \circ \pi$.*

b) $\otimes$ *is a homeomorphism.*

PROOF. Clearly, $\sigma \otimes c_x$ is a group homomorphism from T to $\mathbb{Z}_2$, that takes $-\mathbb{1}$ to -1. Moreover, note that

$$\ker(\sigma \otimes c_x) = \{f \in T : c_x(f) \in \ker \sigma\},$$

and so, if $f \in \ker(\sigma \otimes c_x)$ and $g \in D_T(\mathbb{1}, f)$, then $c_x(g) \in D_T(1, c_x(f))$, which implies $\sigma(c_x(g)) = 1$. Hence, $\sigma \otimes c_x$ is a SG-character of T, i.e., an element of X_T. Now observe that for all $U \in \nu_x$,

$$\sigma \otimes c_x(\eta_U) = \sigma(\eta_{Ux}) = \sigma(-1) = -1,$$

which shows, together with (a) in Fact 1, that $\pi(\sigma \otimes c_x) = x$. It is clear that $p = \otimes \circ \pi$.

$\otimes$ is injective. Let x, y be distinct points in X and select a clopen U such that $x \in U$, $y \notin U$. Then, for all $\sigma \in X_{T_x}$ and $\mu \in X_{T_y}$,

$$\sigma \otimes c_x(\eta_U) = \sigma(-1) = -1 \quad \text{and} \quad \mu \otimes c_y(\eta_u) = \mu(1) = 1,$$

showing that $\sigma \otimes c_x$ is distinct from $\mu \otimes c_y$.

Now suppose that σ, μ are distinct characters of T_x and pick $a \in T_x$ such that $\sigma(a) \neq \mu(a)$. By the definition of stalk and Corollary 6.17.(c), there is $f \in T$ such that $f_x = a$. But then $\sigma \otimes c_x(f) = \sigma(a) \neq \mu(a) = \mu \otimes c_x(f)$.

$\otimes$ is surjective. For $\tau \in X_T$, set $x = \pi(\tau)$. By Fact 1.(b), the value of τ at $f \in T$, depends only on the germ of f at x. Moreover, because the restriction maps γ_{VU} are surjective (Corollary 6.17.(c)), it is easily seen that c_x is surjective, $\forall\, x \in X$ (T is called to be **point-soft**).

Define $\sigma : T_x \longrightarrow \mathbb{Z}_2$ by the following clause : For $a \in T_x$,

$$\sigma(a) = \tau(f), \text{ where } f \text{ is any element of } T \text{ such that } f_x = a.$$

It is straightforward to prove that σ is a group homomorphism into $\mathbb{Z}_2$. To show that it is a SG-character of T_x, let a, $b \in T_x$ be such that $\sigma(a) = 1$ and $b \in D_{T_x}(1, a)$. By Lemma 6.22, there are $U \in \nu_x$ and h, $k \in T_U$, such that $h_x = a$, $k_x = b$ and $k \in D_{T_U}(\mathbb{1}_{|U}, h)$. Let

$$f = h \vee \mathbb{1}_{|-U} \text{ and } g = k \vee \mathbb{1}_{|-U};$$

then f, $g \in T$ and $g \in D_T(\mathbb{1}, f)$. Since τ is a SG-character and $\sigma(a) = \tau(f) = 1$, we get $\sigma(b) = \tau(g) = 1$, as needed. Observe that for all f in T,

$$\sigma(c_x(f)) = -1 \quad \text{iff} \quad \tau(f) = -1, \tag{III}$$

and so τ must be equal to $\sigma \otimes c_x$, proving that $\otimes$ is a bijection.

b) Since $\otimes$ is a bijection, relation (III) above can be rewritten, for an element $h \in \mathcal{D}_T$, to read

$$\otimes(V(H)) = [h = -1],$$

showing that $\otimes$ is open. For continuity, it must be verified that if $f \in T$, then $\otimes^{-1}([f = -1])$ is open in $\mathcal{X}_T$. Let Q be a partition of X such that, for all $q \in Q$, $f_{|q}$ is constant. For each $q \in Q$, let $h_q = f_{|q} \vee \mathbb{1}_{|-q}$; the sequence $\{h_q : q \in Q\}$ satisfies

(i) $h_q \in \mathcal{D}_T$;

(ii) For all $x \in X$, there is a unique $q \in Q$, such that $f_x = h_{qx}$.

We claim that

$$\otimes^{-1}([f = -1]) = \bigcup_{q \in Q} V(h_q).$$

For if $\sigma \otimes c_x(f) = -1$, then $\sigma \in [f_x = -1]$; by (ii), there is $q \in Q$ such that $h_{qx} = f_x$, and so $\sigma \in [h_{qx} = -1]$, that is, $\sigma \otimes c_x \in V(h_q)$. Conversely, it is easily seen that every $V(h_q)$ is contained in $\otimes^{-1}([f = -1])$, ending the proof of Fact 2. $\square$

The last assertion in the statement of the Theorem follows from Fact 2.(b) and Proposition 6.26.(c). $\square$

Theorem 6.27 applies to all clopen U in X. In fact, $p^{-1}(U)$ is naturally homeomorphic to the the space of orders of T_U and to $\pi^{-1}(U)$.

5. The Boolean Hull of SG-Filtered Powers

As a consequence of Theorem 6.27, we may identify $\mathcal{X}_T$ with the space of orders of the filtered power T; with this identification, the Boolean hull of T will be the BA of clopens of $\mathcal{X}_T$, B_T, with $\varepsilon_T : T \longrightarrow B_T$ given by

$$\varepsilon_T(f) = \{(x, \sigma) : \sigma \in [f_x = -1]\} = \otimes^{-1}([f = -1]).$$

This section is devoted to an algebraic description of the Boolean hull a SG-filtered power of a special group G, in which the groups in the filtration are **complete** subgroups of G. This applies, in particular, to Example 6.16, dealing with $\bigoplus_{i \in I}^* G_i$; for if B is the Boolean algebra constructed therein, each G_i is a completely embedded in B. For ease of exposition, we deal with the case where the map λ of Definition 6.8 satisfies condition (*) in the statement of Theorem 6.13, (that appears with the same name in 6.20 and in 6.30, below).

In Remark 6.35, at the end of this section, it is mentioned how the sheaf associated to a pair of Boolean algebras can be used to give an algebraic description of the space of orders of any SG-filtered power.

REMARK 6.28. Let $\Sigma = \{(F_l, G_l) : l \in L\}$ be a continuous L-filtration on (X, G). We show that it is always possible to assume that

$$l \leq k \quad \text{iff} \quad F_l \subseteq F_k, \tag{I}$$

that is, there is a complete lattice Λ and a continuous Λ-filtration Γ satisfying this property, such that $C(X, G : \Sigma)$ **is equal to** $C(X, G; \Gamma)$.

Consider the map $j : L \longrightarrow L$ given by $j(l) = \bigwedge \{k \in L : F_k = F_l\}$. Notice that

(i) $j \circ j = j$

(ii) By continuity, $F_{j(l)} = F_l$.

To show that j is increasing, suppose $l \leq l'$ and consider $k = j(l) \wedge j(l')$; since $k \leq j(l)$ and $F_k = F_l \cap F_{l'} = F_l$, we must have $k = j(l)$, that is, $j(l) \leq j(l')$. By Tarski's fixed point Theorem (Lemma 2.12 in [FS]), the lattice Λ of fixed points of j, is complete. In fact, Λ is the image of j, with inf's computed as in L (but not sup's); moreover,

For all $S \subseteq L$, $\bigwedge_{s \in S} j(s) = j(\bigwedge S)$.

The Λ-filtration $\Gamma = \{(F_\ell, G_\ell) : \ell \in \Lambda\}$, will be continuous and satisfy (I). It is straightforward to verify that $C(X, G; \Sigma) = C(X, G; \Gamma)$. ◇

DEFINITION 6.29. Let X, Y be compact Hausdorff spaces and let L a complete lattice. A **projective L-filtration** on (X, Y) consists of a family $\mathcal{P} = \{(F_l, Y_l) : l \in L\}$, such that :

I. Each Y_l is a compact Hausdorff space and

I.1. For $l \leq k$, there is a continuous map $\pi_{kl} : Y_k \longrightarrow Y_l$, such that for $l \leq k \leq j$, we have $\pi_{kl} \circ \pi_{jk} = \pi_{jl}$;

I.2. $Y_\top = Y$ and $\pi_\top = Id_Y$. We write π_l for $\pi_{\top l} : Y \longrightarrow Y_l$.

II. Each F_l is a closed set in X and

II.1. For all $S \subseteq L$, $F_{\bigwedge S} = \bigcap_{l \in S} F_l$.

II.2. $l \leq k$ iff $F_l \subseteq F_k$.

For $x \in X$, $\lambda(x) = \bigwedge \{l \in L : x \in F_l\}$, exactly as in Definition 6.8.

Given a projective L-filtration $\mathcal{P}$ on (X, Y), define a binary relation $\sim_\mathcal{P}$ on $X \times Y$ by the following clause :

$$(x, y) \sim_\mathcal{P} (t, z) \quad \text{iff} \quad x = t \text{ and } \pi_{\lambda(x)}(y) = \pi_{\lambda(t)}(z).$$

Clearly, $\sim_\mathcal{P}$ is an equivalence relation; write $(X \times Y)/\mathcal{P}$ be the set of its equivalence classes and $\pi_\mathcal{P} : X \times Y \longrightarrow (X \times Y)/\mathcal{P}$ for the canonical quotient map. ◇

PROPOSITION 6.30. *With notation as in 6.29, let* $\mathcal{P} = \{(F_l, Y_l) : l \in L\}$ *be a projective L-filtration on the compact pair* (X, Y). *Assume that λ satisfies*

(∗) *For each $x \in X$, there is $U \in \nu_x$, such that $\lambda(x) \leq \lambda(y)$, for all $y \in U$.*

Then,

a) The following subsets of $X \times Y$ are invariant by $\sim_\mathcal{P}$:

1) $U \times Y$, where U is any open set in X.

2) Sets of the form $U \times V$ where, for some $x \in X$, we have

(i) $U \in \nu_x$ and $\lambda(x)$ is the minimum of $\lambda_{|U}$;

(ii) $V = \pi_{\lambda(x)}^{-1}(W)$, for some open W in $Y_{\lambda(x)}$.

b) With the quotient topology induced by the canonical projection from $X \times Y$, $(X \times Y)/\mathcal{P}$ is a compact Hausdorff space, which is Boolean if the same is true of X, Y and each Y_l.

PROOF. a) If U is open in X, it is clear that $U \times Y$ is invariant by $\sim_\mathcal{P}$ (that is, an equivalence class has non-empty intersection with $U \times Y$ iff it is contained in it). Now suppose that we have a product as in (2) and assume that $(u, v) \in U \times V$ and $(t, z) \sim (u, v)$. Then, $u = t$, while $\pi_{\lambda(u)}(v) = \pi_{\lambda(u)}(z)$. Since x is the minimum of λ in U, this last equality and the composition law I.1 in Definition 6.29, yield $\pi_{\lambda(x)}(v) = \pi_{\lambda(x)}(z)$. Recalling that $v \in V$ and that V is the inverse image by $\pi_{\lambda(x)}$ of an open set in $Y_{\lambda(x)}$, we conclude that $z \in V$, as needed.

b) It is sufficient to verify that the quotient topology on $(X \times Y)/\mathcal{P}$ is Hausdorff. If (x, y) is not $\sim_\mathcal{P}$ equivalent to (t, z) it is because one of the following statements holds true :

(i) $x \neq t$ **or** (ii) $x = t$, but $\pi_{\lambda(x)}(y) \neq \pi_{\lambda(x)}(z)$.

To show that the class of (x, y) can be separated by open sets from (t, z), it is sufficient to find disjoint $\sim_\mathcal{P}$-invariant opens A, B in $X \times Y$, such that $(x, y) \in A$ and $(t, z) \in B$. We shall do this in each of the cases (i) and (ii) mentioned above.

If $x \neq t$, select U, V disjoint opens in X such that $x \in U$ and $t \in V$. Then, $A = U \times Y$ and $B = V \times Y$ will do as the desired disjoint invariant opens (item 1 in (a)). Notice that if X and Y were Boolean, we could have taken A and B clopen.

Now assume that $x = t$ and let U be an open set containing x such that $\lambda(x)$ is the minimum of λ in U. Since $\pi_{\lambda(x)}(y) \neq \pi_{\lambda(x)}(z)$, select disjoint opens W, W' in $Y_{\lambda(x)}$ that separate these points. Then, if $V = \pi_{\lambda(x)}^{-1}(W)$ and $V' = \pi_{\lambda(x)}^{-1}(W')$, item 2 in (a) implies that $A = U \times V$ and $B = U \times V'$ can be taken as the desired disjoint invariants. Again, we could have constructed A, B clopen, if X, Y and each Y_l were Boolean. □

A projective L-filtration $\mathcal{P} = \{(F_l, Y_l) : l \in L\}$ on (X, Y) is **Boolean** if X, Y and all Y_l are Boolean spaces.

NOTATION 6.31. Let $\mathcal{P} = \{(F_l, Y_l) : l \in L\}$ be a Boolean projective L-filtration on (X, Y). For each $l \in L$, let B_{Y_l} be the BA of clopens in Y_l. Write B_Y for the BA of clopens of Y (instead of $B_\top$). For $l \leq k$, let

$$\pi_{kl}^* : B_{Y_l} \longrightarrow B_{Y_k} \quad \text{and} \quad \pi_l^* : B_{Y_l} \longrightarrow B_Y$$

be the Stone dual of π_{kl} and π_l, respectively. For $l \in L$, let B_l be the image of π_l^* in B_Y; B_l is a Boolean <u>subalgebra</u> of B_Y. Now set

$$\mathcal{B} = \{(F_l, B_l) : l \in L\}.$$

To show that $\mathcal{B}$ is a L-filtration, suppose that $F_l \subseteq F_k$; then $l \leq k$ and so $\pi_l = \pi_{kl} \circ \pi_k$; but this implies $\pi_l^* = \pi_k^* \circ \pi_{kl}^*$, and so $B_l \subseteq B_k$ in B_Y.

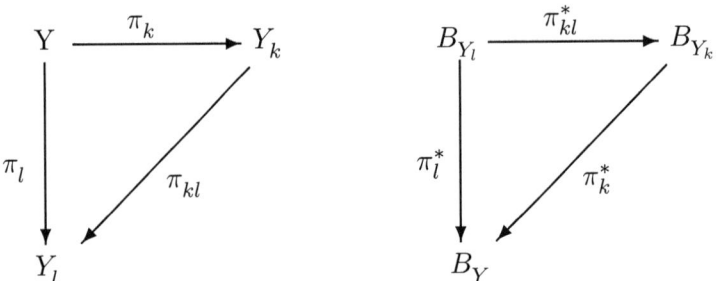

It is clear that $\mathcal{B}$ is a continuous filtration. Moreover, for $U \in B_Y$,

[inv_l] $\qquad\qquad U \in B_l \quad \text{iff} \quad \pi_l^{-1}(\pi_l(U)) = U$,

that is, U is π_l-invariant. If $x \in X$, B_x will stand for $B_{\lambda(x)}$, with λ as in Definition 6.8. ◊

THEOREM 6.32. *Let $\mathcal{P} = \{(F_l, Y_l) : l \in L\}$ be a Boolean projective L-filtration on (X, Y), satisfying condition (*) in 6.30. Let $2 = \{\bot, \top\}$ be the 2-element Boolean algebra. With notation as in 6.31, let*

$$\pi_* : C(X \times Y/\mathcal{P}, 2) \longrightarrow C(X, B_Y),$$

be defined, for $\sigma \in C(X \times Y/\mathcal{P}, 2)$, by

$$\pi_*(\sigma)(x) = \{y \in Y : \sigma(\pi_\mathcal{P}(x, y)) = \top\},$$

where $\pi_\mathcal{P}$ is the canonical projection from $X \times Y$ to $(X \times Y)/\mathcal{P}$. Then π_* is a BA-isomorphism from $C(X \times Y/\mathcal{P}, 2)$ onto $C(X, B_Y; \mathcal{B})$.

PROOF. Let $p_Y : X \times Y \longrightarrow Y$ be the natural projection; write $\mathcal{A}$ for $C(X \times Y/\mathcal{P}, 2)$. For $\sigma \in \mathcal{A}$ and $x \in X$, we have

$$\pi_*(\sigma)(x) = \{y \in Y : \sigma(\pi_\mathcal{P}(x, y) = \top\} = p_Y(\pi_\mathcal{P}^{-1}([\sigma = \top]));$$

since p_Y is an open map, $\pi_*(\sigma)(x)$ is clopen in Y, i.e., $\pi_*(\sigma)(x) \in B_Y$.

The injectivity of π_* is an immediate consequence of the fact that $\pi_\mathcal{P}$ is surjective.

It is clear that the constant functions $\widehat{\top}, \widehat{\bot}$ are taken to the constant functions with value $\top, \bot$, respectively, in $C(X, B_Y)$. For the operations of meet and complementation, we have

$$\pi_*(\sigma \wedge \mu)(x) = \{y \in Y : \sigma \wedge \mu(\pi_\mathcal{P}(x, y)) = \top\}$$
$$= \{y \in Y : \sigma(\pi_\mathcal{P}(x, y)) = \top \text{ and } \mu(\pi_\mathcal{P}(x, y)) = \top\}$$
$$= \pi_*(\sigma)(x) \cap \pi_*(\mu)(x).$$

$$\pi_*(\neg \sigma)(x) = \{y \in Y : \neg \sigma(\pi_\mathcal{P}(x, y)) = \top\} = \{y \in Y : \sigma(\pi_\mathcal{P}(x, y)) = \bot\}$$
$$= Y \setminus \pi_*(\sigma)(x).$$

Thus, π_* is an injective BA-homomorphism. We now prove

Fact. *For all σ in $\mathcal{A}$ and all $x \in X$, $\pi_*(\sigma)(x) \in B_x \ (= B_{\lambda(x)})$.*

PROOF. By condition $[inv_l]$ in 6.31, it is sufficient to verify that $\pi_*(\sigma)(x)$ is $\pi_{\lambda(x)}$-invariant, that is, if $y \in \pi_*(\sigma)(x)$ and $\pi_{\lambda(x)}(y) = \pi_{\lambda(x)}(z)$, then $z \in \pi_*(\sigma)(x)$. But $\pi_{\lambda(x)}(y) = \pi_{\lambda(x)}(z)$ implies $(x, y) \sim_{\mathcal{P}} (x, z)$, and so $\pi_{\mathcal{P}}(x, y) = \pi_{\mathcal{P}}(x, z)$. Hence, $\top = \sigma(\pi_{\mathcal{P}}(x, y)) = \sigma(\pi_{\mathcal{P}}(x, z))$, and z must be in $\pi_*(\sigma)(x)$, as desired. □

It follows easily from Fact 1 that $x \in F_l$ implies $\pi_*(\sigma)(x) \in B_l$, for all $l \in L$; therefore, the image of π_* is contained in $C(X, B_Y; \mathcal{B})$. It remains to verify that π_* is onto $C(X, B_Y; \mathcal{B})$.

Given $f \in C(X, B_Y; \mathcal{B})$, there is a partition P of X such that f has a constant value, say U_p, on each element $p \in P$. Since $\mathcal{B}$ is continuous, Lemma 6.9.(c) implies that $f(x) \in B_x$, for all $x \in X$. Thus, for each $p \in P$ and all $x \in p$, U_p is $\pi_{\lambda(x)}$-invariant. Consider

$$W = \bigcup_{p \in P} p \times U_p \ (\in B_Y).$$

We claim that W is $\sim_{\mathcal{P}}$-invariant. To see this, suppose $(x, y) \in W$ and $(t, z) \sim_{\mathcal{P}} (x, y)$. Then $x = t$ and there is a unique $p \in P$ such that $x \in p$. We have $y \in U_p$ and $\pi_{\lambda(x)}(y) = \pi_{\lambda(x)}(z)$; since U_p is $\pi_{\lambda(x)}$-invariant, we conclude that $z \in U_p$, as needed.

Since W is $\sim_{\mathcal{P}}$-invariant, $\pi_{\mathcal{P}}(W)$ is clopen in $(X \times Y)/\mathcal{P}$. Let σ be the characteristic map of $\pi_{\mathcal{P}}(W)$, i.e., $\sigma(t) = \top$ iff $t \in \pi_{\mathcal{P}}(W)$. Then, for $x \in p \in P$,

$$\pi_*(\sigma)(x) = \{y \in Y : \sigma(\pi_{\mathcal{P}}(x, y)) = \top\} = \{y \in Y : (x, y) \in \pi_{\mathcal{P}}^{-1}(\pi_{\mathcal{P}}(W))\}$$
$$= \{y \in Y : (x, y) \in W\} = U_p,$$

which shows that $\pi_*(\sigma) = f$ and ends the proof. □

We shall apply Theorem 6.32 to get an algebraic description of an important class of SG- filtered powers.

NOTATION 6.33. Let $\Sigma = \{(F_L, G_l) : l \in L\}$ be an **SG**-filtration on (X, G), with G reduced. By Remark 6.28, we may as well assume that for all $l, k \in L$, $F_l \subseteq F_k$ iff $l \leq k$.

Suppose that all G_l are **complete** subgroups of G. It follows immediately that if $l \leq k$ then G_l is a complete subgroup of G_k; by Theorem 5.2, this complete embedding induces a continuous surjection

$$\pi_{kl} : X_{G_k} \longrightarrow X_{G_l}, \ \sigma \mapsto \sigma|_{G_l}.$$

As above, let $\pi_l : X_G \longrightarrow X_{G_l}$ be the continuous surjection associated to the complete embedding $G_l \hookrightarrow G$.

It is clear that $\pi_{jl} = \pi_{lk} \circ \pi_{jk}$, for $l \leq k \leq j$ in L; thus

$$\mathcal{P}_\Sigma = \{(F_l, X_{G_l}) : l \in L\},$$

is a Boolean projective L-filtration on (X, X_G). Let $\mathcal{B}_\Sigma$ be the BA filtration associated to $\mathcal{P}_\Sigma$ as in Notation 6.31.

For each $l \in L$, we have a commuting triangle (A) (on the left)

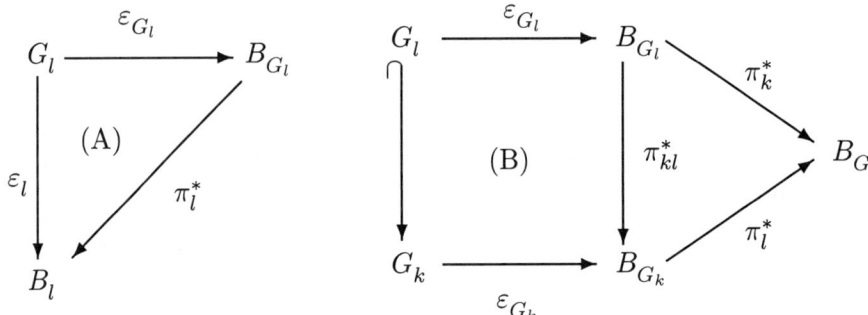

where π_l^* is a BA isomorphism onto $B_l \subseteq B_G$, $\varepsilon_{G_l} : G_l \longrightarrow B_{G_l}$ is the Boolean hull of G_l and $\varepsilon_l = \pi_l^* \circ \varepsilon_{G_l}$. Thus, $\varepsilon_l : G_l \longrightarrow B_l$ can be considered as the Boolean hull of G_l. Moreover, note that since $l \leq k$ implies $B_l \subseteq B_k$, we have, by the commutativity of diagram (B) above,

[rest]   For all $l \leq k$, the restriction of ε_k to G_l coincides with ε_l.

For $x \in X$, write $\varepsilon_x : G_x \longrightarrow B_x$ for $\varepsilon_{\lambda(x)} : G_x \longrightarrow B_{\lambda(x)}$. $\diamond$

THEOREM 6.34. *Let $\Sigma = \{(F_l, G_l) : l \in L\}$ be a **SG** L-filtration on (X, G), such that each G_l is a complete subgroup of the reduced special group G. Assume that Σ satisfies $(*)$ in 6.30. Let $T = C(X, G; \Sigma)$ be the filtered Boolean power determined by Σ. With notation as in 6.33,*

a) *The map $X \times X_G \xrightarrow{\alpha} \mathcal{X}_T$, defined by $\alpha(x, \sigma) = (x, \sigma \circ ev_x)$ is a continuous surjection, that factors through a homeomorphism*

$$h : (X \times X_G)/\mathcal{P}_\Sigma \longrightarrow \mathcal{X}_T,$$

making the following diagram commutative

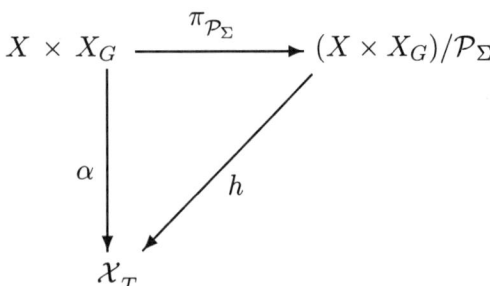

b) *The Boolean hull of T is given by $\varepsilon_T : T \longrightarrow C(X, B_G; \mathcal{B})$, where $\varepsilon_T(f)(x) = \varepsilon_x(f(x))$.*

PROOF. (a) From (I) in Example 6.21, we have $T_x = G_x$ and $c_x = ev_x$. Thus, $\mathcal{X}_T = \bigcup_{x \in X} \{x\} \times X_{G_x}$.

Fact 1. For all $x, y \in X$, for all $\sigma, \mu \in X_G$,
$$\sigma \circ ev_x = \mu \circ ev_y \Leftrightarrow x = y \text{ and } \sigma_{|G_x} = \mu_{|G_x}.$$

PROOF. ($\Leftarrow$) is clear. For the converse, suppose $x \neq y$. Then, there is $U \in \nu_x$ such that $y \notin U$; hence, $\eta_U(x) = -1$, while $\eta_U(y) = 1$ and so, $\sigma \circ ev_x(x) = -1$ and $\mu \circ ev_y(y) = 1$, an impossibility. From the equality of x and y and the fact that T is point-soft, i.e., that all ev_x are onto G_x (see '$\otimes$ is surjective' in the proof of Fact 2 in Theorem 6.27), it is straightforward to see that $\sigma_{|G_x} = \mu_{|G_x}$. □

To show that α is onto, let $(x, \mu) \in \{x\} \times X_{G_x}$; since G_x is a complete subgroup of G, $\pi_{\lambda(x)} : X_G \longrightarrow X_{G_x}$ is surjective. Thus, there is $\sigma \in X_G$ such that $\sigma_{|G_x} = \mu$. Then, $\alpha(x, \sigma) = (x, \sigma \circ ev_x) = (x, \mu)$, as needed. Fact 1 and the surjectivity of α yield a bijective $h : (X \times X_G)/\mathcal{P}_\Sigma \longrightarrow \mathcal{X}_T$ such that $\alpha = h \circ \pi_{\mathcal{P}_\Sigma}$. To verify that h is a homeomorphism it is enough to show that α is continuous. It is easily verified that if $g \in \mathcal{D}_T$ and if a is the value of g in its support, then

$$(x, \sigma \circ ev_x) \in V(g) \quad \text{iff} \quad x \in spt(g) \text{ and } \sigma \in [a = -1],$$

that is $\alpha^{-1}(V(g)) = spt(g) \times [a = -1]$, completing the proof of (a).

(b) Part (a) and Theorem 6.32 yield an isomorphism between the BA of clopens in $\mathcal{X}_T$ and the filtered power $C(X, B_G; \mathcal{B})$. Still to be verified is that if $\varepsilon : T \longrightarrow B_T$ is the Boolean hull of T, then there is an isomorphism $f : B_T \longrightarrow C(X, B_G; \mathcal{B})$, such that the triangle below is commutative :

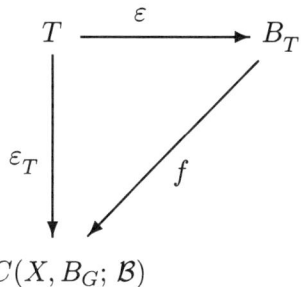

By the universal property of the Boolean hull (Theorem 4.17), for this to be true it is necessary and sufficient that the following conditions be satisfied :

(i) ε_T is a complete embedding;

(ii) For all $\beta \in C(X, B_G; \mathcal{B})$ there are finite sets $F_1, \ldots, F_n \subseteq T$ such that

$$\beta = \bigvee_{j=1}^{n} \bigwedge_{f \in F_j} \varepsilon_T(f).$$

Fact 2. For all $f \in T$, $\varepsilon_T(f) \in C(X, B_G; \mathcal{B})$.

PROOF. Let P be a partition of X such that f is constant in each $p \in P$. Since Σ is **SG**, each p in P has a partition Q_p and, for each $q \in Q_p$, there is a point $x_q \in q$, such that λ attains its minimum λ_q on q at x_q. Set $Q = \bigcup_{p \in P} Q_p$ and $a_q = f(x_q)$. We wish to show that

$$\varepsilon_T(f)|_q = \varepsilon_{\lambda_q}(a_q).$$

If $x \in q$, then $\lambda(x) \geq \lambda_q$ and $f(x) \in G_{\lambda_q}$. Therefore, [rest] in Notation 6.33 yields

$$\varepsilon_x(f(x)) = \varepsilon_x(a_q) = \varepsilon_{\lambda_q}(a_q),$$

verifying $\varepsilon_T(f) \in C(X, B_G)$. For $x \in X$, Lemma 6.9.(c) yields $f(x) \in G_x$ and so $\varepsilon_T(f)(x) = \varepsilon_x(f(x)) \in B_x$. From this it follows easily that $x \in F_l$ implies $\varepsilon_T(f)(x) \in B_l$, for all $l \in L$. $\square$

Fact 3. ε_T is a complete embedding.

PROOF. Clearly, ε_T is injective. Let φ, ψ be n-forms over T. By successive applications of Theorem 6.10 and the fact that each ε_x is a complete embedding we have, with $A = C(X, B_G; \mathcal{B})$

$\varphi \equiv_T \psi$ iff $\forall x \in X, \varphi(x) \equiv_{G_x} \psi(x)$

iff $\forall x \in X, \varepsilon_x \star \varphi(x) \equiv_{B_x} \varepsilon_x \star \psi(x)$

iff $\varepsilon_T \star \varphi \equiv_A \varepsilon_T \star \psi,$

ending the proof of Fact 3. $\square$

To finish the proof, we verify that the image of ε_T generates $C(X, B_G; \mathcal{B})$.

Let $\beta \in C(X, B_G; \mathcal{B})$; we proceed just as in the beginning of the proof of Fact 2 : let P be a partition of X such that for each $p \in P$, β is constant in p. Using **SG**, select a partition Q_p of p such that for each $q \in Q_p$, λ attains its minimum λ_q in $x_q \in q$. Set $Q = \bigcup_{p \in P} Q_p$ and $a_q = \beta(x_q) \in B_{\lambda_q}$.

Since $\varepsilon_{\lambda_q}(G_{\lambda_q})$ generates B_{λ_q}, there are finite sets $K_{q1}, \ldots, K_{qn_q}$ in G_{λ_q} such that

$$a_q = \bigvee_{j=1}^{n_q} \bigwedge_{g \in F_{qj}} \varepsilon_{\lambda_q}(g). \tag{I}$$

For $g \in K_{qj}$, let g_q be the map defined by $g_q(x) = \begin{cases} g & \text{if } x \in q \\ 1 & \text{if } x \notin q. \end{cases}$

By Lemma 6.9.(c), we have $g_q \in T$. Now, because $g \in G_{\lambda_q}$, [rest] in 6.33 entails $\varepsilon_T(g_q)(x) = \begin{cases} \varepsilon_{\lambda_q}(g) & \text{if } x \in q \\ 1 & \text{if } x \notin q. \end{cases}$ Hence, (I) yields

$$\left[\bigvee_{j=1}^{n_q} \bigwedge_{g \in F_{qj}} \varepsilon_T(g_q)\right](x) = \begin{cases} a_q & \text{if } x \in q \\ 1 & \text{if } x \notin q, \end{cases}$$

which implies $\beta = \bigvee_{q \in Q} \bigvee_{j=1}^{n_q} \bigwedge_{g \in F_{qj}} \varepsilon_T(g_q)$, ending the proof. □

REMARK 6.35. If T is a **SG**-filtered power, we describe B_T in terms of a classical construction known as the sheaf associated to a pair of BA's, (A, B), where A is isomorphic to a subalgebra of B (see sections 8.16 and 8.17 of [HBA]). We shall mention the statements of the pertinent results, omitting proofs.

In Proposition 6.26.(b) it was shown that the map $\{x\} \times X_{T_x} \longrightarrow X_{T_x}$, that forgets the first coordinate, is a homeomorphism. This naturally identifies B_x with the BA of clopens in $\{x\} \times X_{T_x}$; we shall use this identification without mention in the sequel.

COROLLARY 6.36. *With notation as in 6.27, let B_T be the BA of clopens in $\mathcal{X}_T$. Let $p^* : B(X) \longrightarrow B_T$ be the injective Stone dual of the continuous projection p. Then,*

a) *For $x \in X$, the filter generated by $\{p^*(U) : U \in \nu_x\}$ in B_T, $\mathcal{F}_x$, is exactly the filter of clopen neighborhoods of $p^{-1}(x)$ in $\mathcal{X}_T$.*

b) *For $x \in X$, the BA-homomorphism $W \mapsto W \cap p^{-1}(x)$ from B_T to B_x, is surjective and factors through the canonical quotient map $B_T \longrightarrow B_T/\mathcal{F}_x$. In particular, B_x is isomorphic to $B_T/\mathcal{F}_x$ and for all $f \in T$,*

$$\varepsilon_T(f)/\mathcal{F}_x = \varepsilon_x(f_x),$$

where $*/cF_x$ denotes the class of $*$ in $B_T/\mathcal{F}_x$ and $\varepsilon_x : T_x \longrightarrow B_x$ is the Boolean hull of T_x.

Observing that Theorem 6.27 holds for all clopen U in X, with the restriction of $\otimes$ to $p^{-1}(U)$ in the role of $\otimes$, the preceding Corollary and Theorem 8.17 in [HBA] yield

THEOREM 6.37. *Let $\mathcal{S}$ be the sheaf of Boolean algebras over X, associated to the pair $(p^*(B(X), B_T)$. Then,*

a) *For each $x \in X$, the stalk of $\mathcal{S}$ at x is B_x.*

b) *If U is clopen U in X and $f \in T_U$, then the map $\varepsilon'_U(f)(x) = \varepsilon_x(f_x)$ is a continuous section of $\mathcal{S}$ over U.*

c) *Let $\mathcal{S}(U)$ be the BA of sections of $\mathcal{S}$ over U and let $\varepsilon'_U : T_U \longrightarrow \mathcal{S}(U)$ be the map in (b). Then, there is a BA-isomorphism $\alpha_U : B_U \longrightarrow \mathcal{S}(U)$, such that the following diagram is commutative.*

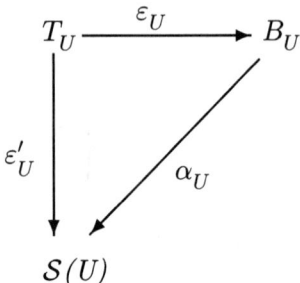

In particular, this holds for the BA of global sections of $\mathcal{S}$, B_T. ◇

It can be shown that in case each G_l is a complete subgroup of G, the BA of global sections of the sheaf in 6.37 is the filtered Boolean power of Theorem 6.34. ◇

CHAPTER 7

Horn-Tarski and Stiefel-Whitney Invariants

Our main result in this chapter, Theorem 7.1, shows that the isometry of quadratic forms of arbitrary dimension n over a reduced special group G is equivalent to the validity, in the Boolean hull B_G of G, of a finite number of (actually n) Boolean identities among their coefficients. The functions of the coefficients occurring in these identities are, in fact, the Boolean analogs of the usual symmetric functions, using join and meet (the Horn-Tarski invariants). On account of the fact that the isometry of forms of dimension $n \geq 3$ is defined in terms of the isometry of forms of dimension 2 by an existential formula of the language of special groups (cf. Notation 1.1.(e) and Chapter 10, below), our main result amounts to a weak form of quantifier-elimination for the isometry relation in terms of the additional operations of meet and join, present in B_G.

We also prove a variant of this result stated in terms of symmetric difference — the natural group operation in B_G — (instead of join), and meet, which are analogous to the Stiefel-Whitney invariants in K-theory and Galois Cohomology. However, owing to the monotonicity of the join operation, the former version is easier to manipulate than the latter.

The first of our characterizations of isometry is inspired by certain results of Horn and Tarski ([HT]) on the extension of measures in Boolean algebras. The Boolean symmetric functions appear there — though in a different way — and that is the reason why we have baptized our necessary and sufficient conditions for isometry after their names. The Boolean functions occuring in the second variant are our own. Our proof of these results depends in a crucial way on the Boolean hull construction, which embedds a reduced special group as a complete subgroup of a BA on which it is dense, originating in the duality theory of Chapter 3.

The Horn-Tarski invariants furnish a powerful conceptual and computational tool for the reduced algebraic theory of quadratic forms. In particular, they seem to put in a proper perspective the algebraic combinatorics pertaining to this theory; this is borne out by some of the examples and applications given here.

In section 1 we prove Theorem 7.1, and in section 2 we establish a number of basic properties of both types of invariants, amongst others, formulas expressing the Stiefel-Whitney invariants in terms of the Horn-Tarski invariants, and viceversa (Theorem 7.6), as well as formulas describing the behaviour of both kinds of invariants under direct sum of forms (Proposition 7.8).

Our applications in section 3 consist, firstly, in a characterization in terms of order of the relation of representation by arbitrary forms in Boolean algebras (Theorem 7.12). Secondly, we characterize the notion of indefiniteness of forms over arbitrary reduced special groups and give necessary and sufficient conditions for a reduced special group to be a Boolean algebra, in terms of the behaviour of certain forms.

In section 4 we compute the Horn-Tarski and Stiefel-Whitney invariants of Pfister forms and their multiples. Some elementary applications illustrate the use of these invariants in quadratic form theory.

In section 5 we describe both kinds invariants for linear combinations of Pfister forms of a fixed degree. As an interesting application we give a Boolean-theoretic proof of the so-called Arason-Pfister Hauptsatz for reduced special groups.

Throughout this chapter we identify a reduced special group G with its image by $\varepsilon_G : G \longrightarrow B_G$, inside its Boolean hull B_G. Thus we consider G as a **complete** subgroup of B_G (Corollary 5.4).

1. The Horn-Tarski conditions

This section is devoted to the proof of the following

THEOREM 7.1. *Let G be a reduced special group, and let $a_1, \ldots, a_n$, $b_1, \ldots, b_n$ be elements of G. For each $1 \leq k \leq n$, let $S^{n,k}$ be the set of all* **strictly** *increasing sequences of length k of elements of $\{1, \ldots, n\}$, denoted $p = (p_1, \ldots, p_k)$. Then, the following are equivalent :*

(1) $\langle a_1, \ldots, a_n \rangle \equiv_G \langle b_1, \ldots, b_n \rangle$.

(2) *For all $1 \leq k \leq n$, the following identities hold in the Boolean hull B_G of G :*

$$[HT_k] \qquad \bigvee_{p \in S^{n,k}} \bigwedge_{i=1}^{k} a_{p_i} = \bigvee_{p \in S^{n,k}} \bigwedge_{i=1}^{k} b_{p_i}.$$

(3) *For all $1 \leq k \leq n$, the following identities hold in the Boolean hull B_G of G :*

$$[SW_k] \qquad \bigtriangleup_{p \in S^{n,k}} \bigwedge_{i=1}^{k} a_{p_i} = \bigtriangleup_{p \in S^{n,k}} \bigwedge_{i=1}^{k} b_{p_i}.$$

Before the proof, we fix notation and establish some simple combinatorial facts.

1. THE HORN-TARSKI CONDITIONS

NOTATION 7.2.

(A) Let n, k be integers such that $1 \leq k \leq n$. Since the elements $p \in S^{n,k}$ are **strictly increasing** sequences with (domain) $Dom(p) = \{1, \ldots, k\}$ and (image) $Im(p) \subseteq \{1, \ldots, n\}$, we have :

(i) $p = q$ iff $Im(p) = Im(q)$.

(ii) If $A \subseteq \{1, \ldots, n\}$ has cardinality k, there is a **unique** $p \in S^{n,k}$ such that $Im(p) = A$.

(B) Let G is a rsg, $\varphi = \langle a_1, \ldots, a_n \rangle$, $\psi = \langle b_1, \ldots, b_n \rangle$ be forms over G, and $\sigma \in X_G$.

(i) We set :

$$p_\sigma(\varphi) = card(\{i \in \{1, \ldots, n\} : \sigma(a_i) = 1\})$$

and

$$n_\sigma(\varphi) = card(\{i \in \{1, \ldots, n\} : \sigma(a_i) = -1\}).$$

Write $\overline{p_\sigma(\varphi)}$ (resp., $\overline{n_\sigma(\varphi)}$) for the unique strictly increasing sequence of length $p_\sigma(\varphi)$ (resp., $n_\sigma(\varphi)$) such that $Im(\overline{p_\sigma(\varphi)}) = \{i : \sigma(a_i) = 1\}$ (resp., $Im(\overline{n_\sigma(\varphi)}) = \{i : \sigma(a_i) = -1\}$).

(ii) For $p = (p_1, \ldots, p_k) \in S^{n,k}$, let $A_p = \bigwedge_{i=1}^{k} a_{p_i}$ (meet in B_G). $\diamond$

FACT 7.3. *With notation as in 7.2 we have,*

(1) $p_\sigma(\varphi) + n_\sigma(\varphi) = n$ *and*

$\sum_{i=1}^{n} \sigma(a_i) = \sum_{i=1}^{n} \sigma(b_i)$ *iff* $p_\sigma(\varphi) = p_\sigma(\psi)$ *iff* $n_\sigma(\varphi) = n_\sigma(\psi)$.

(2) *With notation as in 7.2(B.(ii)),*

$$1 \leq k \leq l \leq n \Rightarrow \bigvee_{p \in S^{n,l}} A_p \leq \bigvee_{q \in S^{n,k}} A_q.$$

(3) *If $p_\sigma(\varphi) = p_\sigma(\psi)$, there is a permutation β of $\{1, \ldots, n\}$ such that for all i, $1 \leq i \leq n$, $\sigma(a_i) = \sigma(b_{\beta(i)})$.*

(4) *Every permutation β of $\{1, \ldots, n\}$ induces a bijection $\beta_k^* : S^{n,k} \longrightarrow S^{n,k}$ ($1 \leq k \leq n$), such that*

(+) $\qquad\qquad Im(\beta_k^*(p)) = Im(\beta \circ p)$.

PROOF. Item (1) is left as an exercise (see Example 1.7). For (2), note that by removing any $l - k$ coordinates from a given $p \in S^{n,l}$ we get a sequence $q \in S^{n,k}$. Clearly, $A_p \leq A_q$.

(3) is an immediate consequence of (1), that is, of the equality of cardinalities of 1's and -1's in $\sigma \star \varphi$ and $\sigma \star \psi$.

(4) For $p \in S^{n,k}$, let β_k^* be the unique strictly increasing sequence of length k enumerating $\beta[Im(p)] = Im(\beta \circ p)$ (see 7.2.(A.ii)). This clearly gives (+). We show (with $\beta^* = \beta_k^*$) :

(a) β^* is injective. For $p, q \in S^{n,k}$, 7.2.(A.i) (used twice),(+), and the bijectivity of β, prove the following sequence of implications :

$$\beta^*(p) = \beta^*(q) \Rightarrow Im(\beta^*(p)) = Im(\beta^*(q)) \Rightarrow \beta[Im(p)] = \beta[Im(q)]$$
$$\Rightarrow Im(p) = Im(q),$$

and the last term implies $p = q$, as needed.

(b) β^* is surjective. Let $q \in S^{n,k}$ and $A = \beta^{-1}[Im(q)]$. Clearly, A has cardinality k; hence there is $p \in S^{n,k}$ such that $Im(p) = A$ (7.2.(A.ii)). Then :

$$Im(\beta^*(p)) = \beta[Im(p)] = \beta[A] = Im(q),$$

and $\beta^*(p) = q$, by 7.2.(A.i). $\square$

Proof of Theorem 7.1. Since G is a complete subgroup of B_G (Corollary 5.4), it suffices to prove the theorem in the case that G is a BA, say B (whence, $B_G = B$). The proof will be divided in two parts, corresponding to (1) $\Leftrightarrow$ (2) and (1) $\Leftrightarrow$ (3).

Let $\varphi = \langle a_1, \ldots, a_n \rangle$, $\psi = \langle b_1, \ldots, b_n \rangle$ be forms over B. Recall that in our notation $1 = \bot$ and $-1 = \top$ (in $\mathbb{Z}_2$ and in B).

Part 1 : (1) and (2) are equivalent.

(1) $\Rightarrow$ (2). Assume that $\varphi \equiv_B \psi$. We prove the identities $[HT_k]$ by induction on $r = n - k$.

$\underline{r = 0.}$ In this case, $k = n$, $S^{n,n} = \{\langle 1, \ldots, n \rangle\}$ and

$$\bigvee_{p \in S^{n,n}} A_p = a_1 \wedge \ldots \wedge a_n \text{ (similarly for the } b_i\text{'s).}$$

It suffices to show that $\sigma(a_1 \wedge \ldots \wedge a_n) = \sigma(b_1 \wedge \ldots \wedge b_n)$, for every $\sigma \in X_B$. Since every such σ is a BA-morphism (Proposition 4.5), what we must prove comes down to

$$\bigwedge_{i=1}^n \sigma(a_i) = \sigma(\bigwedge_{i=1}^n a_i) = \sigma(\bigwedge_{i=1}^n b_i) = \bigwedge_{i=1}^n \sigma(b_i).$$

Our isometry assumption implies $\sum_{i=1}^n \sigma(a_i) = \sum_{i=1}^n \sigma(b_i)$ (Proposition 3.7), which, in turn, implies $p_\sigma(\varphi) = p_\sigma(\psi)$ (see 7.3(1)), i.e. that $\sigma \star \varphi$ and $\sigma \star \psi$ contain the same number of 1's (and -1's). This entails $\bigwedge_{i=1}^n \sigma(a_i) = \bigwedge_{i=1}^n \sigma(b_i)$; indeed, if there is at least a 1 among the $\sigma(a_i)$'s, then there is also one among the $\sigma(b_i)$'s, and so $\bigwedge_{i=1}^n \sigma(a_i) = \bot = \bigwedge_{i=1}^n \sigma(b_i)$. If all

1. THE HORN-TARSKI CONDITIONS

the $\sigma(a_i)$'s are -1, the same is true of the $\sigma(b_i)$'s, and we have $\bigwedge_{i=1}^{n} \sigma(a_i)$
$= \top = \bigwedge_{i=1}^{n} \sigma(b_i)$.

$r-1 \Rightarrow r$. Assume the result true for $r-1$, i.e., for $k+1 \leq n$. By symmetry, it suffices to show :

(*) $$\bigvee_{p \in S^{n,k}} A_p \leq \bigvee_{p \in S^{n,k}} B_p,$$

that is,

(**) For every $\sigma \in X_B$, $\sigma \left(\bigvee_{p \in S^{n,k}} A_p \right) = \top \Rightarrow \sigma \left(\bigvee_{p \in S^{n,k}} B_p \right) = \top$.

Thus, if $\sigma \in X_B$, we may assume $\sigma(A_p) = \top$, for some $p \in S^{n,k}$. We consider two cases:

(i) There is $j \in \{1, \ldots, n\} \setminus Im(p)$ such that $\sigma(a_j) = \top$.

Let p' be the increasing sequence of length $k+1$ enumerating $Im(p) \cup \{j\}$ (7.2.(A.ii)); we have

$$\sigma(A_{p'}) = \sigma(A_p \wedge a_j) = \sigma(A_p) \wedge \sigma(a_j) = \top.$$

By induction hypothesis, $A_{p'} \leq \bigvee_{q \in S^{n,k+1}} B_q$; Fact 7.3(2) gives $A_{p'} \leq \bigvee_{t \in S^{n,k}} B_t$. Hence, $\sigma \left(\bigvee_{t \in S^{n,k}} B_t \right) = \top$.

(ii) $\sigma(a_j) = \bot$, for every $j \in \{1, \ldots, n\} \setminus Im(p)$.

The assumption $\sigma(A_p) = \top$ yields $\sigma(a_{p_i}) = \top$, for $i = 1, \ldots, k$; from the case assumption (ii) it follows that $n_\sigma(\varphi) = k$. Thus, $\sum_{i=1}^{n} \sigma(a_i) = \sum_{i=1}^{n} \sigma(b_i)$ implies $n_\sigma(\psi) = k$ (Fact 7.3(1)). Setting $q = \overline{n_\sigma(\psi)}$, we have $q \in S^{n,k}$ and $\sigma(B_q) = \bigwedge_{i=1}^{k} \sigma(b_{q_i}) = \top$. We conclude, therefore, that $\sigma \left(\bigvee_{t \in S^{n,k}} B_t \right) = \top$.

$(2) \Rightarrow (1)$. Assume the identities $[HT_k]$. In view of Proposition 3.7 and Fact 7.3(1), it suffices to show that $n_\sigma(\varphi) = n_\sigma(\psi)$, for every $\sigma \in X_B$.

Fix $\sigma \in X_B$ and suppose $n_\sigma(\varphi) = d$; let $p = \overline{n_\sigma(\varphi)}$; then $p \in S^{n,d}$. Using $[HT_d]$, we have $A_p \leq \bigvee_{q \in S^{n,d}} B_q$. Since $\sigma(A_p) = \bigwedge_{i=1}^{d} \sigma(a_{p_i}) = \top$, we conclude that $\sigma \left(\bigvee_{q \in S^{n,d}} B_q \right) = \top$. Thus, there is $q \in S^{n,d}$ such that $\sigma(B_q) = \top$, i.e., $\sigma(B_{q_i}) = \top$, for $i = 1, \ldots, d$, and thus $n_\sigma(\varphi) = d \leq n_\sigma(\psi)$. A symmetric argument, starting with the $\sigma(b_i)$'s, proves $n_\sigma(\psi) \leq n_\sigma(\varphi)$ and hence, $n_\sigma(\varphi) = n_\sigma(\psi)$, as required. This ends Part 1 of the proof.

Part 2 : (1) and (3) are equivalent.

(1) ⇒ (3). Assume that $\varphi \equiv_B \psi$; Proposition 3.7 and Fact 7.3(1) show that $p_\sigma(\varphi) = p_\sigma(\psi)$.

We fix $\sigma \in X_B$, an integer $k \leq n$, and with them a permutation β of $\{1, \ldots, n\}$ as in Fact 7.3(3). Let $\beta^* = \beta_k^*$ be the permutation induced by β, as in Fact 7.3(4). Since σ is a Boolean morphism (Proposition 4.5) and β^* is bijective, we obtain :

$$\sigma\left(\triangle_{p \in S^{n,k}} \bigwedge_{i=1}^k a_{p_i}\right) = \triangle_{p \in S^{n,k}} \bigwedge_{i=1}^k \sigma(a_{p_i}) = \triangle_{p \in S^{n,k}} \bigwedge_{i=1}^k \sigma(b_{\beta(p_i)})$$

$$= \triangle_{p \in S^{n,k}} \bigwedge_{i=1}^k \sigma(b_{\beta^*(p)_i}) = \triangle_{q \in S^{n,k}} \bigwedge_{i=1}^k \sigma(b_{q_i})$$

$$= \sigma\left(\triangle_{p \in S^{n,k}} \bigwedge_{i=1}^k b_{p_i}\right).$$

Since σ is arbitrary, this implies the identity $[SW_k]$.

(3) ⇒ (1). Assume that the n identities $[SW_k]$ hold in B. Fix $\sigma \in X_B$ and let $k = n_\sigma(\varphi)$. Observe that for $p \in S^{n,k}$ we have :

$$(*) \qquad \sigma(A_p) = \bigwedge_{i=1}^k \sigma(a_{p_i}) = \begin{cases} \top & \text{if } p = \overline{n_\sigma(\varphi)} \\ \bot & \text{if } p \neq \overline{n_\sigma(\varphi)} \end{cases}$$

The first of these equalities is clear, since $\overline{n_\sigma(\varphi)}$ enumerates $\{i : \sigma(a_i) = \top\}$. For the second, $p \neq \overline{n_\sigma(\varphi)}$ implies $Im(p) \neq Im(\overline{n_\sigma(\varphi)}) = \{i : \sigma(a_i) = \top\}$ (see 7.2.(A.i)); hence, for some $i \leq k$, $\sigma(a_{p_i}) = \bot$ and so, $\sigma(A_p) = \bot$.

Clearly, (*) implies $\sigma\left(\triangle_{p \in S^{n,k}} A_p\right) = \top$, whence, by $[SW_k]$, $\sigma\left(\triangle_{p \in S^{n,k}} B_p\right) = \top$. Then, there is $q \in S^{n,k}$ so that $\sigma(B_q) = \top$, i.e., $\sigma(b_{q_i}) = \top$, for all $i \leq k$, proving that $n_\sigma(\psi) \geq k = n_\sigma(\varphi)$. By symmetry, we conclude $n_\sigma(\varphi) = n_\sigma(\psi)$. Fact 7.3(1) yields $\sum_{i=1}^n \sigma(a_i) = \sum_{i=1}^n \sigma(b_i)$. Since σ is arbitrary, Proposition 3.7 guarantees $\varphi \equiv_B \psi$, ending the proof of Theorem 7.1. □

DEFINITION 7.4. Let G be a reduced special group. We define the **Horn-Tarski** and the **Stiefel-Whitney invariants** of a form $\varphi = \langle a_1, \ldots, a_n \rangle$ over G to be the following elements **of the Boolean hull** B_G of G :

$$\mathcal{HT}_k(\varphi) = \bigvee_{p \in S^{n,k}} \bigwedge_{i=1}^k a_{p_i} \qquad \text{(Horn-Tarski invariants)}$$

and

2. BASIC PROPERTIES

$$\mathcal{SW}_k(\varphi) = \bigtriangleup_{p \in S^{n,k}} \bigwedge_{i=1}^{k} a_{p_i}, \qquad \text{(Stiefel-Whitney invariants)}$$

for every integer k, $1 \leq k \leq n$. ◇

2. Basic Properties

In this section we shall develop some the basic properties of Horn-Tarski and Stiefel-Whitney invariants.

PROPOSITION 7.5. *Let G be a reduced special group, B_G the Boolean hull of G, and $\varphi = \langle a_1, \ldots, a_n \rangle$ be a form of dimension n over G. Then,*

[HT1] $\qquad\qquad\qquad \mathcal{SW}_1(\varphi) = d(\varphi)$, *the discriminant of* φ.

[HT2] $\qquad\qquad\qquad \mathcal{HT}_n(\varphi) = \mathcal{SW}_n(\varphi) = a_1 \wedge \ldots \wedge a_n.$

[HT3] $\qquad$*The Horn-Tarski invariants are decreasing :*

$$\mathcal{HT}_1(\varphi) \geq \mathcal{HT}_2(\varphi) \geq \ldots \geq \mathcal{HT}_n(\varphi).$$

[HT4] $\qquad$*Assume that the sequence of coefficients in φ is decreasing, $a_1 \geq \ldots \geq a_n$ (in the partial order $x \leq y$ iff $x \in D_G(1, y)$, $x, y \in G$). Then, for $1 \leq k \leq n$,*

$$\mathcal{HT}_k(\varphi) = a_k.$$

[HT5] $\qquad\qquad\qquad \varphi \equiv_{B_G} \langle \mathcal{HT}_1(\varphi), \mathcal{HT}_2(\varphi), \ldots, \mathcal{HT}_n(\varphi) \rangle.$

Remarks. (a) By their definition, we have $\mathcal{SW}_k(\varphi) \leq \mathcal{HT}_k(\varphi)$, for all $k \leq n$. But these invariants are not, in general, equal, except for $k = n$.

(b) The invariants $\mathcal{SW}_k(\varphi)$ are not decreasing, nor even pairwise comparable, in general.

(c) The coefficients of the form on the right-hand side of [HT5] do not necessarily belong to G.

(d) Since permutation of the coefficients of a form preserves isometry, Theorem 7.1 implies that [HT4] also holds under the assumption that the a_i's are pairwise comparable in the partial order $\leq$. ◇

PROOF. [HT1] and [HT2] are clear, while [HT3] is Fact 7.3(2).

[HT4]. The assumption implies that $\bigwedge_{i=1}^{k} a_{p_i} = a_{p_k}$, for $p \in S^{n,k}$; hence

$$\mathcal{HT}_k(\varphi) = \bigvee_{p \in S^{n,k}} \bigwedge_{i=1}^{k} a_{p_i} = \bigvee_{p \in S^{n,k}} a_{p_k} \leq a_k,$$

because the sequences $p \in S^{n,k}$ are strictly increasing, and so $p_k \geq k$. On the other hand, k is the last term of a (unique) sequence in $S^{n,k}$, viz.

$p = \{1, \ldots, k\}$, and $a_k = a_{p_k}$ for this sequence. It follows that $\mathcal{HT}_k(\varphi) = a_k$, as claimed.

[HT5] is an immediate consequence of [HT3], [HT4] and Theorem 7.1. $\square$

Our next result gives explicit formulas for both types of invariants in terms of each other, in a way that depends only on k and the dimension of the form, **but not on its coefficients**.

THEOREM 7.6. *With notation as in Proposition 7.5, we have :*

[HT6]   *For* $1 \leq k \leq n$,
$$\mathcal{SW}_k(\varphi) = \mathcal{SW}_k(\langle \mathcal{HT}_1(\varphi), \ldots, \mathcal{HT}_n(\varphi) \rangle) = \bigtriangleup_{p \in S^{n,k}} \mathcal{HT}_{p_k}(\varphi).$$

In particular,

[HT7] $\bigtriangleup_{k=1}^n \mathcal{HT}_k(\varphi) = d(\varphi) \in G.$

[HT8]   *For* $2 \leq k \leq n$,
$$\mathcal{SW}_k(\varphi) = \bigtriangleup_{l=k}^n [\mathcal{HT}_l(\varphi)]^{c_{l,k}},$$

where $c_{l,k}$ is the parity of the binomial coefficient $\binom{l-1}{k-1}$ *(i.e., $c_{l,k} = 0$ (resp., 1) if it is even (resp., odd)).*

In particular,

[HT9]   a) $\mathcal{SW}_2(\varphi) = \bigtriangleup_{j=1}^{[n/2]} \mathcal{HT}_{2j}(\varphi).$

b) $\mathcal{SW}_{n-1}(\varphi) = \begin{cases} \mathcal{HT}_{n-1}(\varphi) & \text{if } n \text{ is odd} \\ \mathcal{HT}_{n-1}(\varphi) \bigtriangleup \mathcal{HT}_n(\varphi) & \text{if } n \text{ is even.} \end{cases}$

[HT10]   *For* $1 \leq k \leq n$,
$$\mathcal{HT}_k(\varphi) = \bigtriangleup_{p=k}^n [\mathcal{SW}_p(\varphi)]^{s(k,p)},$$

where $s(k, k+j)$ is defined by induction on $j \geq 0$ as follows:
$$s(k,k) = 1 \quad \text{and} \quad s(k, k+j) = \sum_{i=0}^{j-1} c_{k+j,k+i} \cdot s(k, k+i).$$

PROOF. [HT6]. The isometry [HT5] (Proposition 7.5) and item 3 in Theorem 7.1 give the first identity. For the second, [HT3] implies $\bigtriangleup_{i=1}^k \mathcal{HT}_{p_i}(\varphi) = \mathcal{HT}_{p_k}(\varphi)$, for every $p \in S^{n,k}$.

[HT7]. Use [HT6] for $k = 1$ and [HT1].

[HT8]. Recall that $p_k \geq k$, for $p \in S^{n,k}$. For all $l \in \{k, \ldots, n\}$, the number of sequences $p \in S^{n,k}$ whose last term is l, equals the number of distinct subsets of $\{1, \ldots, l-1\}$ of cardinality $k-1$, i.e., $\binom{l-1}{k-1}$. Rewriting the last term of [HT6] in this notation yields [HT8].

[HT9]. a) Clearly, $c_{l,2} = \begin{cases} 0 & \text{if } l \text{ odd} \\ 1 & \text{if } l \text{ even} \end{cases}$. Hence, only the terms with even l remain in [HT8]; (a) is just a rewriting of this fact.

b) Since $c_{n-1,n-1} = 1$ and $c_{n,n-1} = \begin{cases} 0 & \text{if } n \text{ odd} \\ 1 & \text{if } n \text{ even} \end{cases}$, [HT8] for the value $k = n-1$ immediately yields (b).

[HT10]. We define the following auxiliary 3-variable function:

(*) $$t(k, k+j, r) = \sum_{i=0}^{j-1} c_{r,k+i} \cdot s(k, k+i),$$

where $j \geq 0$ and $r \geq k+j$. Note that $t(k, k+j, k+j) = s(k, k+j)$. Next, by induction on ℓ we prove: for $k < \ell \leq n$,

(†) $$\triangle_{p=k}^{\ell-1} (\mathcal{SW}_p)^{s(k,p)} = \mathcal{HT}_k \triangle \triangle_{r=\ell}^{n} (\mathcal{HT}_r)^{t(k,\ell,r)},$$

where $\mathcal{HT}_r = \mathcal{HT}_r(\varphi)$, and similarly for $\mathcal{SW}_r$.

Proof of (†). For $\ell = k+1$, the left-hand side of (†) is $\mathcal{SW}_k$. From [HT7] and [HT8] we get:

(**) $$\mathcal{SW}_k = \mathcal{HT}_k \triangle \triangle_{r=k+1}^{n} (\mathcal{HT}_r)^{c_{r,k}}.$$

The definition of the function t gives, for $j = 1$:

$$t(k, k+1, r) = c_{r,k} \cdot s(k, k) = c_{r,k},$$

which yields the desired conclusion.

Now assume the result true for ℓ; we shall verify it for $\ell + 1$. For $k = \ell$ the identity (**) yields

(***) $$(\mathcal{SW}_\ell)^{s(k,\ell)} = (\mathcal{HT}_\ell)^{s(k,\ell)} \triangle \triangle_{r=\ell+1}^{n} (\mathcal{HT}_r)^{c_{r,\ell} \cdot s(k,\ell)}.$$

Multiplying (†) –which holds by induction hypothesis– with (***), we get:

$$\triangle_{p=k}^{\ell} (\mathcal{SW}_p)^{s(k,p)} = \mathcal{HT}_k \triangle (\mathcal{HT}_\ell)^{t(k,\ell,\ell)} \triangle (\mathcal{HT}_\ell)^{s(k,\ell)} \triangle$$
$$\triangle \triangle_{r=\ell+1}^{n} (\mathcal{HT}_r)^{t(k,\ell,r)+c_{r,\ell} \cdot s(k,\ell)}.$$

The two central terms on the right-hand side cancel out, since $s(k, \ell) = t(k, \ell, \ell)$. The identity

$$t(k, \ell+1, r) = t(k, \ell, r) + c_{r,\ell} \cdot s(k, \ell),$$

which follows easily from the definition of t by setting $\ell = k+j$, establishes the formula (†) for $\ell + 1$.

For $\ell = n$ the identity (†) takes the form:

$$\triangle_{p=k}^{n-1} (\mathcal{SW}_p)^{s(k,p)} = \mathcal{HT}_k \triangle (\mathcal{HT}_n)^{t(k,n,n)}.$$

Since $t(k,n,n) = s(k,n)$ and $\mathcal{HT}_n = \mathcal{SW}_n$, we obtain [HT10]. $\square$

The next proposition gives explicit formulas for the Horn-Tarski and Stiefel-Whitney invariants of the sum of two forms in terms of those of the summmands.

NOTATION 7.7. a) For a form φ over G, we set

$$\mathcal{HT}_0(\varphi) = \mathcal{SW}_0(\varphi) = \top \, (= -1).$$

b) Let n, m and k be positive integers such that $k \leq m+n$. We define

$$A^{n,m,k} = \{(s,r) : 0 \leq s \leq \min\{k,n\}, \ 0 \leq r \leq \min\{k,m\}, \text{ and } s+r = k\}.$$

PROPOSITION 7.8. (Addition formulas). *Let φ, ψ be forms over a reduced special group G, of dimensions n, m, respectively. With notation as in 7.7.(b), we have, for $1 \leq k \leq n+m$:*

[HT11] $$\mathcal{HT}_k(\varphi \oplus \psi) = \bigvee_{(s,r) \in A^{n,m,k}} (\mathcal{HT}_s(\varphi) \wedge \mathcal{HT}_r(\varphi)).$$

[HT12] $$\mathcal{SW}_k(\varphi \oplus \psi) = \triangle_{(s,r) \in A^{n,m,k}} (\mathcal{SW}_s(\varphi) \wedge \mathcal{SW}_r(\psi)).$$

PROOF. Let $\varphi = \langle a_1, \ldots, a_n \rangle$, $\psi = \langle b_1, \ldots, b_m \rangle$, and $\varphi \oplus \psi = \langle c_1, \ldots, c_{n+m} \rangle$, where $c_i = \begin{cases} a_i & \text{for } 1 \leq i \leq n \\ b_{i-n} & \text{for } n+1 \leq i \leq n+m. \end{cases}$

We begin by observing that every sequence $p \in S^{n+m,k}$ ($1 \leq k \leq n+m$) can be written as a concatenation of two sequences of lengths $\leq n$ and $\leq m$, respectively. Indeed, since every such p is strictly increasing, there are integers $s(p)$, $r(p)$ — depending on p — such that :

(i) $(s(p), r(p)) \in A^{n,m,k}$ (cf. Notation 7.7.b).

(ii) $\begin{cases} 1 \leq p_i \leq n & \text{if } 1 \leq i \leq s(p) \\ n+1 \leq p_{s(p)+j} \leq n+m & \text{if } 1 \leq j \leq r(p). \end{cases}$

Note that $s(p)$ or $r(p)$ may be zero; for instance, $s(p) = 0$ iff $Im(p) \subseteq \{n+1, \ldots, n+m\}$. Now we define strictly increasing sequences p^1, p^2, as follows :

(iii) length of $p^1 = s(p)$, and $p_i^1 = p_i$ for $1 \leq i \leq s(p)$;

length of $p^2 = r(p)$, and $p_j^2 = p_{s(p)+j}$ for $1 \leq j \leq r(p)$.

Clearly, $p^1 \in S^{n,s(p)}$, $p^2 \in S^{m,r(p)}$ and $p = p^1 {}^\frown p^2$, the concatenation of p^1 and p^2.

Next, we define a map

2. BASIC PROPERTIES

$$\eta : S^{n+m,k} \longrightarrow \coprod_{(s,r)\in A^{n,m,k}} (S^{n,s} \times S^{m,r})$$

by $\eta(p) = (p^1, p^2)$, where $\coprod$ denotes disjoint union. The following is straightforward :

Fact. η is a bijection.

As in 7.2.(B.ii), for $p \in S^{n+m,k}$, $q \in S^{n,s}$, $t \in S^{m,r}$, we set

$$C_p = \bigwedge_{i=1}^{k} c_{p_i}; \quad A_q = \bigwedge_{i=1}^{s} a_{q_i}; \quad B_t = \bigwedge_{i=1}^{r} b_{p_i}.$$

From the definition of p^1, p^2, above, it is clear that $C_p = A_{p^1} \wedge A_{p^2}$, for $p \in S^{n+m,k}$. Since η is a bijection, we have :

$$(+) \qquad \bigvee_{p \in S^{n+m,k}} C_p = \bigvee_{(s,r)\in A^{n,m,k}} \bigvee_{q\in S^{n,s}} \bigvee_{t\in S^{m,r}} (A_q \wedge B_t),$$

and

$$(++) \qquad \bigtriangleup_{p \in S^{n+m,k}} C_p = \bigtriangleup_{(s,r)\in A^{n,m,k}} \bigtriangleup_{q\in S^{n,s}} \bigtriangleup_{t\in S^{m,r}} (A_q \wedge B_r),$$

Clearly, we have

$$(*) \qquad \begin{aligned} \bigvee_{q\in S^{n,s}} \bigvee_{t\in S^{m,r}} (A_q \wedge B_t) &= \bigvee_{q\in S^{n,s}} \left(A_q \wedge \bigvee_{t\in S^{m,r}} B_t \right) \\ &= \left(\bigvee_{q\in S^{n,s}} A_q \right) \wedge \left(\bigvee_{t\in S^{m,r}} B_t \right) \\ &= \mathcal{HT}_s(\varphi) \wedge \mathcal{HT}_r(\varphi). \end{aligned}$$

Since $\wedge$ distributes over $\bigtriangleup$, i.e. $\bigtriangleup_{i=1}^{n}(x \wedge y_i) = x \wedge \bigtriangleup_{i=1}^{n} y_i$, we also get :

$$(**) \qquad \begin{aligned} \bigtriangleup_{q\in S^{n,s}} \bigtriangleup_{t\in S^{m,r}} (A_q \wedge B_t) &= \bigtriangleup_{q\in S^{n,s}} \left(A_q \wedge \bigtriangleup_{t\in S^{m,r}} B_t \right) \\ &= \left(\bigtriangleup_{q\in S^{n,s}} A_q \right) \wedge \left(\bigtriangleup_{t\in S^{m,r}} B_t \right) \\ &= \mathcal{SW}_s(\varphi) \wedge \mathcal{SW}_r(\varphi). \end{aligned}$$

Replacing (*) in (+), and (**) in (++), gives [HT11] and [HT12], respectively. $\square$

A useful case of Proposition 7.8 is :

COROLLARY 7.9. *Let φ be a form of dimension n over a reduced special group G, and $y \in G$. Then, for $1 \leq k \leq n$, we have :*

[HT13]
$$\mathcal{HT}_k(\varphi \oplus \langle y \rangle) = \mathcal{HT}_k(\varphi) \vee (\mathcal{HT}_{k-1}(\varphi) \wedge y)$$
$$\mathcal{SW}_k(\varphi \oplus \langle y \rangle) = \mathcal{SW}_k(\varphi) \triangle (\mathcal{SW}_{k-1}(\varphi) \wedge y)$$
and
$$\mathcal{HT}_{n+1}(\varphi \oplus \langle y \rangle) = \mathcal{SW}_{n+1}(\varphi \oplus \langle y \rangle) = \mathcal{HT}_n(\varphi) \wedge y. \quad \diamond$$

Another natural question is whether the Horn-Tarski and Stiefel-Whitney invariants of a tensor product can be expressed as Boolean functions of those of the factors in a reasonably simple and meaningful way. Proposition 7.10 below gives one such expansion for the Stiefel-Whitney invariants. However, we have not been able to find an expression of this kind for the Horn-Tarski invariants; the difficulty lies in the absence of a tractable distributive law of join over symmetric difference.

PROPOSITION 7.10. *Let G be a reduced special group, $\varphi = \langle a_1, \ldots, a_n \rangle$, $\psi = \langle x_1, \ldots, x_m \rangle$ be forms over G of dimensions n, m, respectively. For $\varepsilon \in \{\pm 1\}$ and $x \in G$, set*

$$\varepsilon x = \begin{cases} x & \text{if } \varepsilon = 1 \\ -x & \text{if } \varepsilon = -1. \end{cases}$$

For integers, k, n, m such that $k \leq m \cdot n$, define

$$F_{k,n}^m = \{(s_1, \ldots, s_m) : 0 \leq s_j \leq n \text{ and } \sum_{j=1}^m s_j = k\}.$$

Then, the following identities hold in B_G, for $1 \leq k \leq n \cdot m$:

[HT14]
$$\mathcal{SW}_k(\varphi \otimes \psi) = \bigtriangleup_{\varepsilon \in 2^m} \left(\bigtriangleup_{s \in F_{k,n}^m} \bigwedge_{j=1}^m \mathcal{SW}_{s_j}(-\varepsilon_j \varphi) \right) \wedge \varepsilon_1 x_1 \wedge \ldots \wedge \varepsilon_m x_m$$

$$= \bigtriangleup_{\eta \in 2^n} \left(\bigtriangleup_{t \in F_{k,m}^n} \bigwedge_{i=1}^n \mathcal{SW}_{t_i}(-\eta_i \psi) \right) \wedge \eta_1 a_1 \wedge \ldots \wedge \eta_n a_n.$$

Remark. In terms of the ring operations of B_G ($\triangle$ and $\wedge$), the right-hand side of, say, the first of these identities, is a polynomial in the variables $x_1, \ldots, x_m, -x_1, \ldots, -x_m$, where the coefficient of the monomial $\varepsilon_1 x_1 \wedge \ldots \wedge \varepsilon_m x_m$ ($\varepsilon \in 2^m$) is the expression in parenthesis. $\quad \diamond$

The case $m = 1$ of Proposition 7.10 is interesting in its own right; it is stated and proven separately.

PROPOSITION 7.11. *Let G be a rsg. Let φ be a form of dimension n over G, and $x \in G$. Then, for $1 \leq k \leq n$:*

[HT15] $\quad \mathcal{SW}_k(x\varphi) = (\mathcal{SW}_k(\varphi) \wedge -x) \triangle (\mathcal{SW}_k(-\varphi) \wedge x).$

[HT16] $\quad \mathcal{HT}_k(x\varphi) = (\mathcal{HT}_k(\varphi) \wedge -x) \triangle (\mathcal{HT}_k(-\varphi) \wedge x).$

2. BASIC PROPERTIES

PROOF. [HT15]. All computations take place in B_G. We use the Boolean identity

(*) $\bigwedge_{i=1}^{r}(x \triangle y_i) = x \triangle (\bigwedge_{i=1}^{r} y_i) \triangle (x \wedge \bigwedge_{i=1}^{r} y_i) \triangle (x \wedge \bigvee_{i=1}^{r} y_i),$

which can be proven by induction, using distributivity of $\wedge$ over $\triangle$. Let $\varphi = \langle a_1, \ldots, a_n \rangle$. Recall that $\mathcal{SW}_k(x\varphi) = \triangle_{p \in S^{n,k}} \bigwedge_{i=1}^{k}(x \triangle a_{p_i})$.

Using (*), we obtain:

(**) $\begin{cases} \triangle_{p \in S^{n,k}} \bigwedge_{i=1}^{k}(x \triangle a_{p_i}) = \left(\triangle_{p \in S^{n,k}} x\right) \triangle \left(\triangle_{p \in S^{n,k}} \bigwedge_{i=1}^{k} a_{p_i}\right) \triangle \\ \triangle \left(\triangle_{p \in S^{n,k}}(x \wedge \bigwedge_{i=1}^{k} a_{p_i})\right) \triangle \left(\triangle_{p \in S^{n,k}}(x \wedge \bigvee_{i=1}^{k} a_{p_i})\right). \end{cases}$

Let $\delta_{n,k}$ denote the parity of $card(S^{n,k}) = \binom{n}{k}$ ($\delta_{n,k} = c_{n+1,k+1}$, cf. Theorem 7.6). Clearly, we have:

(i) $\triangle_{p \in S^{n,k}} x = x^{\delta_{n,k}},$

where $x^\delta = 1$ if $\delta = 0$, and $x^\delta = x$ if $\delta = 1$ (remark the distinction between the $\{0, 1\}$ notation here and the $\{\pm 1\}$ notation in Proposition 7.10).

(ii) $\triangle_{p \in S^{n,k}} \bigwedge_{i=1}^{k} a_{p_i} = \mathcal{SW}_k(\varphi).$

(iii) $\triangle_{p \in S^{n,k}} (x \wedge \bigwedge_{i=1}^{k} a_{p_i}) = x \wedge \triangle_{p \in S^{n,k}} \bigwedge_{i=1}^{k} a_{p_i} = x \wedge \mathcal{SW}_k(\varphi).$

(Distributivity of $\wedge$ over $\triangle$).

Next, we compute the last term on the right-hand side of (**):

(iv) $\triangle_{p \in S^{n,k}} (x \wedge \bigvee_{i=1}^{k} a_{p_i}) = x \wedge (-1)^{\delta_{n,k}} \mathcal{SW}_k(-\varphi).$

Indeed, using distributivity and the De Morgan laws, we obtain:

$$\triangle_{p \in S^{n,k}}(x \wedge \bigvee_{i=1}^{k} a_{p_i}) = x \wedge \left(\triangle_{p \in S^{n,k}}\left(-\bigwedge_{i=1}^{k} -a_{p_i}\right)\right)$$

$$= x \wedge \left[\triangle_{p \in S^{n,k}}(-1) \triangle \triangle_{p \in S^{n,k}} \bigwedge_{i=1}^{k} -a_{p_i}\right]$$

$$= x \wedge (-1)^{\delta_{n,k}} \mathcal{SW}_k(-\varphi).$$

Substituting (i) – (iv) on the right-hand side of (**) gives:

$$\mathcal{SW}_k(x\,\varphi) = [\mathcal{SW}_k(\varphi) \triangle (x \wedge \mathcal{SW}_k(\varphi))] \triangle$$
$$\triangle\, [x^{\delta_{n,k}} \triangle (x \wedge (-1)^{\delta_{n,k}}\,\mathcal{SW}_k(-\varphi))].$$

The first term on the right-hand side of this expression is $\mathcal{SW}_k(\varphi) \wedge -x$. We now compute the second term :

a) If $\delta_{n,k}$ is even, the second term is clearly $x \wedge \mathcal{SW}_k(-\varphi)$.

b) If $\delta_{n,k}$ is odd, then :

$$x^{\delta_{n,k}} \triangle (x \wedge (-1)^{\delta_{n,k}}\,\mathcal{SW}_k(-\varphi)) = x \triangle (x \wedge -\mathcal{SW}_k(-\varphi))$$
$$= x \wedge \mathcal{SW}_k(-\varphi).$$

Thus, the second term is always $x \wedge \mathcal{SW}_k(-\varphi)$, which proves [HT15]. Identity [HT16] follows easily from [HT15], using [HT10]. □

Proof of Proposition 7.10. The second identity in [HT14] follows from the first applied to the form $\psi \otimes \varphi$ ($\equiv_G \varphi \otimes \psi$), and Theorem 7.1. The first identity is proved by induction on m; the case $m = 1$ is Proposition 7.11. We introduce the following notation :

For integers $n, l \geq 1$, $\varepsilon \in 2^l$, and $0 \leq s_1, \ldots, s_l \leq n$,

(i) $a_{\varepsilon,\langle s_1,\ldots,s_l\rangle} = \left(\bigwedge_{j=1}^{l} \mathcal{SW}_{s_j}(-\varepsilon_j\,\varphi)\right) \wedge \varepsilon_1 x_1 \wedge \ldots \wedge \varepsilon_l x_l;$

(ii) $a_{\langle s_1,\ldots,s_l\rangle} = \underset{\varepsilon \in 2^l}{\triangle}\, a_{\varepsilon,\langle s_1,\ldots,s_l\rangle}.$

Further, given an integer s, $0 \leq s \leq n \cdot l$, we set

(iii) $a_{\varepsilon,s,n,l} = \underset{\rho \in F^l_{k,n}}{\triangle}\, a_{\varepsilon,\rho},$

where $F^l_{k,n}$ is as in the statement of Proposition 7.10.

By induction hypothesis, for $1 \leq s \leq n(m-1)$, we have :

(1) $\mathcal{SW}_s(\langle x_1,\ldots,x_{m-1}\rangle \otimes \varphi) = \underset{\varepsilon \in 2^{m-1}}{\triangle}\, \underset{\rho \in F^{m-1}_{s,n}}{\triangle}\, a_{\varepsilon,\rho} =$

$= \underset{\varepsilon \in 2^{m-1}}{\triangle}\, a_{\varepsilon,s,n,m-1} = \underset{\rho \in F^{m-1}_{s,n}}{\triangle}\, a_\rho.$

From [HT15], Proposition 7.11, we have, for $0 \leq r \leq n$, $\mathcal{SW}_r(x_m\,\varphi) = b \triangle b^-$, where,

$$b = \mathcal{SW}_r(\varphi) \wedge -x_m \quad \text{and} \quad b^- = \mathcal{SW}_r(-\varphi) \wedge x_m.$$

Distributing $\wedge$ over $\triangle$ gives :

$$\mathcal{SW}_s(\langle x_1,\ldots,x_{m-1}\rangle \otimes \varphi) \wedge \mathcal{SW}_r(x_m\,\varphi) = a \wedge (b \triangle b^-)$$
$$= (a \wedge b) \triangle (a \wedge b^-),$$

where $a = \mathcal{SW}_s(\langle x_1,\ldots,x_{m-1}\rangle \otimes \varphi)$; using (1) we get

$$a \wedge b = \triangle_{\rho \in F_{s,n}^{m-1}} (a_\rho \wedge b) \quad \text{and} \quad a \wedge b^- = \triangle_{\rho \in F_{s,n}^{m-1}} (a_\rho \wedge b^-),$$

whence

(2) $$\mathcal{SW}_s(\langle x_1,\ldots,x_{m-1}\rangle \otimes \varphi) \wedge \mathcal{SW}_r(x_m\,\varphi) = \triangle_{\rho \in F_{s,n}^{m-1}} a_{\rho\widehat{\,}\langle r\rangle},$$

where $\rho\widehat{\,}\langle r\rangle$ denotes concatenation of the displayed sequences. Taking symmetric difference over all pairs of integers $(s,r) \in A^{n(m-1),n,k}$, (2) yields:

(3) $$\begin{cases} \triangle_{(s,r) \in A^{n(m-1),n,k}} \mathcal{SW}_s(\langle x_1,\ldots,x_{m-1}\rangle \otimes \varphi) \wedge \mathcal{SW}_r(x_m\varphi) = \\ \\ = \triangle_{(s,r) \in A^{n(m-1),n,k}} \triangle_{\rho \in F_{s,n}^{m-1}} a_{\rho\widehat{\,}\langle r\rangle} = \triangle_{\delta \in F_{k,n}^m} a_\delta. \end{cases}$$

Unravelling notation through (i) – (iii), we see that the last term in (3) is just the right-hand side of the first identity in [HT14]. On the other hand, the addition formula [HT12] shows that the first term in (3) equals $\mathcal{SW}_k(\psi \otimes \varphi)$, ending the proof. $\square$

3. Some applications

.

We shall give here some applications of the Horn-Tarski conditions; further applications will be presented in sections 3 and 4, below. Our first result is a characterization of the set of elements represented by an arbitrary form in a BA, in terms of order.

THEOREM 7.12. *Let B be a BA and $a_1,\ldots,a_n \in B$. Then,*

$$D_B(\langle a_1,\ldots,a_n\rangle) = \left[\bigwedge_{i=1}^n a_i,\ \bigvee_{i=1}^n a_i\right]$$
$$= \{b \in B : \bigwedge_{i=1}^n a_i \leq b \leq \bigvee_{i=1}^n a_i\}.$$

In particular, we have :

COROLLARY 7.13. *Let G be a reduced special group, $a_1,\ldots,a_n \in G$, and let B_G be the Boolean hull of G. Then*

$$D_G(\langle a_1,\ldots,a_n\rangle) \subseteq \left[\bigwedge_{i=1}^n a_i,\ \bigvee_{i=1}^n a_i\right] \cap G. \quad \diamond$$

The following Lemma will be used in the proof of Theorem 7.12.

LEMMA 7.14. *Let B be a BA, and $b, a_1, \ldots, a_n \in B$. The following are equivalent:*

(1) $b \in D_B(\langle a_1, \ldots, a_n \rangle)$.

(2) *There are elements $x_{n-1} \leq x_{n-2} \leq \ldots \leq x_1$ in B such that*

$$\mathcal{HT}_1(\langle a_1, \ldots, a_n \rangle) = b \vee x_1,$$
$$\mathcal{HT}_k(\langle a_1, \ldots, a_n \rangle) = (b \wedge x_{k-1}) \vee x_k \quad \text{for } 2 \leq k \leq n-1,$$
$$\mathcal{HT}_n(\langle a_1, \ldots, a_n \rangle) = b \wedge x_{n-1}.$$

PROOF. (1) $\Rightarrow$ (2). Let $b_2, \ldots, b_n \in B$ be such that $\langle a_1, \ldots, a_n \rangle \equiv_B \langle b, b_2, \ldots, b_n \rangle$. By Theorem 7.1,

(*) $\quad \mathcal{HT}_k(\langle a_1, \ldots, a_n \rangle) = \mathcal{HT}_k(\langle b, b_2, \ldots, b_n \rangle) \quad$ for $1 \leq k \leq n$.

For $0 \leq k \leq n-1$, let $x_k = \mathcal{HT}_k(b_2, \ldots, b_n)$. By [HT3] (Proposition 7.5), the x_k's form a decreasing sequence. The identities [HT13] (Corollary 7.9) and (*) yield

$$\mathcal{HT}_k(\langle a_1, \ldots, a_n \rangle) = \mathcal{HT}_k(\langle b \rangle \oplus \langle b_2, \ldots, b_n \rangle) = x_k \vee (x_{k-1} \wedge b)$$

for $1 \leq k \leq n-1$; in particular, since $x_0 = \top$ (see Notation 7.7.(a)), for $k = 1$ we obtain the first identity of (2). Also:

$$\mathcal{HT}_n(\langle a_1, \ldots, a_n \rangle) = b \wedge \mathcal{HT}_{n-1}(\langle b_2, \ldots, b_n \rangle) = b \wedge x_{n-1}.$$

(2) $\Rightarrow$ (1). Given a decreasing sequence as in (2), [HT4] (Proposition 7.5) shows that $\mathcal{HT}_k(\langle x_1, \ldots, x_{n-1} \rangle) = x_k$, for $1 \leq k \leq n-1$. Using [HT13] (Corollary 7.9) we obtain

$$\mathcal{HT}_k(\langle b, x_1, \ldots, x_{n-1} \rangle) = (b \wedge \mathcal{HT}_{k-1}(\langle x_1, \ldots, x_{n-1} \rangle)) \vee$$
$$\vee \mathcal{HT}_k(\langle x_1, \ldots, x_{n-1} \rangle) = (b \wedge x_{k-1}) \vee x_k,$$

which, by assumption (2), equals $\mathcal{HT}_k(\langle a_1, \ldots, a_n \rangle)$, for $1 \leq k \leq n-1$. For $k = n$, by the assumptions in (2), we have:

$$\mathcal{HT}_n(\langle a_1, \ldots, a_n \rangle) = b \wedge x_{n-1} = b \wedge \bigwedge_{i=1}^{n-1} x_i = \mathcal{HT}_n(\langle b, x_1, \ldots, x_{n-1} \rangle).$$

Thus, the forms $\langle a_1, \ldots, a_n \rangle$ and $\langle b, x_1, \ldots, x_{n-1} \rangle$ have the same Horn-Tarski invariants and hence, are isometric (in B) by Theorem 7.1, i.e., $b \in D_B(\langle a_1, \ldots, a_n \rangle)$. $\square$

Proof of Theorem 7.12. The inclusion $\subseteq$ is easy: assuming $b \in D_B(\langle a_1, \ldots, a_n \rangle)$, let $b_2, \ldots, b_n \in B$ be such that $\langle a_1, \ldots, a_n \rangle \equiv_B \langle b, b_2, \ldots, b_n \rangle$. By Theorem 7.1 these forms have the same Horn-Tarski invariants $\mathcal{HT}_k$, for $1 \leq k \leq n$. For $k = 1, n$ we get:

$$\bigvee_{i=1}^n a_i = b \vee \bigvee_{j=2}^n b_j \geq b \quad \text{and} \quad \bigwedge_{i=1}^n a_i = b \wedge \bigwedge_{j=2}^n b_j \leq b.$$

3. SOME APPLICATIONS

For the other inclusion, we assume $\bigwedge_{i=1}^{n} a_i \leq b \leq \bigvee_{i=1}^{n} a_i$, and verify condition (2) of Lemma 7.14. Let

$$x_k = (\mathcal{HT}_k(\langle a_1, \ldots, a_n \rangle) \wedge -b) \vee (b \wedge \mathcal{HT}_{k+1}(\langle a_1, \ldots, a_n \rangle)),$$

for $1 \leq k \leq n-1$. We have :

(i) The sequence x_k is decreasing.

Indeed, let $2 \leq k \leq n-1$; [HT3] (Proposition 7.5) yields :

$$\mathcal{HT}_k(\langle a_1, \ldots, a_n \rangle) \wedge -b \leq \mathcal{HT}_{k-1}(\langle a_1, \ldots, a_n \rangle) \wedge -b,$$
$$\mathcal{HT}_{k+1}(\langle a_1, \ldots, a_n \rangle) \wedge b \leq \mathcal{HT}_k(\langle a_1, \ldots, a_n \rangle) \wedge b,$$

which clearly imply $x_k \leq x_{k-1}$.

(ii) $\mathcal{HT}_1(\langle a_1, \ldots, a_n \rangle) = b \vee x_1$.

We have,

$$b \vee x_1 = b \vee (b \wedge \mathcal{HT}_2(\langle a_1, \ldots, a_n \rangle)) \vee (\mathcal{HT}_1(\langle a_1, \ldots, a_n \rangle) \wedge -b) =$$
$$= b \vee \mathcal{HT}_1(\langle a_1, \ldots, a_n \rangle).$$

Since $b \leq \bigvee_{i=1}^{n} a_i = \mathcal{HT}_1(\langle a_1, \ldots, a_n \rangle)$, (ii) follows at once.

(iii) $\mathcal{HT}_k(\langle a_1, \ldots, a_n \rangle) = (b \wedge x_{k-1}) \vee x_k,$ for $2 \leq k \leq n-1$.

By the definition of x_{k-1}, $b \wedge x_{k-1} = b \wedge \mathcal{HT}_k(\langle a_1, \ldots, a_n \rangle)$; hence,

$$(b \wedge x_{k-1}) \vee x_k =$$
$$= (b \wedge \mathcal{HT}_k(\langle a_1, \ldots, a_n \rangle)) \vee (-b \wedge \mathcal{HT}_k(\langle a_1, \ldots, a_n \rangle)) \vee$$
$$\vee (\mathcal{HT}_{k+1}(\langle a_1, \ldots, a_n \rangle) \wedge b) =$$
$$= \mathcal{HT}_k(\langle a_1, \ldots, a_n \rangle) \vee (\mathcal{HT}_{k+1}(\langle a_1, \ldots, a_n \rangle) \wedge b) = \mathcal{HT}_k(\langle a_1, \ldots, a_n \rangle),$$

where the last equality comes from [HT3].

(iv) $\mathcal{HT}_n(\langle a_1, \ldots, a_n \rangle) = b \wedge x_{n-1}$.

As in (iii), $b \wedge x_{n-1} = b \wedge \mathcal{HT}_n(\langle a_1, \ldots, a_n \rangle)$. But, $b \geq \bigwedge_{i=1}^{n} a_i = \mathcal{HT}_n(\langle a_1, \ldots, a_n \rangle)$, and this clearly implies (iv). □

Our next series of results deal with the notion of indefinite form :

DEFINITION 7.15. Let G be a special group, and $\varphi = \langle a_1, \ldots, a_n \rangle$ be a form over G. We say that φ is **indefinite (in G)** iff $|\sum_{i=1}^{n} \sigma(a_i)| < n$, for every $\sigma \in X_G$ (the sum takes place in $\mathbb{Z}$). ◊

A special feature of the **reduced** theory of quadratic forms is the equivalence of the notions of isotropic and universal forms (Definition 1.3); see [L2; Cor. 1.20, p. 8], for the field case. Notice that in the present context, axiom [SG2] (Definition 1.2) gives at once that isotropic implies universal. The converse follows from Proposition 1.6.(e.3) : if φ is universal, then $1, -1 \in D_G(\varphi)$; hence $2 \cdot \varphi$ is isotropic, and then so is φ.

The following Proposition gives several characterizations of indefinite forms over reduced special groups.

PROPOSITION 7.16. *Let G be a reduced special group and $\varphi = \langle a_1, \ldots, a_n \rangle$ be a form over G. The following are equivalent :*

(1) *φ is indefinite in G.*

(2) *For every $\sigma \in X_G$ there are indices $i, j \in \{1, \ldots, n\}$ such that $\sigma(a_i) = 1$ and $\sigma(a_j) = -1$.*

(3) *In B_G, $\mathcal{HT}_1(\varphi) = \bigvee_{i=1}^{n} a_i = \top$ and $\mathcal{HT}_n(\varphi) = \bigwedge_{i=1}^{n} a_i = \bot$.*

(4) *The ideals of B_G generated by $\{a_1, \ldots, a_n\}$ and $\{-a_1, \ldots, -a_n\}$ are improper.*

(5) *φ is isotropic in B_G.*

(6) *Both the Pfister forms $\bigotimes_{i=1}^{n} \langle 1, a_i \rangle$ and $\bigotimes_{i=1}^{n} \langle 1, -a_i \rangle$ represent -1 in G (i.e., they are isotropic, or, equivalently, hyperbolic over G).*

PROOF. The equivalence of (1) and (2) is clear from Definition 7.15.

(2) $\Rightarrow$ (3). Let $\gamma \in S(B_G)$ be a Boolean character of B_G. Then, $\sigma = \gamma_{|G} \in X_G$ (Proposition 4.6); (2) implies :

$$\gamma(\bigvee_{i=1}^{n} a_i) = \bigvee_{i=1}^{n} \gamma(a_i) = \bigvee_{i=1}^{n} \sigma(a_i) = \top;$$

similarly, $\gamma(\bigwedge_{i=1}^{n} a_i) = \bot$. Since γ is arbitrary, we conclude $\bigvee_{i=1}^{n} a_i = \top$ and $\bigwedge_{i=1}^{n} a_i = \bot$.

(3) $\Leftrightarrow$ (4). Clear.

(4) $\Rightarrow$ (5). From [HT5] in Proposition 7.5 and (4) comes

$$\varphi \equiv_{B_G} \langle \mathcal{HT}_1(\varphi), \ldots, \mathcal{HT}_n(\varphi) \rangle \equiv_{B_G} \langle 1, -1 \rangle \oplus \langle \mathcal{HT}_2(\varphi), \ldots, \mathcal{HT}_{n-1}(\varphi) \rangle,$$

and hence φ is isotropic in B_G.

(5) $\Rightarrow$ (3). Since φ is isotropic in B_G, both $\bot = 1$ and $\top = -1$ are in $D_{B_G}(\varphi)$. It follows from Theorem 7.12 and [HT3] in Proposition 7.5 that the equalities in (3) are satisfied.

(3) $\Rightarrow$ (2). Every SG-character $\sigma \in X_G$ extends to a BA-character $\gamma \in S(B_G)$ (Corollary 5.4.(b)). Clearly, (3) implies that $\gamma(a_j) = \top$, i.e., $\sigma(a_j) = -1$, for some $j \leq n$, and $\gamma(a_i) = \bot$, i.e., $\sigma(a_i) = 1$, for some $i \leq n$.

(3) $\Leftrightarrow$ (6). This comes from Proposition 4.11.(b), if we recall that we are identifying $\varepsilon_G(a)$ with a, for all $a \in G$. Thus, we have

$\bigvee_{i=1}^{n} a_i = \top$ iff $\bigcup_{i=1}^{n} \varepsilon_G(a_i) = \varepsilon_G(-1) = X_G$ iff $-1 \in D_G(\bigotimes_{i=1}^{n} \langle 1, a_i \rangle)$,

as well as

$\bigwedge_{i=1}^{n} a_i = \bot$ iff $\bigvee_{i=1}^{n} -a_i = \top$ iff $\bigcup_{i=1}^{n} \varepsilon_G(-a_i) = \varepsilon_G(-1)$ iff

iff $-1 \in D_G(\bigotimes_{i=1}^{n} \langle 1, -a_i \rangle)$,

which prove the asserted equivalence. $\square$

Our next result characterizes BA's in terms of the notions introduced above. The result is well-known for fields; cf. [L2] (Thms. 16.2, p. 119, and 17.12, p. 131).

PROPOSITION 7.17. *Let G be a reduced special group. The following are equivalent:*

(1) *G is a Boolean algebra.*

(2) *Every indefinite form over G is isotropic.*

(3) *The form $\langle 1, a, b, -ab \rangle$ is isotropic for every $a, b \in G$.*

(4) *For all $a, b \in G$, there is $c \in G$ such that $\langle 1, a, b, -ab \rangle \equiv_G \langle 1, -1, c, c \rangle$.*

(5) *For all $a, b \in G$, there is $c \in G$ such that $D_G(\langle 1, a \rangle \otimes \langle 1, b \rangle) = D_G(1, c)$.*

PROOF. (1) $\Rightarrow$ (2). The equivalence of items (1) and (5) in Proposition 7.16 shows that 'indefinite $\Leftrightarrow$ isotropic' in any BA.

(2) $\Rightarrow$ (3) is clear, since the form $\langle 1, a, b, -ab \rangle$ is indefinite.

(3) $\Rightarrow$ (4). Let $a, b \in G$; since the discriminant of $\langle 1, a, b, -ab \rangle$ is -1 and this form is isometric over G to $\langle 1, -1 \rangle \oplus \langle x, y \rangle$, then $x = y$, and there is $c \in G$ such that:

$$\langle 1, a, b, -ab \rangle \equiv_G \langle 1, -1, c, c \rangle.$$

(4) $\Rightarrow$ (5). Given $a, b \in G$, let $-c \in G$ be the element such that

$$\langle 1, -a, -b, -ab \rangle \equiv_G \langle 1, -1, -c, -c, \rangle.$$

This isometry implies $\langle -1, -a, -b, -ab \rangle \equiv_G \langle -1, -1, -c, -c \rangle$, which, when multiplied by -1 on both sides, gives

(*) $\qquad \langle 1, a \rangle \otimes \langle 1, b \rangle \equiv_G \langle 1, c \rangle \oplus \langle 1, c \rangle.$

By Proposition 1.6.(e.1), we have $D_G(\langle 1, c \rangle) = D_G(\langle 1, c \rangle \oplus \langle 1, c \rangle)$. It now follows directly from (*) that $D_G(\langle 1, a \rangle \otimes \langle 1, b \rangle) = D_G(\langle 1, c \rangle)$.

(5) $\Rightarrow$ (1). For $a, b \in G$, let $c \in G$ be such that $D_G(\langle 1, a \rangle \otimes \langle 1, b \rangle) = D_G(\langle 1, c \rangle)$. Theorem 7.12 assures that, **in $\mathbf{B_G}$**, we have $c = a \vee b$. Thus, G contains the join **in $\mathbf{B_G}$** of any pair of its elements. Hence, the same is true of any finite subset $\{a_1, \ldots, a_n\} \subseteq G$. Taking into account the De Morgan

laws in a BA, as well as that G is closed under complements, we conclude that both the join and the meet, **in B_G**, of any finite subset of G, is in G. From Proposition 4.10.(b), we get $G = B_G$, or, more precisely, that ε_G is a bijective complete SG-morphism. It is readily verified that this implies that ε_G is a SG-isomorphism. Thus, G is a Boolean algebra. □

Remark. It is **not** true that a reduced special group that has meets and joins of every pair of its elements, in the representation partial order, is a BA. In fact, all fans F have this property: in the partial order $a \leq b$ iff $a \in D_F(\langle 1, b \rangle)$, we have

$$\inf\{a, b\} = \begin{cases} 1 & \text{if } a \neq b \text{ and } -1 \notin \{a, b\} \\ a & \text{if } a = b \text{ or } b = -1, \end{cases}$$

and

$$\sup\{a, b\} = \begin{cases} -1 & \text{if } a \neq b \text{ and } 1 \notin \{a, b\} \\ a & \text{if } a = b \text{ or } b = 1. \end{cases}$$

(The distributive laws fail in this example.)

The proof of Proposition 7.17 shows the existence of meets and joins **in B_G**, not solely in the representation partial order in G. It is also possible to prove directly that a reduced special group satisfying (4) (or (5)) is a BA, but this proof is much more involved, although elementary. ◇

4. The invariants of Pfister forms

The main results of this paragraph, Theorems 7.18 and 7.24, give exact and explicit computations of the Horn-Tarski and Stiefel-Whitney invariants of Pfister forms and their multiples. As an illustration of the use of these invariants we prove — in some cases, reprove — certain results concerning Pfister forms, well-known in the theory of quadratic forms.

THEOREM 7.18. *Let G be a reduced special group. Let $a \in G$, and $\varphi = \bigotimes_{i=1}^{n} \langle 1, a_i \rangle$ be a Pfister form over G of degree $n \geq 1$. Then:*

$$\mathcal{HT}_k(a\varphi) = \begin{cases} a \vee \bigvee_{i=1}^{n} a_i = \mathcal{HT}_1(a\varphi) & \text{for } 1 \leq k \leq 2^{n-1} \\ a \wedge -\bigvee_{i=1}^{n} a_i = \mathcal{HT}_{2^n}(a\varphi) & \text{for } 2^{n-1} + 1 \leq k \leq 2^n. \end{cases}$$

In particular,

$$\mathcal{HT}_k(\varphi) = \begin{cases} \bigvee_{i=1}^{n} a_i = \mathcal{HT}_1(\varphi) & \text{for } 1 \leq k \leq 2^{n-1} \\ \bot = \mathcal{HT}_{2^n}(\varphi) & \text{for } 2^{n-1} + 1 \leq k \leq 2^n. \end{cases}$$

Before proving this result, we shall give some consequences in order to illustrate its use. As a matter of fact, most of the standard algebraic results concerning Pfister forms can be derived from Theorem 7.18 by simple Boolean computations.

COROLLARY 7.19. *Let G be a rsg. Let $a \in G$, and φ be a Pfister form over G. The following are equivalent :*

(1) $a\varphi$ *is indefinite in G.*   (2) $a\varphi$ *is isotropic in G.*

(3) $a\varphi$ *is hyperbolic in G.*

PROOF. The implications (3) $\Rightarrow$ (2) $\Rightarrow$ (1) are obvious.
(1) $\Rightarrow$ (3). By Proposition 7.16.(3), assumption (1) implies $\mathcal{HT}_1(a\varphi) = \top$ and $\mathcal{HT}_{2^n}(a\varphi) = \bot$ (in B_G). By Theorem 7.18,

$$\mathcal{HT}_k(a\ \varphi) = \begin{cases} -1 & \text{if } 1 \leq k \leq 2^{n-1} \\ 1 & \text{if } 2^{n-1}+1 \leq k \leq 2^n. \end{cases}$$

Using [HT5], Proposition 7.5, we get

$$a\varphi \equiv_{B_G} \langle \mathcal{HT}_1(a\varphi),\ldots,\mathcal{HT}_n(a\varphi) \rangle = \langle \underbrace{-1,\ldots,-1}_{2^{n-1}}, \underbrace{1,\ldots,1}_{2^{n-1}} \rangle.$$

Both terms of this isometry have coefficients in G; since G is a complete subgroup of B_G (Corollary 5.4.(a)), this isometry holds in G. □

COROLLARY 7.20. *Let G be a rsg. Let $a, b \in G$, and $\varphi = \bigotimes_{i=1}^n \langle 1, a_i \rangle$, $\psi = \bigotimes_{i=1}^n \langle 1, b_i \rangle$ be Pfister forms of the same dimension over G. Then :*

i) $D_G(\varphi) = \left[\bot, \bigvee_{i=1}^n a_i\right]_{B_G} \cap G = \{g \in G : \bot \leq g \leq \bigvee_{i=1}^n a_i\}$.

ii) $a \in D_G(\varphi) \Leftrightarrow a\varphi \equiv_G \varphi$.

iii) $\varphi \equiv_G \psi \Leftrightarrow \mathcal{HT}_1(\varphi) = \mathcal{HT}_1(\psi) \Leftrightarrow \bigvee_{i=1}^n a_i = \bigvee_{i=1}^n b_i$ *(in B_G).*

iv) *If $a \in D_G(\varphi)$, then* $\langle 1, b \rangle \otimes \varphi \equiv_G \langle 1, ab \rangle \otimes \varphi$.

v) $a \in D_G(\langle 1, -a \rangle \otimes \varphi) \Rightarrow a \in D_G(\varphi)$.

Remark. (ii) and (v) are (*f*) and (*l*.3) of Proposition 2.2, respectively. For a proof of (iv) in the case of fields, see [L2; Cor. X.1.8, p. 280]. ◇

PROOF. (i) is a direct consequence of Theorems 7.12 and 5.2.(6) (the latter applied to $f = \varepsilon_G$).

(ii) ($\Rightarrow$). Assume that $a \in D_G(\varphi)$; from (i) we get $a \leq \bigvee_{i=1}^n a_i = \mathcal{HT}_1(\varphi)$. Then, Theorem 7.18 implies $\mathcal{HT}_1(a\varphi) = a \vee \mathcal{HT}_1(\varphi) = \mathcal{HT}_1(\varphi)$, $\mathcal{HT}_{2^n}(a\varphi) = a \wedge -\mathcal{HT}_1(\varphi) = \bot$, and hence $\mathcal{HT}_k(a\varphi) = \mathcal{HT}_k(\varphi)$, for all $1 \leq k \leq 2^n$. By Theorem 7.1, $a\varphi \equiv_G \varphi$.

($\Leftarrow$). The assumption $a\varphi \equiv_G \varphi$ and Theorems 7.1 and 7.18 imply $\mathcal{HT}_1(a\varphi) = a \vee \mathcal{HT}_1(\varphi) = \mathcal{HT}_1(\varphi)$, whence $a \leq \mathcal{HT}_1(\varphi) = \bigvee_{i=1}^{n} a_i$; by (i), $a \in D_G(\varphi)$.

(iii) is clear from Theorems 7.18 and 7.1 : $\mathcal{HT}_1(\varphi) = \mathcal{HT}_1(\psi)$ implies that φ and ψ have the same Horn-Tarski invariants for all $k \leq n$; hence $\varphi \equiv_G \psi$.

(iv) By (iii), it suffices to prove that $\mathcal{HT}_1(\langle 1, b \rangle \otimes \varphi) = \mathcal{HT}_1(\langle 1, ab \rangle \otimes \varphi)$. We have :

$$\mathcal{HT}_1(\langle 1, ab \rangle \otimes \varphi) = (a \triangle b) \vee \mathcal{HT}_1(\varphi).$$

The following Boolean identity is easily checked :

(*) $\qquad (a \triangle b) \vee c = (a \vee c) \triangle (b \vee c) \triangle c.$

Since $a \in D_G(\varphi)$, (i) yields $a \leq \bigvee_{i=1}^{n} a_i = \mathcal{HT}_1(\varphi)$. Using (*) with $c = \mathcal{HT}_1(\varphi)$, we obtain :

$$(a \triangle b) \vee \mathcal{HT}_1(\varphi) = b \vee \mathcal{HT}_1(\varphi),$$

that is,

$$\mathcal{HT}_1(\langle 1, ab \rangle \otimes \varphi) = \mathcal{HT}_1(\langle 1, b \rangle \otimes \varphi).$$

(v) The assumption and (i) give $a \leq -a \vee \mathcal{HT}_1(\varphi)$, which clearly yields $a \leq \mathcal{HT}_1(\varphi)$; whence $a \in D_G(\varphi)$. $\square$

COROLLARY 7.21. *Let G be a reduced special group, and φ' be the pure subform of a Pfister form φ over G, i.e., $\varphi = \langle 1 \rangle \oplus \varphi'$ and, say, $dim(\varphi') = 2^n - 1$. Then,*

$$\mathcal{HT}_k(\varphi') = \begin{cases} \mathcal{HT}_1(\varphi) & \text{if } 1 \leq k \leq 2^{n-1} \\ \bot & \text{if } 2^{n-1} + 1 \leq k \leq 2^n - 1. \end{cases}$$

PROOF. Left to the reader, using formula [HT13] of Corollary 7.9, and Theorem 7.18. $\square$

Remark . Theorem 7.18 makes it possible to compute some of the Horn-Tarski invariants of certain subforms of Pfister forms called Pfister neighbours (cf. [KS], §13, pp. 41 ff.). However, this requires more delicate combinatorial arguments that those employed above. $\diamond$

Now we turn to :

Proof of Theorem 7.18. By induction on $n = deg(\varphi)$. The case $n = 1$ is clear :

$\mathcal{HT}_1(a\varphi) = a \vee (a \triangle a_1) = a \vee a_1$ and $\mathcal{HT}_2(a\varphi) = a \wedge (a \triangle a_1) = a \wedge -a_1$.

$n \Rightarrow n+1$. Let $\psi = \langle 1, x \rangle \otimes \varphi$, with φ a Pfister form of degree n. Clearly, $a\psi \equiv_G a\varphi \oplus ax\varphi$. The addition formula [HT11] in Proposition 7.8 gives, for $1 \leq k \leq 2^{n+1}$ and with $A^{2^n, 2^n, k} = A^{2^n, k}$ (Notation 7.7) :

(+) $$\mathcal{HT}_k(a\psi) = \bigvee_{(s,r)\in A^{2^n,k}} (\mathcal{HT}_s(a\varphi) \wedge \mathcal{HT}_r(ax\varphi)).$$

In order to ease notation, we set
$$\alpha_{s,r} = \mathcal{HT}_s(a\varphi) \wedge \mathcal{HT}_r(ax\varphi),$$
for $(s,r) \in A^{2^n,k}$. We split the proof in three cases:

Case I. $1 \leq k \leq 2^{n-1}$.

Under this assumption, $k = s + r$ implies that $s, r \leq 2^{n-1}$; further, both indices may be 0. Separating the terms $s = 0$ and $r = 0$ in (+), we have:

(*) $$\mathcal{HT}_k(a\psi) = \mathcal{HT}_k(a\varphi) \vee \mathcal{HT}_k(ax\varphi) \vee \bigvee_{\substack{1 \leq s,r \leq k \\ s+r=k}} \alpha_{s,r}.$$

The induction hypothesis gives, for $1 \leq s, r \leq k$:
$$\mathcal{HT}_s(a\varphi) = \mathcal{HT}_1(a\varphi) \quad \text{and} \quad \mathcal{HT}_r(ax\varphi) = \mathcal{HT}_1(ax\varphi);$$
hence,
$$\alpha_{s,r} = \mathcal{HT}_1(a\varphi) \wedge \mathcal{HT}_1(ax\varphi).$$

Substituting this in (*), we get
$$\mathcal{HT}_k(a\psi) = \mathcal{HT}_1(a\varphi) \vee \mathcal{HT}_1(ax\varphi) = [a \vee \mathcal{HT}_1(\varphi)] \vee [(a \triangle x) \vee \mathcal{HT}_1(\varphi)].$$

Since $a \vee (a \triangle x) = a \vee x$, we obtain
$$\mathcal{HT}_k(a\psi) = a \vee x \vee \mathcal{HT}_1(\varphi) = a \vee \mathcal{HT}_1(\psi).$$

Case II. $2^{n-1} + 1 \leq k \leq 2^n$.

In this case, $k = s + r$ implies that either s or r is $\leq 2^{n-1}$. Separating in (+) the terms in which both s and r are $\leq 2^{n-1}$ from those in which one of the indices is $> 2^{n-1}$, we get:

(**) $$\begin{cases} \mathcal{HT}_k(a\psi) = \alpha \vee \beta \vee \gamma, \text{ where} \\[4pt] \alpha = \bigvee_{\substack{0 \leq s,r \leq 2^{n-1} \\ s+r=k}} \alpha_{s,r} \;; \quad \beta = \bigvee_{\substack{0 \leq s \leq 2^{n-1} \\ 2^{n-1} < r \leq k \\ s+r=k}} \alpha_{s,r} \quad \text{and} \\[4pt] \gamma = \bigvee_{\substack{0 \leq r \leq 2^{n-1} \\ 2^{n-1} < s \leq k \\ s+r=k}} \alpha_{s,r} \end{cases}$$

A computation similar to that of case I yields:

(1) $\alpha = [a \vee \mathcal{HT}_1(\varphi)] \wedge [(a \triangle x) \vee \mathcal{HT}_1(\varphi)] = (a \wedge (a \triangle x)) \vee \mathcal{HT}_1(\varphi)$.

In order to compute β we observe that, by the induction hypothesis :
$$\mathcal{HT}_r(ax\varphi) = \mathcal{HT}_k(ax\varphi) \qquad \text{for all } 2^{n-1} < r \leq k,$$
whence,
$$\alpha_{s,r} \leq \mathcal{HT}_k(ax\varphi) \qquad \text{for all } s, r \text{ so that } 1 \leq s \leq 2^{n-1} < r \leq k.$$

Separating the term in $s = 0$ in β gives :

(2) $\qquad\qquad \beta = \mathcal{HT}_k(ax\varphi) = (a \triangle x) \wedge - \mathcal{HT}_1(\varphi).$

As for the term γ in (**), the induction hypothesis implies
$$\mathcal{HT}_s(a\varphi) = \mathcal{HT}_k(a\varphi) \qquad \text{for } 2^{n-1} < s \leq k,$$
whence,
$$\alpha_{s,r} \leq \mathcal{HT}_k(a\varphi) \qquad \text{for } s, r \text{ so that } 1 \leq r \leq 2^{n-1} < s \leq k.$$

Separating the term $r = 0$ gives :

(3) $\qquad\qquad \gamma = \mathcal{HT}_k(a\psi) = a \wedge - \mathcal{HT}_1(\varphi).$

Substituting in (**) the values obtained in (1), (2) and (3), we get :
$$\mathcal{HT}_k(a\psi) = [(a \wedge (a \triangle x)) \vee \mathcal{HT}_1(\varphi)] \vee [(a \triangle x) \wedge - \mathcal{HT}_1(\varphi)] \vee$$
$$\vee [a \wedge - \mathcal{HT}_1(\varphi)]$$
$$= [(a \wedge (a \triangle x)) \vee \mathcal{HT}_1(\varphi)] \vee [(a \vee (a \triangle x)) \wedge - \mathcal{HT}_1(\varphi)]$$
$$= (a \wedge (a \triangle x)) \vee (a \vee (a \triangle x)) \vee \mathcal{HT}_1(\varphi) = a \vee x \vee \mathcal{HT}_1(\varphi)$$
$$= a \vee \mathcal{HT}_1(\psi).$$

Case III. $2^n + 1 \leq k \leq 2^{n+1}$.

In this case, $k = r + s$ and $s, r \leq 2^n$, imply that at least one of the indices s, r is $> 2^{n-1}$, and both must be ≥ 1. Separating in (+) the terms where both s, r are $> 2^{n-1}$ from those in which one of these indices is $\leq 2^{n-1}$, we get :

(***) $\begin{cases} \mathcal{HT}_k(a\psi) = \alpha' \vee \beta' \vee \gamma', \text{ where} \\[1em] \alpha' = \displaystyle\bigvee_{\substack{2^{n-1}<s,r\leq 2^n \\ s+r=k}} \alpha_{s,r} \; ; \quad \beta' = \displaystyle\bigvee_{\substack{1\leq s\leq 2^{n-1} \\ 2^{n-1}<r\leq 2^n \\ s+r=k}} \alpha_{s,r} \quad \text{and} \\[2em] \gamma' = \displaystyle\bigvee_{\substack{1\leq r\leq 2^{n-1} \\ 2^{n-1}<s\leq 2^n \\ s+r=k}} \alpha_{s,r} \end{cases}$

Note that the index sets in each of α', β' and γ' are non-empty. The induction hypothesis gives, for all s, r such that $2^{n-1} < s, r \leq 2^n$:

$$\mathcal{HT}_s(a\varphi) = a \wedge -\mathcal{HT}_1(\varphi) \quad \text{and} \quad \mathcal{HT}_r(ax\varphi) = (a \triangle x) \wedge -\mathcal{HT}_1(\varphi).$$

Hence,

(1') $\alpha' = (a \wedge -\mathcal{HT}_1(\varphi)) \wedge ((a \triangle x) \wedge -\mathcal{HT}_1(\varphi)) = a \wedge -x \wedge -\mathcal{HT}_1(\varphi).$

Similarly, for s, r such that $1 \leq s \leq 2^{n-1} < r \leq 2^n$, we have :

$$\mathcal{HT}_s(a\varphi) = a \vee \mathcal{HT}_1(\varphi) \quad \text{and} \quad \mathcal{HT}_r(ax\varphi) = (a \triangle x) \wedge -\mathcal{HT}_1(\varphi),$$

whence,

(2') $\beta' = (a \vee \mathcal{HT}_1(\varphi)) \wedge ((a \triangle x) \wedge -\mathcal{HT}_1(\varphi))$
$= a \wedge (a \triangle x) \wedge -\mathcal{HT}_1(\varphi) = a \wedge -x \wedge -\mathcal{HT}_1(\varphi).$

Finally, if $1 \leq r \leq 2^{n-1} < s \leq 2^n$, we have

$$\mathcal{HT}_s(a\varphi) = a \wedge -\mathcal{HT}_1(\varphi) \quad \text{and} \quad \mathcal{HT}_r(ax\varphi) = (a \triangle x) \vee \mathcal{HT}_1(\varphi).$$

This yields :

(3') $\gamma' = (a \wedge -\mathcal{HT}_1(\varphi)) \wedge ((a \triangle x) \vee \mathcal{HT}_1(\varphi))$
$= a \wedge (a \triangle x) \wedge -\mathcal{HT}_1(\varphi) = a \wedge -x \wedge -\mathcal{HT}_1(\varphi).$

The equalities (1'), (2') and (3') show that $\alpha' = \beta' = \gamma'$; hence in (***) we have :

$$\mathcal{HT}_k(a\psi) = a \wedge -x \wedge -\mathcal{HT}_1(\varphi) = a \wedge -\mathcal{HT}_1(\psi). \qquad \square$$

The remainder of this section is devoted to the computation of the Stiefel-Whitney invariants of Pfister forms and their multiples. We begin with the Pfister forms.

THEOREM 7.22. *Let G be a reduced special group, and φ be a Pfister form of degree $n \geq 1$ over G. Then*

$$\mathcal{SW}_k(\varphi) = \begin{cases} 1 & \text{if } 1 \leq k \leq 2^n, \ k \neq 2^{n-1}, \\ \mathcal{HT}_1(\varphi) & \text{if } k = 2^{n-1}. \end{cases}$$

PROOF. The result holds for $n = 1$, by straightforward checking. Henceforth, we suppose $n \geq 2$. By [HT1], Proposition 7.5, $\mathcal{SW}_1(\varphi) = d(\varphi) = 1$ (discriminant of φ); hence, we may assume $2 \leq k \leq 2^n$. Formula [HT8] of Proposition 7.6 shows :

$$\mathcal{SW}_k(\varphi) = \triangle_{l=k}^{2^n} \mathcal{HT}_l(\varphi)^{c_{l,k}}.$$

Since $\mathcal{HT}_l(\varphi) = 1$ for $2^{n-1} + 1 \leq l \leq 2^n$, and $\mathcal{HT}_l(\varphi) = \mathcal{HT}_1(\varphi)$ for $1 \leq l \leq 2^{n-1}$ (Theorem 7.18), the preceding identity becomes

$$(*) \quad \mathcal{SW}_k(\varphi) = \begin{cases} \triangle_{l=k}^{2^{n-1}} \mathcal{HT}_1(\varphi)^{c_{l,k}} = \mathcal{HT}_1(\varphi)^{d_k} & \text{for } 2 \leq k \leq 2^{n-1} \\ 1 & \text{for } 2^{n-1}+1 \leq k \leq 2^n, \end{cases}$$

where d_k = parity of $\sum_{l=k}^{2^{n-1}} c_{k,l}$

$$= \text{parity of } card(\{l \in \{k, \ldots, 2^{n-1}\} : \binom{l-1}{k-1} \text{ is odd}\}).$$

It is clear that $d_{2^{n-1}} = 1$ ($c_{r,r} = 1$ for $r \geq 2$), which shows that $\mathcal{SW}_{2^{n-1}}(\varphi) = \mathcal{HT}_1(\varphi)$. In order to compute the value of d_k for $2 \leq k \leq 2^{n-1} - 1$, we use the following classical result, a consequence of the so-called Glaisher's Theorem (cf. [Di]; Ch. IX, p. 273]) :

Proposition. *For integers k, l, such that $2 \leq k \leq l$, let $l - 1 = \sum_i \alpha_i \, 2^i$, $k - 1 = \sum_i \beta_i \, 2^i$, be the developments of $l - 1$ and $k - 1$ in base 2. Then,*

$$\binom{l-1}{k-1} \text{ is odd} \Leftrightarrow \alpha_i \geq \beta_i, \text{ for all } i. \qquad \diamond$$

Let r be the integer such that $2^{r-1} \leq k < 2^r$; since $2 \leq k < 2^{n-1}$, we have $2 \leq r \leq n-1$. In the binary expansion of $k-1$, $\beta_i = 0$ for $i \geq r$, i.e., $k - 1 = \beta_0 + \beta_1 \, 2 + \ldots + \beta_{r-1} \, 2^{r-1}$.

Let $\{i_1, \ldots, i_s\} = \{i \in \{0, \ldots, r-1\} : \beta_i = 1\}$. Since $k - 1 \leq l - 1 < 2^{n-1}$, we have

$$l - 1 = \alpha_0 + \alpha_1 \, 2 + \alpha_{r-1} \, 2^{r-1} + \alpha_r \, 2^r + \ldots + \alpha_{n-2} \, 2^{n-2}.$$

From the Proposition above, we get

$$\binom{l-1}{k-1} \text{ is odd iff } \alpha_i \geq \beta_i, \text{ for } 0 \leq i \leq r-1 \text{ iff } \alpha_{i_1} = \cdots = \alpha_{i_s} = 1.$$

Hence,

$$(**) \quad card(\{l \in \{k, \ldots, 2^{n-1}\} : \binom{l-1}{k-1} \text{ is odd}\}) = 2^p,$$

where $p = card(\{0, \ldots, n-2\} \setminus \{i_1, \ldots, i_s\}) = n - 1 - s$. Next, $k < 2^r$ entails $\beta_j = 0$ for some index $j \in \{0, \ldots, r-1\}$ (otherwise, $k - 1 = 1 + 2 + \ldots + 2^{r-1} = 2^r - 1$), which shows that $s < r \leq n-1$, i.e., $p > 0$.

Thus, we have proved that $d_k = 0$ for $2 \leq k < 2^{n-1}$, which, in view of (*), proves $\mathcal{SW}_k(\varphi) = 1$, for $2 \leq k < 2^{n-1}$. $\square$

In order to compute the Stiefel-Whitney invariants of a multiple of a Pfister form, we will use formula [HT15] of Proposition 7.11. This requires

PROPOSITION 7.23. *Let G be a reduced special group, and φ be a Pfister form over G of degree $n \geq 1$. Then, for $1 \leq k \leq 2^n$,*

$$\mathcal{SW}_k(-\varphi) = \begin{cases} 1 & \text{if } k \neq 2^{n-1},\ 2^n \\ \mathcal{HT}_1(\varphi) & \text{if } k = 2^{n-1} \\ -\mathcal{HT}_1(\varphi) & \text{if } k = 2^n \end{cases}$$

PROOF. Note that since $2^n \cdot \langle 1, -1 \rangle \equiv_G \bigotimes_{i=1}^{n+1} \langle 1, -1 \rangle$ is a Pfister form of degree $n+1$, Theorem 7.22 proves :

$$\mathcal{SW}_l(2^n \cdot \langle 1, -1 \rangle) = \begin{cases} 1 & \text{if } 1 \leq l \leq 2^{n+1},\ l \neq 2^n \\ -1 & \text{if } l = 2^n \end{cases}$$

For the proof of the Proposition, straightforward checking shows that the result holds for $n = 1$. So we assume $n \geq 2$. Since $\varphi \oplus -\varphi \equiv_G 2^n \cdot \langle 1, -1 \rangle$, the addition formula [HT12] of 7.8 applied on the left-hand side of this isometry, and the computation above, yield, for $1 \leq k \leq 2^n$,

$$(+) \quad \mathcal{SW}_k(\varphi) \triangle \mathcal{SW}_k(-\varphi) \triangle \bigtriangleup_{\substack{1 \leq s,r \leq k-1 \\ k=s+r}} (\mathcal{SW}_s(\varphi) \wedge \mathcal{SW}_r(-\varphi)) =$$

$$= \begin{cases} 1 & \text{if } k \neq 2^n \\ -1 & \text{if } k = 2^n. \end{cases}$$

Now, for $n \geq 2$ fixed, we use induction on k.

$\underline{k=1}$. In this case the last factor of $(+)$ does not occur. Hence, $\mathcal{SW}_1(-\varphi) = \mathcal{SW}_1(\varphi) = 1$ (7.22).

Induction step. Assume the statement true for all $r \leq k-1$. We analyse several cases :

Case I. k is odd.

If $k = s + r$, then one of s, r must be odd. If s, $r \geq 1$, then s, $r \leq k-1$, and the induction hypothesis applies. If s is odd, then $\mathcal{SW}_s(\varphi) = 1$ by Theorem 7.22. If r is odd, then $\mathcal{SW}_r(-\varphi) = 1$, by the induction hypothesis. Hence, the last factor in $(+)$ vanishes $(= 1)$, and so $\mathcal{SW}_k(\varphi) = \mathcal{SW}_k(-\varphi)$. Since k is odd, Theorem 7.22 implies that $\mathcal{SW}_k(-\varphi) = 1$.

Case II. k is even, $k \neq 2^{n-1},\ 2^n$.

If $k = s+r$, then either both s, r are odd or both are even. By Theorem 7.22 and the induction hypothesis (for $r \leq k-1$) all factors in the third term of $(+)$ with s, r odd, vanish $(= 1)$. Thus, let both s, r be even; we consider two subcases :

(i) $k < 2^{n-1}$.

Then, s, $r < k < 2^{n-1}$. By the induction hypothesis, $\mathcal{SW}_r(-\varphi) = 1$ and the last factor in $(+)$ vanishes.

(ii) $k > 2^{n-1}$.

By the induction hypothesis, all terms in the third factor of (+) with $r \neq 2^{n-1}$ vanish (since $\mathcal{SW}_r(-\varphi) = 1$). So, let $r = 2^{n-1}$ and $s = k - 2^{n-1}$; since $k < 2^n$, we have $s < 2^{n-1}$; by Theorem 7.22, $\mathcal{SW}_s(\varphi) = 1$. Then, $\mathcal{SW}_{k-2^{n-1}}(\varphi) \wedge \mathcal{SW}_{2^{n-1}}(\varphi) = 1$, and the last factor of (+) also vanishes in this subcase.

Thus, for both (i) and (ii), $\mathcal{SW}_k(-\varphi) = \mathcal{SW}_k(\varphi)$, which equals 1 by Theorem 7.22.

Case III. $k = 2^{n-1}$.

By the induction hypothesis, $\mathcal{SW}_r(-\varphi) = 1$ for $r < 2^{n-1}$. Hence, $\mathcal{SW}_s(\varphi) \wedge \mathcal{SW}_r(-\varphi) = 1$, for all $s, r \geq 1$ such that $2^{n-1} = s + r$, and the third factor in (+) vanishes. By Theorem 7.22 we have $\mathcal{SW}_{2^{n-1}}(-\varphi) = \mathcal{SW}_{2^{n-1}}(\varphi) = \mathcal{HT}_1(\varphi)$.

Case IV. $k = 2^n$.

If $2^n = s + r$, $s, r \geq 1$ and $r \neq 2^{n-1}$, the induction hypothesis yields $\mathcal{SW}_r(-\varphi) = 1$ and $\mathcal{SW}_s(\varphi) \wedge \mathcal{SW}_r(-\varphi) = 1$. Consider the case $r = s = 2^{n-1}$; by Theorem 7.22, $\mathcal{SW}_{2^{n-1}}(\varphi) = \mathcal{HT}_1(\varphi)$ and, by case III, $\mathcal{SW}_{2^{n-1}}(-\varphi) = \mathcal{HT}_1(\varphi)$. Hence, the third factor of (+) equals $\mathcal{HT}_1(\varphi)$. Again, by Theorem 7.22, $\mathcal{SW}_{2^n}(\varphi) = 1$. Since the product (+) equals -1 in this case, we conclude

$$\mathcal{SW}_{2^n}(-\varphi) = -\mathcal{HT}_1(\varphi),$$

as asserted. This ends the proof. $\square$

Now we prove

THEOREM 7.24. *Let G be a reduced special group. Let $a \in G$, and φ be a Pfister form of degree $n \geq 1$ over G. Then, for $1 \leq k \leq 2^n$,*

$$\mathcal{SW}_k(a\varphi) = \begin{cases} 1 & \text{if } k \neq 2^{n-1}, 2^n \\ \mathcal{HT}_1(\varphi) & \text{if } k = 2^{n-1} \\ \mathcal{HT}_{2^n}(a\varphi) & \text{if } k = 2^n. \end{cases}$$

PROOF. Recall formula [HT15], Proposition 7.11 :

(+) $\mathcal{SW}_k(a\varphi) = (\mathcal{SW}_k(\varphi) \wedge -a) \triangle (\mathcal{SW}_k(-\varphi) \wedge a)$, for $1 \leq k \leq 2^n$.

Case I. $k \neq 2^{n-1}, 2^n$.

By Theorem 7.22 and Proposition 7.23 we have $\mathcal{SW}_k(\varphi) = 1$ and $\mathcal{SW}_k(-\varphi) = 1$. Replacing these values in (+), we obtain $\mathcal{SW}_k(a\varphi) = 1$.

Case II. $k = 2^{n-1}$.

In this case, Theorem 7.22 and Proposition 7.23 give

$$\mathcal{SW}_{2^{n-1}}(\varphi) = \mathcal{HT}_1(\varphi) \quad \text{and} \quad \mathcal{SW}_{2^{n-1}}(-\varphi) = \mathcal{HT}_1(\varphi).$$

Replacing in (+), we get

$$\mathcal{SW}_{2^{n}-1}(a\varphi) = (\mathcal{HT}_{1}(\varphi) \wedge -a) \bigtriangleup (\mathcal{HT}_{1}(\varphi) \wedge a)$$
$$= (\mathcal{HT}_{1}(\varphi) \wedge -a) \vee (\mathcal{HT}_{1}(\varphi) \wedge a) = \mathcal{HT}_{1}(\varphi).$$

Case III. $k = 2^n$.

By 7.22 and 7.23, $\mathcal{SW}_{2^n}(\varphi) = 1$ and $\mathcal{SW}_{2^n}(-\varphi) = -\mathcal{HT}_{1}(\varphi)$. Replacing these values in (+), we have

$$\mathcal{SW}_{2^n}(a\varphi) = 1 \bigtriangleup (-\mathcal{HT}_{1}(\varphi) \wedge a) = \mathcal{HT}_{2^n}(a\varphi). \qquad \Box$$

COROLLARY 7.25. *Let G be a reduced special group. Let φ' be the pure subform of a Pfister form φ of degree $n \geq 1$ over G. Then, for $1 \leq k \leq 2^n - 1 \; (= dim(\varphi'))$, we have*

$$\mathcal{SW}_k(\varphi') = \mathcal{SW}_k(\varphi) = \begin{cases} 1 & \text{if } k \neq 2^{n-1} \\ \mathcal{HT}_1(\varphi) & \text{if } k = 2^{n-1} \end{cases}$$

PROOF. Use formula [HT13] of Corollary 7.9 on $\varphi = \langle 1 \rangle \oplus \varphi'$, and Theorem 7.22. $\qquad \Box$

5. The invariants of linear combinations of Pfister forms.

Using the foregoing results we shall now prove that the Horn-Tarski and Stiefel-Whitney invariants of forms of the type $\bigoplus_{i=1}^{r} a_i \varphi_i$ over an arbitrary reduced special group G — where the φ_i's are Pfister forms of the same degree $n \geq 1$ and the a_i's are in G — can be expressed as Horn-Tarski or Stiefel-Whitney invariants, respectively, of simpler forms **over $\mathbf{B_G}$**, whose coefficients are the Horn-Tarski invariants of the forms $a_i \varphi_i$ (see Theorems 7.26 and 7.29, below).

In the terminology of Witt rings, these results make it possible to obtain Boolean expressions for the Horn-Tarski invariants of any form in the nth power of the fundamental ideal $I(G)$ of the Witt ring associated to the special group G, in terms of the invariants of Pfister forms of degree n and coefficients in G (for the construction of the Witt ring $W(G)$ of a special group G, see [D]; however, this construction will not be used here). As an interesting application we give a new, Boolean-theoretic proof of the Arason-Pfister Hauptsatz in the case of reduced special groups (Theorem 7.31).

We deal first with the Horn-Tarski invariants.

THEOREM 7.26. *Let G be a reduced special group. Let $\varphi_1, \ldots, \varphi_r$ ($r \geq 1$) be Pfister forms over G of the same degree $n \geq 1$. Let $a_1, \ldots, a_r$ be elements of G. Given an integer m, $1 \leq m \leq r \cdot 2^n$, let k be the unique integer such that $(k-1) \cdot 2^{n-1} + 1 \leq m \leq k \cdot 2^{n-1}$. Then*

$$\mathcal{HT}_m(\bigoplus_{i=1}^{r} a_i \varphi_i) =$$
$$= \mathcal{HT}_k(\langle \mathcal{HT}_1(a_1 \varphi_1), \ldots, \mathcal{HT}_1(a_r \varphi_r), \mathcal{HT}_{2^n}(a_1 \varphi_1), \ldots, \mathcal{HT}_{2^n}(a_r \varphi_r) \rangle).$$

In order to ease the proof of this and later results, we introduce

NOTATION 7.27.   (a) The degree n of the forms φ_i remains fixed throughout; hence, we omit it in the notation.

(b) We set

— $\psi_r = \bigoplus_{i=1}^{r} a_i\varphi_i$; this is a form of dimension $r \cdot 2^n$ over G.

— $\eta_r = \langle \mathcal{HT}_1(a_1\varphi_1), \ldots, \mathcal{HT}_1(a_r\varphi_r), \mathcal{HT}_{2^n}(a_1\varphi_1), \ldots, \mathcal{HT}_{2^n}(a_r\varphi_r) \rangle$; this is a form of dimension $2r$ over $\mathbf{B_G}$.

— For $1 \leq k \leq 2r$, we write : $\chi_k^r = \mathcal{HT}_k(\eta_r)$; χ_k^r is an element of B_G.

(c) We split the interval $[1, r \cdot 2^n]$ in $\mathbb{Z}$ in $2r$ consecutive intervals of length 2^{n-1}, namely,

$$I_k = \{l \in \mathbb{Z} : (k-1) \cdot 2^{n-1} + 1 \leq l \leq k \cdot 2^{n-1}\} \qquad \text{for } 1 \leq k \leq 2r. \quad \Diamond$$

Proof of Theorem 7.26. Induction on r. The case $r = 1$ is Theorem 7.18.

$r \Rightarrow r+1$.   With notation as in 7.27.(b), we have $\psi_{r+1} \equiv_G \psi_r \oplus a_{r+1}\varphi_{r+1}$. The addition formula [HT11], Proposition 7.8, gives, for $1 \leq m \leq (r+1)2^n$:

$$(+) \qquad \mathcal{HT}_m(\psi_{r+1}) = \bigvee_{(s,t) \in A^{r2^n, 2^n, m}} \left(\mathcal{HT}_s(\psi_r) \wedge \mathcal{HT}_t(a_{r+1}\varphi_{r+1}) \right).$$

Remark that, in this formula, $t \in \{0\} \cup I_1 \cup I_2$, and either $s = 0$ or $s \in I_l$, for some l belonging to $\{1, \ldots, 2r\}$, depending on the values of m.

For the sake of readability, we introduce the following abbreviations :

$$(*) \qquad x_i = \begin{cases} \mathcal{HT}_1(a_i\varphi_i) & \text{for } 1 \leq i \leq r \\ \mathcal{HT}_{2^n}(a_j\varphi_j) & \text{for } i = r + j, 1 \leq j \leq r, \end{cases}$$

and $x_{2r+1} = \mathcal{HT}_1(a_{r+1}\varphi_{r+1})$, $x_{2r+2} = \mathcal{HT}_{2^n}(a_{r+1}\varphi_{r+1})$. Thus, $\eta_r = \langle x_1, \ldots, x_r, x_{r+1}, \ldots, x_{2r} \rangle$ and $\eta_{r+1} \equiv_{BG} \eta_r \oplus \langle x_{2r+1}, x_{2r+2} \rangle$.

$$(**) \qquad \alpha_{s,t} = \mathcal{HT}_s(\psi_r) \wedge \mathcal{HT}_t(a_{r+1}\varphi_{r+1}),$$

for $(s, t) \in A^{r2^n, 2^n, m}$ as in $(+)$.

$(***)$ For $j = 1, 2$, and $l \in \{1, \ldots, 2r\}$, we set

$$\zeta_{l,j} = \bigvee_{s \in I_l, t \in I_j} \alpha_{s,t}.$$

The proof falls into five cases, according to whether m belongs to the intervals I_1, I_2, $[2^n + 1, r2^n] = I_3 \cup \ldots \cup I_{2r}$, I_{2r+1}, I_{2r+2}, respectively.

Case I. $m \in I_1$.   In this case, if $m = s + t$, then $s, t \in \{0\} \cup I_1$, and $(+)$ takes the form :

5. THE INVARIANTS OF LINEAR COMBINATIONS OF PFISTER FORMS.

(1) $\quad \mathcal{HT}_m(\psi_{r+1}) = \mathcal{HT}_m(\psi_r) \vee \mathcal{HT}_m(a_{r+1}\varphi_{r+1}) \vee \zeta_{1,1};$

by induction hypothesis, $\mathcal{HT}_s(\psi_r) = \chi_1^r$ for $1 \leq s \leq m$; by Theorem 7.18, we have $\mathcal{HT}_t(a_{r+1}\varphi_{r+1}) = \mathcal{HT}_1(a_{r+1}\varphi_{r+1})$ for $1 \leq t \leq m$; hence, $\zeta_{1,1} = \chi_1^r \wedge \mathcal{HT}_1(a_{r+1}\varphi_{r+1})$ and (1) boils down to

$$\mathcal{HT}_m(\psi_{r+1}) = \chi_1^r \vee \mathcal{HT}_1(a_{r+1}\varphi_{r+1}).$$

On the other hand, we have, $\chi_1^{r+1} = \bigvee_{i=1}^{r+1} \mathcal{HT}_1(a_i\varphi_i) = \chi_1^r \vee \mathcal{HT}_1(a_{r+1}\varphi_{r+1})$, and we conclude

$$\mathcal{HT}_m(\psi_{r+1}) = \chi_1^{r+1},$$

as required.

Case II. $m \in I_2$. In this case, if $m = s + t$, then either s or t belong to $\{0\} \cup I_1$, and if one of these indices is in I_1, the other is in $I_1 \cup I_2$. Separating terms according to these alternatives, the addition formula (+) becomes :

(2) $\quad \mathcal{HT}_m(\psi_{r+1}) = \mathcal{HT}_m(\psi_r) \vee \mathcal{HT}_m(a_{r+1}\varphi_{r+1}) \vee \zeta_{1,1} \vee \zeta_{1,2} \vee \zeta_{2,1},$

where the disjuncts $\zeta_{1,2}, \zeta_{2,1}$ <u>do not occur</u> if $m = 2^{n-1} + 1$ (but <u>all terms occur</u> if $m \geq 2^{n-1} + 2$). By induction hypothesis, $\mathcal{HT}_s(\psi_r) = \chi_j^r$, for $s \in I_j$ ($j = 1, 2$). From Theorem 7.18, we get $\alpha_{s,t} = \chi_j^r \wedge \mathcal{HT}_q(a_{r+1}\varphi_{r+1})$ for $s \in I_j$, while $q = 1$ or 2^n according to whether $t \in I_1$ or $t \in I_2$, respectively. It follows that

$\zeta_{1,1} = \chi_1^r \wedge \mathcal{HT}_1(a_{r+1}\varphi_{r+1}), \quad \zeta_{1,2} = \chi_1^r \wedge \mathcal{HT}_{2^n}(a_{r+1}\varphi_{r+1})$ and
$\zeta_{2,1} = \chi_2^r \wedge \mathcal{HT}_1(a_{r+1}\varphi_{r+1}).$

From [HT3] (Proposition 7.5) we get that $\zeta_{1,2}, \zeta_{2,1}$ are both $\leq \zeta_{1,1}$; then the identity (2) reduces to

(2') $\quad \mathcal{HT}_m(\psi_{r+1}) = \chi_2^r \vee \mathcal{HT}_{2^n}(a_{r+1}\varphi_{r+1}) \vee (\chi_1^r \wedge \mathcal{HT}_1(a_{r+1}\varphi_{r+1})),$

in <u>both the cases</u> $m = 2^{n-1} + 1$ and $m \geq 2^{n-1} + 2$.

Next, the addition formula [HT11] (Proposition 7.8) applied to the isometry

$$\eta_{r+1} \equiv_{BG} \eta_r \oplus \langle x_{2r+1}, x_{2r+2}\rangle \qquad (\text{see } (*) \text{ above})$$

yields :

$$\chi_2^{r+1} = \mathcal{HT}_2(\eta_{r+1}) = \mathcal{HT}_2(\eta_r) \vee \mathcal{HT}_2(\langle x_{2r+1}, x_{2r+2}\rangle) \vee$$
$$\vee (\mathcal{HT}_1(\eta_r) \wedge \mathcal{HT}_1(\langle x_{2r+1}, x_{2r+2}\rangle))$$
$$= \chi_2^r \vee x_{2r+2} \vee (\chi_1^r \wedge x_{2r+1}),$$

which is the right-hand side of (2'). This proves that

$$\mathcal{HT}_m(\psi_{r+1}) = \chi_2^{r+1} \qquad \text{for } m \in I_2,$$

as required.

Case III. $2^n + 1 \leq m \leq r \cdot 2^n$, i.e., $m \in I_3 \cup \ldots \cup I_{2r}$.

Let k be the unique integer, $3 \leq k \leq 2r$, such that $m \in I_k$. Thus, $m = s + t$ implies

$t = 0$ (and $s = m$), **or** $t \in I_1$ and $s \in I_{k-1} \cup I_k$, **or**

$t \in I_2$ and $s \in I_{k-2} \cup I_{k-1}$.

The addition formula (+) takes the form :

(3) $\mathcal{HT}_m(\psi_{r+1}) = \mathcal{HT}_m(\psi_r) \vee \zeta_{k-1,1} \vee \zeta_{k,1} \vee \zeta_{k-2,2} \vee \zeta_{k-1,2}$,

where the disjuncts $\zeta_{k,1}$ and $\zeta_{k-1,2}$ <u>do not occur</u> for $m = (k-1)2^{n-1} + 1$ (but all the others do), while all five disjuncts occur if $m \geq (k-1)2^{n-1} + 2$. By the induction hypothesis and Theorem 7.18 we obtain

$$\alpha_{s,t} = \chi_l^r \wedge \mathcal{HT}_q(a_{r+1}\varphi_{r+1}) \quad \text{for } s \in I_l \ (l \in \{k-2, k-1, k\}),$$

$$\text{and } t \in I_j \ (j = 1, 2),$$

with $q = 1$ or 2^n, according to whether $j = 1$ or 2, respectively. It follows that $\zeta_{l,j}$ takes on this value, as well. From [HT3] (Proposition 7.5) we have

$$\zeta_{k,1} \leq \zeta_{k-1,1} \quad \text{and} \quad \zeta_{k-1,2} \leq \zeta_{k-2,2},$$

and the identity (3) reduces to :

(3') $\mathcal{HT}_m(\psi_{r+1}) = \chi_k^r \vee (\chi_{k-1}^r \wedge \mathcal{HT}_1(a_{r+1}\varphi_{r+1})) \vee$

$\vee (\chi_{k-2}^r \wedge \mathcal{HT}_{2^n}(a_{r+1}\varphi_{r+1}))$,

<u>in both the cases</u> $m = (k-1)2^{n-1} + 1$ and $m \geq (k-1)2^{n-1} + 2$, i.e., for all $m \in I_k$.

As in case II, the addition formula [HT11] (Proposition 7.8) applied to the isometry $\eta_{r+1} \equiv_{BG} \eta_r \oplus \langle x_{2r+1}, x_{2r+2} \rangle$, gives :

$$\chi_k^{r+1} = \chi_k^r \vee (\chi_{k-1}^r \wedge x_{2r+1}) \vee (\chi_{k-2}^r \wedge x_{2r+2}),$$

which coincides with the right-hand side of (3'). Thus, we have shown

$$\mathcal{HT}_m(\psi_{r+1}) = \chi_k^{r+1} \qquad \text{for } m \in I_k \ (3 \leq k \leq 2r),$$

as asserted.

Case IV. $m \in I_{2r+1}$. Under this assumption, $m = s + t$ with $s \leq r \cdot 2^n$, $t \leq 2^n$, imply

$t \in I_1$ and $s \in I_{2r}$, **or** $t \in I_2$ and $s \in I_{2r-1} \cup (I_{2r} \setminus \{r \cdot 2^n\})$.

In particular, both s and t are ≥ 1. Note also that the set $I_{2r} \setminus \{r \cdot 2^n\}$ is empty if $n = 1$, but non-empty if $n \geq 2$. The formula (+) comes down to :

(4) $\mathcal{HT}_m(\psi_{r+1}) = \zeta_{2r,1} \vee \zeta_{2r-1,2} \vee \zeta_{2r,2}$,

5. THE INVARIANTS OF LINEAR COMBINATIONS OF PFISTER FORMS.

where the disjunct $\zeta_{2r,2}$ does not occur if $m = r\cdot 2^n + 1$, but does occur if $m \geq r\cdot 2^n + 2$ (this being possible only if $n \geq 2$).

As in the previous cases, the induction hypothesis and Theorem 7.18 yield :

$$\zeta_{2r,1} = \chi^r_{2r} \wedge \mathcal{HT}_1(a_{r+1}\varphi_{r+1}), \quad \zeta_{2r-1,2} = \chi^r_{2r-1} \wedge \mathcal{HT}_{2^n}(a_{r+1}\varphi_{r+1}), \text{ and}$$

$$\zeta_{2r,2} = \chi^r_{2r} \wedge \mathcal{HT}_{2^n}(a_{r+1}\varphi_{r+1}).$$

Then, $\zeta_{2r,2} \leq \zeta_{2r,1}$, and, the identity (4) becomes

(4') $\mathcal{HT}_m(\psi_{r+1}) = (\chi^r_{2r} \wedge \mathcal{HT}_1(a_{r+1}\varphi_{r+1})) \vee (\chi^r_{2r-1} \wedge \mathcal{HT}_{2^n}(a_{r+1}\varphi_{r+1})),$

for both $m = r\cdot 2^n + 1$ and $m \geq r\cdot 2^n + 2$, that is, for arbitrary $m \in I_{2r+1}$.

As in the preceding cases, from [HT11] (Proposition 7.8) and $\eta_{r+1} \equiv_{B_G} \eta_r \oplus \langle x_{2r+1}, x_{2r+2} \rangle$, we get :

$$\chi^{r+1}_{2r+1} = (\chi^r_{2r} \wedge x_{2r+1}) \vee (\chi^r_{2r-1} \wedge x_{2r+2}),$$

which equals the right-hand side of (4') and shows that

$$\mathcal{HT}_m(\psi_{r+1}) = \chi^{r+1}_{2r+1} \qquad \text{for } m \in I_{2r+1}.$$

Case V. $m \in I_{2r+2}$. Here, the condition $m = s + t$, with $s \leq r\cdot 2^n$, $t \leq 2^n$, implies $t \in I_2$ and $s \in I_{2r}$. The formula (+) takes the form

(5) $$\mathcal{HT}_m(\psi_{r+1}) = \zeta_{2r,2}.$$

The induction hypothesis and 7.18 entail $\zeta_{2r,2} = \chi^r_{2r} \wedge \mathcal{HT}_{2^n}(a_{r+1}\varphi_{r+1})$, whence

$$\mathcal{HT}_m(\psi_{r+1}) = \chi^r_{2r} \wedge \mathcal{HT}_{2^n}(a_{r+1}\varphi_{r+1}) = \bigwedge_{i=1}^{r+1} \mathcal{HT}_{2^n}(a_i\varphi_i) = \chi^{r+1}_{2r+2}. \qquad \square$$

As an application of Theorem 7.26 we prove

PROPOSITION 7.28. *Let G be a reduced special group. With notation as in 4.14, the following are equivalent :*

(1) *G is a Boolean algebra.*

(2) *$\mathcal{I} \circ \Sigma = id_{\mathcal{I}(B_G)}$.*

PROOF. We identify G with its image by ε_G in the Boolean hull B_G. It is clear that (1) implies (2). For the converse, we shall prove that for all $a, b \in G$, $a \wedge b$ (in B_G) is actually in G. Hence, the de Morgan Laws imply that G satisfies 7.17.(5), and so must be a Boolean algebra. The reasoning is an adaptation of that in Theorem 17.6 in [L2].

For a, b in G, let I be the principal ideal generated by $a \wedge b$ in B_G. Since $I = \mathcal{I}(\Sigma(I))$, there are $c_1, \ldots, c_n$ in $G \cap I$ such that, in B_G,

$$a \wedge b = \bigvee\nolimits_{i=1}^{n} c_i. \tag{A}$$

We may assume that $n \geq 2$. Set $\theta = \varphi \oplus \psi$, where $\varphi = \bigotimes_{i=1}^{n} \langle 1, c_i \rangle$, $\psi = 2^{n-2}(\langle 1, -a \rangle \otimes \langle 1, -b \rangle)$. Write

$$\eta = \langle \mathcal{HT}_1(\varphi), \mathcal{HT}_1(\psi), \mathcal{HT}_{2^n}(\varphi), \mathcal{HT}_{2^n}(\psi) \rangle$$

$$= \langle \bigvee\nolimits_{i=1}^{n} c_i, (-a \vee -b), \bot, \bot \rangle.$$

Note that $\bigvee_{i=1}^{n} c_i \vee (-a \vee -b) = \top$, while $\bigvee_{i=1}^{n} c_i \wedge (-a \vee -b) = \bot$. By Theorem 7.26, we have, for $1 \leq m \leq 2 \cdot 2^n$

$$\mathcal{HT}_m(\theta) = \mathcal{HT}_k(\eta),$$

where k is the unique integer such that $(k-1)2^{n-1} + 1 \leq m \leq k 2^{n-1}$, that is,

$$\mathcal{HT}_m(\theta) = \begin{cases} \top & \text{if } k = 1 \\ \bot & \text{if } k = 2, 3, 4 \end{cases}$$

Hence, Theorem 7.1 entails $\theta \equiv_G 2^{n-1}\langle 1, -1 \rangle \oplus 2^n \langle 1 \rangle$. Since $n \geq 2$, we conclude that θ is isotropic in G. By Proposition 1.6.(d), there is $x \in G$, such that $x \in D_G(\varphi)$ and $-x \in D_G(\psi)$. But then, Theorem 7.12 yields

$$x \leq \bigvee\nolimits_{i=1}^{n} c_i \quad \text{and} \quad -x \leq (-a \vee -b),$$

which together with the equality in (A) implies $a \wedge b = x$, ending the proof. $\square$

Next we compute the Stiefel-Whitney invariants for linear combinations of Pfister forms. The result is a symmetric-difference version of Theorem 7.26 and the proof follows a similar plan, although the computations to be carried out are different. We use notation as in 7.27 and in the proof of Theorem 7.26.

THEOREM 7.29. *Let G be a reduced special group. Let $\varphi_1, \ldots, \varphi_r$ ($r \geq 1$) be Pfister forms over G of the same degree $n \geq 1$. Let $a_1, \ldots, a_r$ be elements of G. Then, for $1 \leq m \leq r \cdot 2^n$, we have :*

$$\mathcal{SW}_m(\bigoplus\nolimits_{i=1}^{r} a_i \varphi_i) = \begin{cases} 1 & \text{if } m \notin \{k \cdot 2^{n-1} : 1 \leq k \leq 2r\} \\ \mathcal{SW}_k(\eta_r) & \text{if } m = k \cdot 2^{n-1}, \text{ for some } k, 1 \leq k \leq 2r, \end{cases}$$

where $\eta_r = \langle \mathcal{HT}_1(a_1\varphi_1), \ldots, \mathcal{HT}_1(a_r\varphi_r), \mathcal{HT}_{2^n}(a_1\varphi_1), \ldots, \mathcal{HT}_{2^n}(a_r\varphi_r) \rangle$.

Remark. This result sharpens, in the reduced case, Corollary 3.3 of [Mi].

We shall need

LEMMA 7.30. *Let G be a reduced special group, $a \in G$, and φ be a Pfister form over G. Then :*

5. THE INVARIANTS OF LINEAR COMBINATIONS OF PFISTER FORMS. 169

a) $\mathcal{HT}_1(a\varphi) \triangle \mathcal{HT}_{2^n}(a\varphi) = \mathcal{HT}_1(\varphi)$.

Hence,

b)   For $1 \leq k \leq 2^n = dim(\varphi)$, $\mathcal{HT}_k(a\varphi) \triangle \mathcal{HT}_{2^n-k}(a\varphi) = \mathcal{HT}_1(\varphi)$.

c) $\mathcal{SW}_1(\langle \mathcal{HT}_1(a\varphi), \mathcal{HT}_{2^n}(a\varphi) \rangle) = \mathcal{HT}_1(\varphi)$;

     $\mathcal{SW}_2(\langle \mathcal{HT}_1(a\varphi), \mathcal{HT}_{2^n}(a\varphi) \rangle) = \mathcal{HT}_{2^n}(a\varphi)$.

PROOF. (b) follows from (a) and Theorem 7.18; (c) is straightforward, using (a).

(a) By Theorem 7.18, using $x \vee y = x \triangle y \triangle (x \wedge y)$:

$$\mathcal{HT}_1(a\varphi) \triangle \mathcal{HT}_{2^n}(a\varphi) = (a \vee \mathcal{HT}_1(\varphi)) \triangle (a \wedge -\mathcal{HT}_1(\varphi)) =$$
$$= a \triangle \mathcal{HT}_1(\varphi) \triangle (a \wedge \mathcal{HT}_1(\varphi)) \triangle (a \wedge -\mathcal{HT}_1(\varphi))$$
$$= a \triangle \mathcal{HT}_1(\varphi) \triangle (a \wedge [\mathcal{HT}_1(\varphi) \triangle -\mathcal{HT}_1(\varphi)])$$
$$= a \triangle \mathcal{HT}_1(\varphi) \triangle (a \wedge \top) = a \triangle \mathcal{HT}_1(\varphi) \triangle a$$
$$= \mathcal{HT}_1(\varphi). \qquad \square$$

Proof of Theorem 7.29. Induction on r.

$r = 1$. The values of $\mathcal{SW}_m(a_1\varphi_1)$, for $1 \leq m \leq 2^n$, are given by 7.24.

On the other hand, $\eta_1 = \langle \mathcal{HT}_1(a_1\varphi_1), \mathcal{HT}_{2^n}(a_1\varphi_1) \rangle$. By Lemma 7.30.(c), if $m = 2^{n-1}$ (i.e., $k = 1$), $\mathcal{SW}_1(\eta_1) = \mathcal{HT}_1(\varphi_1) = \mathcal{SW}_{2^{n-1}}(a_1\varphi_1)$; if $m = 2^n$ (i.e., $k = 2$), $\mathcal{SW}_2(\eta_1) = \mathcal{HT}_{2^n}(a_1\varphi_1) = \mathcal{SW}_{2^n}(a_1\varphi_1)$.

$r \Rightarrow r + 1$. Using the addition formula [HT12], Proposition 7.8, on the isometry $\psi_{r+1} \equiv_G \psi_r \oplus a_{r+1}\varphi_{r+1}$, we get, for $1 \leq m \leq (r+1) \cdot 2^n$,

$$\mathcal{SW}_m(\psi_{r+1}) = \bigtriangleup_{(s,t) \in A^{r \cdot 2^n, 2^n, m}} (\mathcal{SW}_s(\psi_r) \wedge \mathcal{SW}_t(a_{r+1}\varphi_{r+1})).$$

This formula admits some simplification. Indeed, by Theorem 7.24, $\mathcal{SW}_t(a_{r+1}\varphi_{r+1}) = 1$ for all t, except when $t = 2^{n-1}$ or 2^n. Then the preceding identity reduces to

$$(+) \qquad \mathcal{SW}_m(\psi_{r+1}) = \bigtriangleup_{\substack{t \in \{0, 2^{n-1}, 2^n\} \\ 0 \leq m-t \leq r \cdot 2^n}} (\mathcal{SW}_{m-t}(\psi_r) \wedge \mathcal{SW}_t(a_{r+1}\varphi_{r+1})).$$

We consider two cases :

Case I. $m \notin \{k \cdot 2^{n-1} : 1 \leq k \leq 2r + 2\}$.

In this case, $m - t$ is not a multiple of 2^{n-1} for any of the relevant values of t in $(+)$. By the induction hypothesis, $\mathcal{SW}_{m-t}(\psi_r) = 1$ for $t \in \{0, 2^{n-1}, 2^n\}$; hence $\mathcal{SW}_m(\psi_{r+1}) = 1$.

Case II. $m = k \cdot 2^{n-1}$, for some k, $1 \leq k \leq 2r+2$.

We have
$$\eta_{r+1} \equiv_{BG} \eta_r \oplus \langle \mathcal{HT}_1(a_{r+1}\varphi_{r+1}), \mathcal{HT}_{2^n}(a_{r+1}\varphi_{r+1}) \rangle.$$

Theorem 7.1 and the addition formula [HT12] (Proposition 7.8) applied to this isometry give, for $1 \leq k \leq 2r+2$,

$$\mathcal{SW}_k(\eta_{r+1}) =$$
$$= \bigtriangleup_{(s,t) \in A^{2r,2,k}} (\mathcal{SW}_s(\eta_r) \wedge \mathcal{SW}_t(\langle \mathcal{HT}_1(a_{r+1}\varphi_{r+1}), \mathcal{HT}_{2^n}(a_{r+1}\varphi_{r+1}) \rangle)).$$

Using Lemma 7.30.(c) and Theorem 7.24, we obtain for the different values of k :

(1) $\mathcal{SW}_1(\eta_{r+1}) = \mathcal{SW}_1(\eta_r) \bigtriangleup \mathcal{HT}_1(\varphi_{r+1})$.

(2) $\mathcal{SW}_2(\eta_{r+1}) = \mathcal{SW}_2(\eta_r) \bigtriangleup [\mathcal{SW}_1(\eta_r) \wedge \mathcal{HT}_1(\varphi_{r+1})] \bigtriangleup \mathcal{HT}_{2^n}(a_{r+1}\varphi_{r+1})$.

(3) For $2 < k \leq 2r$:
$$\mathcal{SW}_k(\eta_{r+1}) = \mathcal{SW}_k(\eta_r) \bigtriangleup [\mathcal{SW}_{k-1}(\eta_r) \wedge \mathcal{HT}_1(\varphi_{r+1})] \bigtriangleup$$
$$\bigtriangleup [\mathcal{SW}_{k-2}(\eta_r) \wedge \mathcal{HT}_{2^n}(a_{r+1}\varphi_{r+1})].$$

(4) $\mathcal{SW}_{2r+1}(\eta_{r+1}) = [\mathcal{SW}_{2r}(\eta_r) \wedge \mathcal{HT}_1(\varphi_{r+1})] \bigtriangleup$
$$\bigtriangleup [\mathcal{SW}_{2r-1}(\eta_r) \wedge \mathcal{HT}_{2^n}(a_{r+1}\varphi_{r+1})].$$

(5) $\mathcal{SW}_{2r+2}(\eta_{r+1}) = \mathcal{SW}_{2r}(\eta_r) \wedge \mathcal{HT}_{2^n}(a_{r+1}\varphi_{r+1})$.

An analysis of the five cases, corresponding to the values of k given in (1) – (5), using (+), the induction hypothesis, and Theorem 7.24, proves the identity claimed, namely
$$\mathcal{SW}_{k \cdot 2^{n-1}}(\psi_{r+1}) = \mathcal{SW}_k(\eta_{r+1}).$$

In order to illustrate the argument, we do the case $2 < k \leq 2r$, corresponding to (3) above. In this situation all three values $t = 0$, 2^{n-1}, 2^n in (+) are possible, and (+) takes the form
$$\mathcal{SW}_{k \cdot 2^{n-1}}(\psi_{r+1}) = \mathcal{SW}_{k \cdot 2^{n-1}}(\psi_r) \bigtriangleup [\mathcal{SW}_{(k-1) \cdot 2^{n-1}}(\psi_r) \wedge \mathcal{SW}_{2^{n-1}}(a_{r+1}\varphi_{r+1})]$$
$$\bigtriangleup [\mathcal{SW}_{(k-2) \cdot 2^{n-1}}(\psi_r) \wedge \mathcal{SW}_{2^n}(a_{r+1}\varphi_{r+1})].$$

The induction hypothesis and Theorem 7.24 give :

$$\mathcal{SW}_{k \cdot 2^{n-1}}(\psi_{r+1}) = \mathcal{SW}_k(\eta_r) \bigtriangleup [\mathcal{SW}_{k-1}(\eta_r) \wedge \mathcal{HT}_1(\varphi_{r+1})] \bigtriangleup$$
$$\bigtriangleup [\mathcal{SW}_{k-2}(\eta_r) \wedge \mathcal{SW}_{2^n}(a_{r+1}\varphi_{r+1})].$$

Comparing with (3) shows
$$\mathcal{SW}_{k \cdot 2^{n-1}}(\psi_{r+1}) = \mathcal{SW}_k(\eta_{r+1}),$$

5. THE INVARIANTS OF LINEAR COMBINATIONS OF PFISTER FORMS. 171

as asserted. The other four cases can be handled using essentially the same argument and their proofs will be omitted. □

As an application of the results in this paragraph we present a new proof of the following fundamental result in the theory of quadratic forms.

THEOREM 7.31. (The Arason-Pfister Hauptsatz). *Let G be a reduced special group. Fix an integer $n \geq 2$. Assume that ψ is a form over G of dimension $m < 2^n$, Witt equivalent to a linear combination of Pfister forms of degree n, over G. Then, ψ is hyperbolic over G.*

Remarks. In the language of Witt rings, the assumption means that ψ belongs to the nth power of $I(G)$, the ideal of even dimensional forms of the Witt ring $W(G)$ of the special group G; see 1.25 (or [D]). Marshall ([Ma1], Ch.III, p. 61; Ch.IV, §4, p. 80 and §8, pp. 88-91) proves the result for abstract Witt rings. Lam ([L1], Ch. X, pp. 289-292) contains a proof for the case of fields. ◇

PROOF. Notation is as in 7.27 and the proof of Theorem 7.26.

Since $m < 2^n$, our assumption means

$$\psi \oplus s\langle 1, -1 \rangle \equiv_G \bigoplus_{i=1}^{r} a_i \varphi_i = \psi_r,$$

where the φ_i are Pfister forms of degree n over G, and $a_i \in G$. Note that $m + 2s = r \cdot 2^n$. Since $n, r \geq 1$, m is even and $s = r \cdot 2^{n-1} - \frac{m}{2} > 1$. This isometry implies that the Horn-Tarski invariants of ψ_r are as follows:

$$(+) \quad \mathcal{HT}_l(\psi_r) = \begin{cases} \top & \text{if } 1 \leq l \leq s \\ \mathcal{HT}_j(\psi) & \text{if } l = s+j, 1 \leq j \leq m \\ \bot & \text{if } s+m+1 \leq l \leq r \cdot 2^n. \end{cases}$$

The assumption $m < 2^n$ implies:

$$(*) \quad (r-1) \cdot 2^{n-1} < s \quad \text{and} \quad (**) \quad s, s+1, \ldots, s + \tfrac{m}{2} \leq r \cdot 2^{n-1}.$$

Hence, the integers $s, s+1, \ldots, s + \frac{m}{2}$ lie all in the interval $I_r = [(r-1) \cdot 2^{n-1} + 1, r 2^{n-1}]$. By Theorem 7.26, the invariants $\mathcal{HT}_k(\psi_r)$ are equal for $t \in \{s, s+1, \ldots, s + \frac{m}{2}\}$. Since, $\mathcal{HT}_s(\psi_r) = \top$, $(+)$ entails

$$(++) \quad \mathcal{HT}_j(\psi) = \mathcal{HT}_{s+j}(\psi_r) = \top \quad \text{for } 1 \leq j \leq \tfrac{m}{2}.$$

From $m < 2^n$ we also get

$$(***) \quad s + m + 1 \leq (r+1) \cdot 2^{n-1}.$$

Since $s + \frac{m}{2} = r \cdot 2^{n-1}$, it follows that $s + \frac{m}{2} + 1, \ldots, s + m + 1$ all belong to the interval I_{r+1}, and the invariants $\mathcal{HT}_l(\psi_r)$ are equal for $l \in \{s + \frac{m}{2} + 1, \ldots, s + m + 1\}$. Since $\mathcal{HT}_{s+m+1}(\psi_r) = \bot$ (see $(+)$), we conclude

$$(+++) \quad \mathcal{HT}_j(\psi) = \mathcal{HT}_{s+j}(\psi_r) = \bot \quad \text{for } \tfrac{m}{2} + 1 \leq j \leq m.$$

Equations $(++)$ and $(+++)$ prove ([H5], Proposition 7.5) :
$$\psi \equiv_{B_G} \langle \underbrace{\top,\ldots,\top}_{\frac{m}{2}}, \underbrace{\bot,\ldots,\bot}_{\frac{m}{2}} \rangle.$$

Since G is a complete subgroup of B_G and this last isometry has coefficients in G, ψ is hyperbolic in G, completing the proof. $\square$

CHAPTER 8

Algebraic K-theory of Fields and Special Groups

This Chapter is an exposition of the contents of [DM2]. If F is a field of characteristic $\neq 2$, let K_*F and k_*F be the K-theory and the mod 2 K-theory of F, set down by J. Milnor in [Mi]. The main result of this Chapter, describes, when F is formally real, graded ring homomorphisms from K_*F and k_*F to the Boolean hull of $G_{red}(F)$ (Theorem 8.7). Moreover, the graded ring homomorphism from k_*F to $G_{red}(F)$ takes the K-theoretic Stiefel-Whitney invariants of a form over F, to the Boolean-theoretic Stiefel-Whitney invariants of the image form in $G_{red}(F)$. As an application, we give a new, simpler proof of a result of R. Elman and T.Y. Lam, namely Theorem 2.15 in [EL2]. Theorem 8.7 will also be important in Chapter 9.

Although it is assumed that the reader is familiar with the presentation in [Mi], whose notation shall be followed closely, section 1 contains a brief overview of the main points we need.

The results proven below exhibit an application of Chapter 7 to the non-reduced theory of quadratic forms. Corollary 8.10 shows that for Pythagorean fields, the K-theoretic Stiefel-Whitney invariants contain the same information as the Boolean-theoretic Stiefel-Whitney invariants and the Horn-Tarski invariants. However, as is shown by properties developed in Chapter 7, the latter, which have no analog in K-theory or cohomology, are easier to compute.

Notation is as in section 3 of Chapter 1. For a formally real field F, $G(F) = \dot{F}/\dot{F}^2$ is the special group of squares classes of F and $G_{red}(F)$ is the reduced special group of classes modulo sums of squares. Write $\overline{a}$ for class of $a \in \dot{F}$ in $G(F)$. By section 3 in Chapter 1, $G(F)$ faithfully codes classical isometry of non-zero forms over F, that is, if $\langle a_1, \ldots, a_n \rangle$, $\langle b_1, \ldots, b_n \rangle$ are forms over $\dot{F}$, then

$$\langle a_1, \ldots, a_n \rangle \equiv_{\dot{F}} \langle b_1, \ldots, b_n \rangle \quad \text{iff} \quad \langle \overline{a}_1, \ldots, \overline{a}_n \rangle \equiv_{G(F)} \langle \overline{b}_1, \ldots, \overline{b}_n \rangle.$$

Thus, we may work in $G(F)$ to obtain results on the isometry of forms over F.

Let $\pi : G(F) \longrightarrow G_{red}(F)$ be the surjective SG-morphism that associates to a square class $\overline{a} \in G(F)$, its class modulo sums of squares.

Let $\varepsilon_{red} : G_{red}(F) \longrightarrow B(F)$ be the boolean hull of $G_{red}(F)$.

Set $\mathfrak{p} = \varepsilon_{red} \circ \pi$, that is, $\mathfrak{p}$ is the composition

$$G(F) \xrightarrow{\pi} G_{red}(F) \xrightarrow{\varepsilon_{red}} B(F).$$

Thus, for all $a, c \in \dot{F}$, we have $\mathfrak{p}(\bar{a} \cdot \bar{c}) = \mathfrak{p}\bar{a} \triangle \mathfrak{p}\bar{c}$, where $\triangle$ is symmetric difference in $B(F)$ (Proposition 4.10).

PROPOSITION 8.1. *Let F be a formally real field. With notation as above, we have*

a) *If φ, ψ are n-forms in $G(F)$, then*

$$\pi \star \varphi \equiv_{G_{red}(F)} \pi \star \psi \quad \text{iff} \quad \exists\, l \geq 1 \text{ such that } 2^l \otimes \varphi \equiv_{G(F)} 2^l \otimes \psi.$$

b) *For n-forms φ, ψ over $G(F)$, the following are equivalent:*

(1) $\forall\, 1 \leq k \leq n$, $\mathcal{HT}_k(\mathfrak{p} \star \varphi) = \mathcal{HT}_k(\mathfrak{p} \star \psi)$.

(2) $\forall\, 1 \leq k \leq n$, $\mathcal{SW}_k(\mathfrak{p} \star \varphi) = \mathcal{SW}_k(\mathfrak{p} \star \psi)$.

(3) $\exists\, l \geq 1$ such that $2^l \otimes \varphi \equiv_{G(F)} 2^l \otimes \psi$.

PROOF. a) Since F is formally real, we have $1 \neq -1$ in $G(F)$ and in $G_{red}(F)$. By Example 2.24, $G_{red}(F) = G(F)/Sat(G(F))$, and so the conclusion follows from Proposition 2.21.

b) It follows from Theorem 7.1, applied to the forms $\pi \star \varphi$, $\pi \star \psi$, that (1) and (2) are equivalent to $\pi \star \varphi \equiv_{G_{red}(F)} \pi \star \psi$; by (a), condition (3) must be equivalent to the other two. $\square$

1. Milnor's Algebraic K-theory

Let F be a field, as always of characteristic $\neq 2$. We briefly recall the algebraic K-theory set down by J. Milnor in [Mi], together with its mod 2 counterpart. Let $K_1 F$ be the multiplicative group $\dot{F}$ written additively. 'To keep notation straight' (dixit Milnor), we **fix an isomorphism**

$$l : \dot{F} \longrightarrow K_1 F, \qquad \text{('logarithm')}$$

that is, a bijection such that $l(ab) = l(a) + l(b)$, for all $a, b \in \dot{F}$. Notice that $l(1/a) = -l(a)$ and that $l(1)$ is the zero of $K_1 F$.

Let $\underbrace{K_1 F \otimes \ldots \otimes K_1 F}_{n\, times}$ be the n-fold tensor product of $K_1 F$ over $\mathbb{Z}$. Then,

$$A = (\mathbb{Z}, K_1 F, \ldots, \bigotimes_{i=1}^{n} K_1 F, \ldots),$$

is a graded ring under tensor product. Define

$$K_* F = (\mathbb{Z}, K_1 F, K_2 F, \ldots, K_n F, \ldots)$$

as the graded ring determined by the quotient of A by the ideal generated by the set

$$\{l(a) \otimes l(1-a) : a \neq 1 \text{ in } \dot{F}\}.$$

Thus, $K_n F$ is the quotient of $\underbrace{K_1 F \otimes \ldots \otimes K_1 F}_{n \, times}$ by the subgroup consisting of finite sums of elements $l(a_1) \otimes \ldots \otimes l(a_n)$, such that $a_i + a_{i+1} = 1$, for some i between 1 and $n-1$.

Remark. In [Mi], Milnor states that the idea for the definition of K_2 comes from the work of Calvin Moore, Steinberg and Matsumoto on algebraic groups, while the definition of K_n, $n \geq 3$, is *ad hoc*.

Just as in [Mi], we shall write tensors and their classes in $K_* F$ by superposition, i.e.,

$$l(a_1) l(a_2) \ldots l(a_n) \text{ stands for } l(a_1) \otimes \ldots \otimes l(a_n),$$

as well as its class in $K_n F$.

PROPOSITION 8.2. (Basic properties of $K_* F$, [Mi]) *With notation as above,*

a) $\forall\, a \in \dot{F}$, $l(a)l(-a) = 0$.

b) $\forall\, \xi \in K_n F$, $\forall\, \eta \in K_m F$, $\xi \eta = (-1)^{nm} \eta \xi$.

c) $\forall\, a \in \dot{F}$, $l(a)^2 = l(a)l(-1)$.

d) $\forall\, a_1, \ldots, a_n \in \dot{F}$, $\sum_{i=1}^n a_i = \begin{cases} 1 \\ 0 \end{cases}$ *implies* $l(a_1) \ldots l(a_n) = 0$. $\diamond$

We now describe the mod 2 K-theory of fields. Let

$$k_* F = (\mathbb{Z}/2\mathbb{Z}, k_1 F, \ldots, k_n F, \ldots)$$

be the quotient of $K_* F$ by the **ideal**

$$2K_* F = (2\mathbb{Z}, 2K_1 F, \ldots, 2K_n F, \ldots),$$

where $2K_n F$ is the subgroup consisting of all finite sums of elements $l(a_1) \ldots l(a_n)$, such that $a_i \in \dot{F}^2$, for some $1 \leq i \leq n$. Note that $k_* F$ is a graded algebra over $\mathbb{Z}/2\mathbb{Z}$. We will still denote a generator in the quotient $k_n F$ by $l(a_1) \ldots l(a_n)$.

PROPOSITION 8.3. *Let* $a_1, \ldots, a_n, x \in \dot{F}$; *let σ be a permutation of* $\{1, \ldots, n\}$. *In $k_n F$, we have*

a) $l(a_1) \ldots l(x^2 a_i) \ldots l(a_n) = l(a_1) \ldots l(a_n)$.

b) $l(a_1) \ldots l(a_n) = l(a_{\sigma(1)}) \ldots l(a_{\sigma(n)})$.

PROOF. a) Follows immediately from the fact that the product is multilinear and that in $k_1 F$ we have

$$l(x^2 a_i) = 2\, l(x) + l(a_i) = l(a_i).$$

b) By 8.2.(b), in $k_2 F$, we have $l(a)l(b) - l(b)l(a) = 2l(a)l(b) = 0$ and so we may exchange any two successive terms in $l(a_1) \ldots l(a_n)$, without altering its value. Since permutations of the form $(i, i+1)$ generate the group of permutations of $\{1, \ldots, n\}$, (b) is proven. $\square$

It is clear that the map (square class of a) $\in \dot{F}/\dot{F}^2 \mapsto l(a) \in k_1 F$, is an isomorphism between $G(F) = \dot{F}/\dot{F}^2$ and $k_1 F$.

Let $k_\pi F$ be the $\mathbb{Z}/2\mathbb{Z}$-algebra of all formal series $\xi_0 + \xi_1 + \ldots$, with $\xi_i \in k_i F$. Additively, $k_\pi F$ is isomorphic to the product $k_0 F \times k_1 F \times \ldots$.

For a form $\varphi = \langle a_1, \ldots, a_n \rangle$ over $\dot{F}$, Milnor defines the **total Stiefel-Whitney invariant** of φ as the following element of $k_\pi F$:

$$w(\varphi) = (1 + l(a_1))(1 + l(a_2)) \ldots (1 + l(a_n)).$$

For φ as above, set

$$w_i(\varphi) = \sum_{p \in S^{n,i}} l(a_{p_1}) \ldots l(a_{p_i}),$$

called the ith **(K-theoretic) Stiefel-Whitney invariant** of φ, an element of $k_n F$. The sets $S^{n,i}$ are defined in the statement of Theorem 7.1.

PROPOSITION 8.4. *With notation as above, for n-forms φ, ψ over $\dot{F}$,*

a) $w(\varphi) = 1 + \sum_{i=1}^n w_i(\varphi)$.

b) $w(\varphi)$ *is a unit in* $k_\pi F$.

c) $w(\varphi) = w(\psi)$ *iff* $\forall\ 1 \leq i \leq n$, $w_i(\varphi) = w_i(\psi)$.

d) $\varphi \equiv_{G(F)} \psi$ *implies* $w(\varphi) = w(\psi)$.

PROOF. Items (a) - (c) are straightforward. The proof of (d) will proceed by verifying the statement for $n = 1, 2$ and then using Witt's celebrated chain equivalence Theorem. Item (c) above tells us that it is enough to verify that the w_i's are the same for φ and ψ. One should keep in mind that $\dot{F}/\dot{F}^2$ is isomorphic, via l, to $k_1 F$.

It follows readily from Proposition 8.3.(a) that the statement holds for forms of dimension 1, because in this case isometry is equivalent to equality of square classes. For forms of dimension 2, let a, b, c, d be elements of $\dot{F}$, such that $\langle a, b \rangle \equiv_{\dot{F}} \langle c, d \rangle$, or equivalently (see section 3, Chapter 1, right after proof of 1.30),

$$\overline{ab} = \overline{cd} \quad \text{and} \quad D_T(a, b) = D_T(c, d),$$

where $T = \dot{F}^2$. From Proposition 8.3.(a), we get $l(ab) = l(cd)$ and so

$$w_1(\langle a, b \rangle) = l(a) + l(b) = l(c) + l(d) = w_1(\langle c, d \rangle). \tag{$*$}$$

By the definition of representation in section 3 of Chapter 1, there are $x, y \in F$ such that $a = x^2c + y^2d$. Hence, $x^2\dfrac{c}{a} + y^2\dfrac{d}{a} = 1$. The definition of $K_2 F$ and 8.3.(a) yield $l(c/a)l(d/a) = 0$. So, using (**), we get

$$\begin{aligned} 0 = l(c/a)l(d/a) &= (l(c) - l(a))\,(l(d) - l(a)) \\ &= l(c)l(d) - l(a)(l(c) + l(d)) + l(a)^2 \\ &= l(c)l(d) - l(a)(l(a) + l(b)) + l(a)^2 \\ &= l(c)l(d) - l(a)l(b), \end{aligned}$$

proving that $w_2(\langle a, b \rangle) = w_2(\langle c, d \rangle)$.

We now show that if φ is simply equivalent to ψ, then $w_i(\varphi) = w_i(\psi)$, $1 \leq i \leq dim(\varphi) = dim(\psi)$. The definition of simple equivalence appears just before the statement of Lemma 1.22. If $\varphi = \langle a_1, \ldots, a_n \rangle$ is simply equivalent to $\psi = \langle b_1, \ldots, b_n \rangle$, then are k, l such that $\langle a_k, a_l \rangle \equiv_{G(F)} \langle b_k, b_l \rangle$, while $a_j = b_j$, for all j distinct from k and l.

Since the discriminants of φ and ψ are in the same square class, Proposition 8.3.(a) and what has been proven for 2-forms, yield

$$w_1(\varphi) = \sum_{i=1}^n l(a_i) = \sum_{i=1}^n l(b_i) = w_1(\psi).$$

Now fix i between 2 and n; consider the sets

$A_1 = \{p \in S^{n,i} : \{1, 2\} \cap p = \emptyset\}; \quad A_2 = \{p \in S^{n,i} : \{1, 2\} \subseteq p\};$

$A_3 = \{p \in S^{n,i} : 1 \in p \text{ and } 2 \notin p\}; \quad A_4 = \{p \in S^{n,i} : 2 \in p \text{ and } 1 \notin p\},$

with $S^{n,i}$ as the statement of Theorem 7.1. Clearly, $S^{n,i}$ is the union of the A_i's. Moreover, if

$C = \{q = (q_1, \ldots, q_{i-1}) : q_1 < \ldots < q_{i-1} \text{ and } q_k \in \{1, \ldots, n\} \setminus \{j, k\}\},$

then $A_3 = \{q \cup \{j\} : q \in C\}$ and $A_4 = \{q \cup \{k\} : q \in C\}$. Since $\otimes$ is commutative in $k_* F$ and induction guarantees that

$$l(a_j) + l(a_k) = l(b_j) + l(b_k) \quad \text{and} \quad l(a_j)l(a_k) = l(b_j)l(b_k),$$

we have, with $l(p, \varphi) = l(a_{p_1}) \ldots l(a_{p_i})$,

$$\begin{aligned} w_i(\varphi) &= \sum_{p \in A_1} l(p, \varphi) + \sum_{p \in A_2} l(p, \varphi) + \sum_{p \in A_3} l(p, \varphi) + \sum_{p \in A_4} l(p, \varphi) \\ &= \sum_{p \in A_1} l(p, \psi) + \sum_{p \in A_2} l(p, \psi) + l(a_j) \sum_{q \in C} l(q, \varphi) + l(a_k) \sum_{q \in C} l(q, \varphi) \\ &= \sum_{p \in A_1 \cup A_2} l(p, \psi) + [l(a_j) + l(a_k)] \sum_{q \in C} l(q, \varphi) \\ &= \sum_{p \in A_1 \cup A_2} l(p, \psi) + [l(b_j) + l(b_k)] \sum_{q \in C} l(q, \psi) = w_i(\psi). \end{aligned}$$

The above argument shows that w does change if φ is chain equivalent to ψ. By Witt's chain-equivalence Theorem (or Theorem 1.23), $\varphi \equiv_{G(F)} \psi$ implies the equality of all their Stiefel-Whitney classes. □

REMARK 8.5. Using Theorem 1.23, the same argument as above will give a another proof that isometry of forms in a reduced special group implies the equality of their Horn-Tarski and Boolean-theoretic Stiefel-Whitney invariants. ◇

With notation as in 1.25, Milnor proves (Corollary 3.3 and Theorem 4.1 in [Mi]) :

PROPOSITION 8.6. *For $i = 1, 2$, the K-theoretic Stiefel-Whitney invariant $w_i(\varphi)$ induces an isomorphism $w_i : I^i F/I^{i+1}F \longrightarrow k_i F$.* ◇

2. Isometry and K-Theoretic Stiefel-Whitney Invariants

In this section we show that if F is a formally real field, then the K-theoretic Stiefel-Whitney invariants determine isometry class iff $I^3 F$ is torsion-free (as a group). This appears in [EL2], but depends on a result of Scharlau involving the Galois cohomology ring of F. Our proof will, instead, use the Boolean-theoretic Stiefel-Whitney invariants, together with

THEOREM 8.7. *Let F be a formally real field and let $B(F)$ be the Boolean hull of G_{red}. Then, there are graded ring homomorphisms*

$$g = (g_n)_{n \geq 0} : K_* F \longrightarrow B(F) \quad \text{and} \quad \mathfrak{e} = (\varepsilon_n)_{n \geq 0} : k_* F \longrightarrow B(F),$$

such that

a) *The following diagram is commutative :*

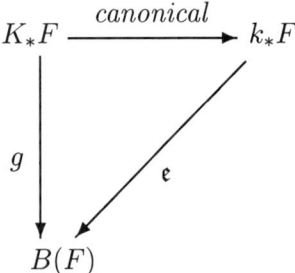

b) *For $n \geq 1$, let $B_F(n)$ be the subgroup of $B(F)$ generated by meets of at most n elements in the image of $\mathfrak{p}$. Then the $Im(g_n) = Im(\varepsilon_n) = B_F(n)$.*

c) *For all $1 \leq i \leq n$, $\varepsilon_i(w_i(\overline{\varphi})) = \mathcal{SW}_i(\mathfrak{p} \star \overline{\varphi})$, where w_i is the ith K-theoretic Stiefel-Whitney invariant of φ.*

PROOF. For $n = 0$, there is a natural map $f_0 : \mathbb{Z}/2\mathbb{Z} \longrightarrow B(F)$, given by $0 \mapsto \perp$ and $1 \mapsto \top$. Let g_0 be the composition $\mathbb{Z} \xrightarrow{can.} \mathbb{Z}/2\mathbb{Z} \xrightarrow{f_0} B(F)$.

2. ISOMETRY AND K-THEORETIC STIEFEL-WHITNEY INVARIANTS

For $n \geq 1$, define a map α_n from $(K_1 F)^n$ to $B(F)$ by the rule:
$$\alpha_n(l(a_1), \ldots, l(a_n)) = \mathfrak{p}\bar{a}_1 \wedge \mathfrak{p}\bar{a}_2 \wedge \ldots \wedge \mathfrak{p}\bar{a}_n.$$
To show that α_n is n-linear, we use distributivity of $\triangle$ over meets and $\mathfrak{p}(\bar{a} \cdot \bar{c}) = \mathfrak{p}\bar{a} \triangle \mathfrak{p}\bar{c}$. Indeed,
$$\alpha_n(l(a_1), \ldots, l(a_i) + l(c_i), \ldots, l(a_n)) = \alpha_n(l(a_1), \ldots, l(a_i \cdot c_i), \ldots, l(a_n))$$
$$= \mathfrak{p}\bar{a}_1 \wedge \ldots \wedge [\mathfrak{p}(\bar{a}_i \cdot \bar{c}_i)] \wedge \ldots \wedge \mathfrak{p}\bar{a}_n$$
$$= \mathfrak{p}\bar{a}_1 \wedge \ldots \wedge [\mathfrak{p}\bar{a}_i \triangle \mathfrak{p}\bar{c}_i)] \wedge \ldots \wedge \mathfrak{p}\bar{a}_n$$
$$= (\mathfrak{p}\bar{a}_1 \wedge \ldots \wedge \mathfrak{p}\bar{a}_i \wedge \ldots \wedge \mathfrak{p}\bar{a}_n) \triangle (\mathfrak{p}\bar{a}_1 \wedge \ldots \wedge \mathfrak{p}\bar{c}_i \wedge \ldots \wedge \mathfrak{p}\bar{a}_n)$$
$$= \alpha_n(l(a_1), \ldots, l(a_i), \ldots, l(a_n)) \triangle \alpha_n(l(a_1), \ldots, l(c_i), \ldots, l(a_n)),$$
verifying the n-linearity of α_n. Thus, α_n induces a homomorphism $\widehat{\alpha}_n$ from the n-fold tensor product $\bigotimes_{i=1}^{n} K_1 F$ to $B(F)$, such that $l(a_1) l(a_2) \ldots l(a_n)$ is taken to $\mathfrak{p}\bar{a}_1 \wedge \mathfrak{p}\bar{a}_2 \wedge \ldots \wedge \mathfrak{p}\bar{a}_n$. To show that $\widehat{\alpha}_n$ factors through $K_n F$, it must be verified that if $\eta = l(a_1) l(a_2) \ldots l(a_n)$, with $a_i + a_{i+1} = 1$ for some $i \leq n-1$, then $\widehat{\alpha}_n(\eta) = \bot$ (1 in $B(F)$).

But if $a_i = 1 - a_{i+1}$, then we have $\bar{a}_i \in D_G(1, -\bar{a}_{i+1})$; since π is a morphism of special groups, we conclude $\pi(\bar{a}_i) \in D_{G_{red}(F)}(1, -\pi(\bar{a}_{i+1}))$. Consequently, Proposition 4.10 and Corollary 4.12 yield
$$\mathfrak{p}\bar{a}_i \subseteq \mathfrak{p}(-\bar{a}_{i+1}) = -\mathfrak{p}\bar{a}_{i+1},$$
where $-x$ denotes the complement of x in $B(F)$. Hence, $\mathfrak{p}\bar{a}_i \wedge \mathfrak{p}\bar{a}_{i+1} = \bot$, which in turn implies $\widehat{\alpha}_n(\eta) = \bot$, as needed.

Hence, $\widehat{\alpha}_n$ induces a homomorphism $g_n : K_n F \longrightarrow B(F)$, taking a generator $l(a_1) l(a_2) \ldots l(a_n)$ to the element $\mathfrak{p}\bar{a}_1 \wedge \mathfrak{p}\bar{a}_2 \wedge \ldots \wedge \mathfrak{p}\bar{a}_n$ in $B(F)$. Clearly, we have that for all $\eta \in K_n F$, $\zeta \in K_m F$, $n, m \geq 0$,
$$g_{n+m}(\eta \zeta) = g_n(\eta) \wedge g_m(\zeta),$$
and so $g = (g_n) : K_* F \longrightarrow B(F)$ is a homomorphism of graded rings.

Now observe that if $\eta = l(a_1) l(a_2) \ldots l(a_n)$ is such that $a_i = c^2 \in \dot{F}^2$, then
$$g_n(\eta) = \mathfrak{p}\bar{a}_1 \wedge \ldots \wedge [\mathfrak{p}\bar{c} \triangle \mathfrak{p}\bar{c}] \wedge \ldots \wedge \mathfrak{p}\bar{a}_n = \bot,$$
which shows that $2K_n F \subseteq \ker g_n$, for all $n \geq 0$. Since $k_* F = K_* F / 2 K_* F$, g_n induces a homomorphism ε_n from $k_n F$ to $B(F)$, that maps a generator $l(a_1) l(a_2) \ldots l(a_n)$ to $\mathfrak{p}\bar{a}_1 \wedge \mathfrak{p}\bar{a}_2 \ldots \wedge \mathfrak{p}\bar{a}_n$ in $B(F)$. By construction, the diagram in (a) is commutative. Moreover, since $g = (g_n)$ is a graded ring homomorphism, the same is true of $\mathfrak{e} = (\varepsilon_n)$. It is straightforward to check that the image of ε_n (or of g_n) is $B_F(n)$ (item (b)).

It is clear from the values of ε_i on the generators of $k_i F$, that $\varepsilon_i(w_i(\varphi)) = \mathcal{SW}_i(\mathfrak{p} \star \overline{\varphi})$, verifying (c) and ending the proof. $\square$

We apply the preceding constructions to get a new proof of a result in [EL2]. We state it as

THEOREM 8.8. *Let F be a formally real field such that I^3F is torsion-free. For n-forms φ, ψ over $\dot{F}$, the following are equivalent*

(1) $w(\varphi) = w(\psi)$.

(2) $\varphi \equiv_{G(F)} \psi$.

(3) $\forall\ 1 \leq i \leq n,\ w_i(\varphi) = w_i(\psi)$.

PROOF. By Proposition 8.4, it is enough to verify that (3) implies (2). This is the main simplification of the original proof, using the Boolean-theoretic Stiefel-Whitney invariants in place of Scharlau's result on Delzant's Stiefel-Whitney classes in Galois cohomology.

Item (3) and the homomorphisms ε_n in Theorem 8.7 yield $\mathcal{SW}_i(\mathfrak{p} \star \overline{\varphi}) = \mathcal{SW}_i(\mathfrak{p} \star \overline{\psi})$, for all $1 \leq i \leq n$. By Proposition 8.1, there is $l \geq 1$ such that $2^l \otimes \overline{\varphi} \equiv_{G(F)} 2^l \otimes \overline{\psi}$. Since isometry in $G(F)$ of the image forms $\overline{\varphi}, \overline{\psi}$ is equivalent to isometry in $\dot{F}$, we conclude that $2^l \otimes \varphi \equiv_{\dot{F}} 2^l \otimes \psi$, that is, $\varphi - \psi$ is torsion in $W(F)$. Note that $\varphi - \psi$ is in IF, because $dim(\varphi) = dim(\psi) = n$. To show that it is in fact in I^3F, we reason just as Elman and Lam in [EL2].

Since $w_i(\varphi) = w_i(\psi)$, $i = 1, 2$, it is readily verified that $\varphi - \psi$ must be in the kernel of both maps in Proposition 8.6. Hence, $\varphi - \psi \in I^3F$; since we already know that this difference is torsion, we conclude that $\varphi = \psi$ in $W(F)$. Taking into account that φ and ψ are of the same dimension, we get $\varphi \equiv_{\dot{F}} \psi$, as desired. □

REMARK 8.9. In [EL2], Thm. 2.15, the following result is proven :

Theorem. *Let F be a field. Then, dimension and total Stiefel-Whitney class characterize isometry of quadratic forms in F iff I^3F is torsion-free.* ◇

We have presented above an alternative for the "if" direction of this result, in the case of formally real fields. It can also be shown, using the Boolean-theoretic invariants, that if isometry is determined by total Stiefel-Whitney class and dimension, then I^3F is torsion-free. However, the computations for this are not as conceptually different from the original proof as those presented above and will be omitted. ◇

If F is a Pythagorean field, then $G_{red}(F) = G(F)$ and the map $\mathfrak{p}$ is just the Boolean hull of the reduced special group $G(F)$. As a corollary of the preceding results we get that, for forms over formally real Pythagorean fields, the Horn-Tarski invariants carry the same information as the total Stiefel-Whitney class :

COROLLARY 8.10. *Let F be a formally real Pythagorean field and φ, ψ be n-forms over F. The following are equivalent :*

(1) $\varphi \equiv_{G(F)} \psi$.
(2) $w(\varphi) = w(\psi)$.
(3) $\forall\ 1 \leq i \leq n,\ w_i(\varphi) = w_i(\psi)$.
(4) $\forall\ 1 \leq i \leq n,\ \mathcal{SW}_i(\mathfrak{p} \star \overline{\varphi}) = \mathcal{SW}_i(\mathfrak{p} \star \overline{\psi})$. $\diamond$

CHAPTER 9

Marshall's Conjecture for Pythagorean Fields

This Chapter is an exposition of the contents of [DM1], wherein we give an affirmative answer to the following problem, posed by M. Marshall in 1974 :

Let F be a formally real, Pythagorean field. Is it true that for every quadratic form φ over F and any integer $n \geq 1$, the following holds :

[MC] If $sgn_P(\varphi) \equiv 0 \pmod{2^n}$ for every order P of F, then $\varphi \in I^n(F)$.

Here $sgn_P(\varphi)$ denotes the signature of φ under the order P and $I^n(F)$ is the nth power of the fundamental ideal $I(F)$ of the Witt ring $W(F)$ of F; cf. 3.5, 1.25, and 1.26. As far as we know, the above question first appeared in print in Marshall's 1977 paper ([Ma6]; Open question 2; p. 575) where it is posed for the more general context of abstract spaces of orders, as well as in Lam's lectures [L3] in the same volume, where it occurs for the field case, in a more general form, as Open Problem B, p. 49. The aforementioned date (1974) was communicated to us by M. Marshall.

Our proof is divided in two parts, whose main points are outlined below.

Part I. In the framework of special groups and the related Boolean-theoretic techniques developed above, it is shown that [MC] is equivalent to

[WMC]   If $2\varphi \in I^{n+1}(F)$, then $\varphi \in I^n(F)$, for every integer $n \geq 1$.

As a matter of fact, both [MC] and [WMC] make sense in the context of formally real special groups. Here [MC] takes the following form : for every integer $n \geq 1$ and every quadratic form φ over such a group G

[MC]   If $sgn_\sigma(\varphi) \equiv 0 \pmod{2^n}$ for every $\sigma \in X_G$, then $\varphi \in I^n(G)$,

where X_G denotes the space of orders of G (Chapter 3) and $I^n(G)$ has been defined in 1.26.

We prove the equivalence of [MC] and [WMC] for formally real special groups G with the property that $2^n \langle 1 \rangle \notin I^{n+1}(G)$, for every integer $n \geq 1$; we call these groups $\mathcal{AP}$ (for Arason-Pfister). By the Arason-Pfister Hauptsatz ([AP]; see also [L1], Thm. X.3.1, p. 289 ff), the special groups of formally real fields are $\mathcal{AP}$ groups, and so are the reduced special groups by

Theorem 7.31; see Lemma 9.6 below. Since reduced special groups and abstract order spaces are (dually) equivalent notions (Theorem 3.19), this part of the proof also works for Marshall's original formulation of the problem.

The reduction of $[MC]$ to $[WMC]$ proceeds by the following steps :

(a) First we show that, whenever G is $\mathcal{AP}$, the **inductive limit** $\mathcal{W}(G)$ of the groups $\overline{I^n}(G) = I^n(G)/I^{n+1}(G)$ (under the operation induced by the sum of forms in $I^n(G)$), with transition functions induced by multiplication (i.e., tensor product) by $2 = \langle 1,1 \rangle$, is a non-trivial Boolean ring. This is also the case, when F is a formally real field, for the **inductive limit** $k(F)$ of Milnor's mod 2 K-theoretic groups $k_n F$, with transition functions induced by multiplication by $l(-1)$ (Theorem 9.13.(a)). As Boolean algebras, these inductive limits possess a natural structure of reduced special groups, cf. Chapter 4.

(b) Furthermore, the natural group homomorphisms $\rho : G \longrightarrow \mathcal{W}(G)$ and $\kappa : G(F) \longrightarrow k(F)$, are, in addition, morphisms of special groups, which are injective iff G is reduced and F is Pythagorean, respectively (Theorem 9.13.(b), (c)). Owing to the universal property of the Boolean hull B_G of G (Theorem 4.17), these maps extend to Boolean algebra morphisms $B(\rho) : B_G \longrightarrow \mathcal{W}(G)$ and $B(\kappa) : B(F) \longrightarrow k(F)$, respectively (Corollary 9.15).

(c) The next step is the proof that these Boolean homomorphisms are **isomorphisms**. Indeed, they have inverses $M : \mathcal{W}(G) \longrightarrow B_G$ and $\varepsilon : k(F) \longrightarrow B(F)$, respectively.

The homomorphism M is defined on the image of $\overline{I^n}$ in $\mathcal{W}(G)$ (by the homomorphism α_n given by the inductive limit construction) to be the clopen in X_G

$$\{\sigma \in X_G : sgn_\sigma(\varphi) \not\equiv 0 \pmod{2^{n+1}}\} \qquad (\varphi \in I^n(G)).$$

M is a Boolean analogue of a map considered by Milnor (§ 3 in [Mi]), induced by the Stiefel-Whitney invariant of order 2^{n-1} on forms in I^n. The map ε, defined on the image of $k_n F$ in $k(F)$ sends (the image of) a generator $l(a_1)l(a_2)\ldots l(a_n)$ to the element $a_1 \wedge a_2 \wedge \ldots \wedge a_n$ in $B(F)$.

(d) It follows from the definition of the map M that $[MC]$ holds for a given integer $n \geq 1$ if and only if the map $M \circ \alpha_n : \overline{I^n} \longrightarrow B_G$ is injective. And this is the case, if and only if, $[WMC]$ holds at level n (equivalently, the transition homomorphism $\overline{I^n}(G) \xrightarrow{2 \cdot} \overline{I^{n+1}}(G)$ induced by multiplication by 2 is injective).

Part II. In the second step of the proof we deal with (formally real) Pythagorean fields properly. We establish the injectivity of the map $\overline{I^n}(G) \xrightarrow{2 \cdot} \overline{I^{n+1}}(G)$ as follows :

(a) If F is such a field and $n \geq 1$, write $H^n(F)$ for the nth Galois cohomology group of F, with coefficients in $\mathbb{Z}/2\mathbb{Z}$. In Proposition 9.27 we show that the map $(-1) \cup : H^n(F) \longrightarrow H^{n+1}(F)$ is injective. This comes from, among other things, the long exact sequence in cohomology associated to quadratic extensions of F, due to Arason [A], together with Voevodsky's recent and celebrated result ([V]) that the Milnor homomorphisms $h_n^F : k_n F \longrightarrow H^n(F)$ are isomorphisms for all $n \geq 1$.

(b) The commutativity of the diagram

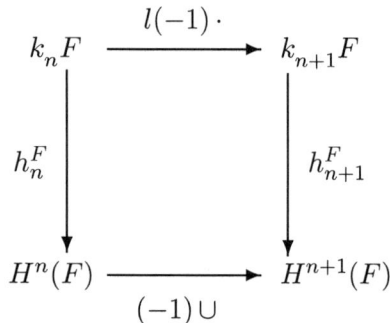

where the map in the upper row is induced by multiplication by $l(-1)$, yields at once that this homomorphism is injective for all $n \geq 1$ and every formally real, Pythagorean field F (Corollary 9.28). It follows that every such field verifies Milnor's conjecture for the graded Witt ring, i.e., the homomorphism $s_n : k_n F \longrightarrow \overline{I^n}(F)$ is an isomorphism for all $n \geq 1$.

(c) Finally, the commutativity of the diagram

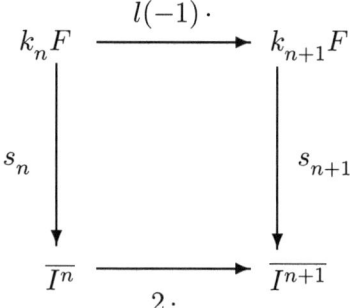

proves [WMC], and hence [MC], for all formally real Pythagorean fields.

1. Introduction

9.1. NOTATION AND PRELIMINARIES

a) In the sequel **all Pythagorean fields are considered to be formally real**.

b) We shall use notation concerning the special groups $G(F)$ and $G_{red}(F)$ associated to a field F, as set down in section 3 of Chapter 1. The class of $a \in \dot{F}$ modulo $\dot{F}^2$ will be denoted by $\bar{a}$.

c) The signature of a form at a character was defined in 3.5. Since the signature commutes with sum and product of forms, and Witt-equivalent forms have the same signature, we get:

FACT 9.2. *Let G be a formally real special group and $n \geq 1$.*

i) *If φ is a Pfister form of degree n over G and $\sigma \in X_G$, then $sgn_\sigma(\varphi)$ is either 0 or 2^n. Furthermore,*

ii) *If $\varphi = \otimes_{i=1}^{n} \langle 1, a_i \rangle$, then $sgn_\sigma(\varphi) = 0$ iff $\sigma(a_i) = -1$ for at least one i, $1 \leq i \leq n$.*

iii) *If ψ is a linear combination of Pfister forms of degree n, then $sgn_\sigma(\psi)$ is a multiple of 2^n. The same holds for every $\psi \in I^n(G)$.* ◇

The last assertion in (iii) tells us that the signature defines a map of $I^n(G)$ into $\mathbb{Z}$ whose values are multiples of 2^n.

In the terminology of the Boolean hull (Chapter 4), item (ii) can be restated as follows:

d) If $\varphi = \otimes_{i=1}^{n} \langle 1, a_i \rangle$, then

— $\{\sigma \in X_G : sgn_\sigma(\varphi) = 0\} = \bigcup_{i=1}^{n} [a_i = -1] = \bigvee_{i=1}^{n} \varepsilon_G(a_i),$

— $\{\sigma \in X_G : sgn_\sigma(\varphi) = 2^n\} = -\bigvee_{i=1}^{n} \varepsilon_G(a_i).$

Both these sets are clopen in X_G i.e., they belong to B_G.

In terms of the Horn-Tarski invariants, using Theorem 7.18, these equalities take the following form:

e) If $\varphi = \otimes_{i=1}^{n} \langle 1, a_i \rangle$, then

— $\{\sigma \in X_G : sgn_\sigma(\varphi) = 0\} = \mathcal{HT}_1(\varphi) = \bigvee_{i=1}^{n} a_i,$ and

— $\{\sigma \in X_G : sgn_\sigma(\varphi) = 2^n\} = -\mathcal{HT}_1(\varphi) = \bigwedge_{i=1}^{n} -a_i.$

f) The **Weak Marshall Conjecture** is the following statement :

[WMC] For all forms φ and integers $n \geq 1$, $2\varphi \in I^{n+1}$ implies $\varphi \in I^n$. ◇

LEMMA 9.3. *For a formally real special group G, the following conditions are equivalent :*

(1) *G verifies [WMC].*

(2) *For all forms φ and all integers k, $n \geq 1$,*
$$2^k \varphi \in I^{n+k} \quad implies \quad \varphi \in I^n.$$

(3) *For all forms φ over G and all integers $n \geq 1$,*
$$\varphi \in I^{n-1} \ and \ 2\varphi \in I^{n+1} \quad implies \quad \varphi \in I^n.$$

PROOF. It is clear that $(1) \Leftrightarrow (2) \Rightarrow (3)$. It remains to verify that $(3) \Rightarrow (1)$. Suppose that φ is a form over G such that $2\varphi \in I^{n+1}$, for some $n \geq 1$. Our first observation is

Fact. *If $2\varphi \in I^{n+1}$ for some $n \geq 1$, then $dim(\varphi)$ is even, that is, $\varphi \in I$.*

Proof. For integers p, $q \geq 0$, we have
$$2\varphi \oplus p\langle 1, -1 \rangle \equiv_G \psi \oplus q\langle 1, -1 \rangle, \tag{I}$$
where ψ is a linear combination of Pfister forms of degree $n+1$ over G. Thus, applying any SG-character σ on both sides of (I) and recalling Fact 9.2.(iii), yields
$$2 \, sgn_\sigma(\varphi) = sgn_\sigma(\psi) = \alpha \, 2^{n+1},$$
for some integer $\alpha \in \mathbb{Z}$. Thus, $sgn_\sigma(\varphi) = \alpha \, 2^n$ is even, since $n \geq 1$. Write $\varphi = \langle a_1, \ldots, a_m \rangle$ and set β = the cardinal of $\{j \leq m : \sigma(a_j) = -1\}$. Then,
$$sgn_\sigma(\varphi) = \sum_{i=1}^m sgn_\sigma(a_i) = -\beta + (m - \beta) = m - 2\beta,$$
which clearly forces $m = dim(\varphi)$ to be even. $\diamond$

To end the proof that $(3) \Rightarrow (1)$, we shall verify by induction on $1 \leq k \leq n$, that $\varphi \in I^k$. The Fact takes care of the case $k = 1$. Suppose our contention is true for $k < n$. Then, $k + 2 \leq n + 1$ and so $\varphi \in I^k$, with $2\varphi \in I^{k+2}$. By (3), we get $\varphi \in I^{k+1}$, completing the induction step. $\square$

If G is a formally real group, just as in Chapter 8, write $G \xrightarrow{\pi} G_{red}$ for the canonical SG-quotient map.

LEMMA 9.4. *If a formally real special group, G, verifies [WMC], then G_{red} verifies [WMC].*

PROOF. Since π is a SG-morphism, $\varphi \in I^n(G)$ implies $\pi \star \varphi \in I^n(G_{red})$. Since it is surjective, every form ψ over G_{red} <u>lifts</u> to G, i.e., there is a form φ over G such that $\pi \star \varphi = \psi$.

Assume that $2\psi \in I^{n+1}(G_{red})$, ψ a form over G_{red}. Thus, there are integers p, q, ≥ 0 and a linear combination, θ, of Pfister forms of degree $n + 1$ over G_{red}, such that

2. THE INDUCTIVE LIMIT OF GRADED RINGS OF EXPONENT TWO

$$2\psi \oplus p\langle 1, -1 \rangle \equiv_{G_{red}} \theta \oplus q\langle 1, -1 \rangle. \tag{I}$$

Thus, we may consider a lifting φ of ψ to G, and a linear combination τ of Pfister forms of degree $n+1$ over G that is a lifting of θ. By Proposition 2.21, (I) implies that there is $k \geq 0$ such that

$$2^k(2\varphi \oplus p\langle 1, -1 \rangle) \equiv_G 2^k(\tau \oplus q\langle 1, -1 \rangle),$$

and so $2^{k+1}\varphi \in I^{k+n+1}(G)$. Since G verifies $[WMC]$, we conclude that $\varphi \in I^n(G)$, which in turn implies $\pi \star \varphi = \psi \in I^n(G_{red})$, as desired. □

DEFINITION 9.5. A special group G is said to be $\mathcal{AP}$ if it is formally real and for all integers $k \geq 1$, $2^k \notin I^{k+1}$. ◇

LEMMA 9.6. *If G is reduced, or is formally real and verifies $[WMC]$, or is the special group of a formally real field, then G is $\mathcal{AP}$.*

PROOF. If $2^k \in I^{k+1}$ and G is the special group of a formally real field or a reduced special group, we apply the classical version of the Arason-Pfister Hauptsatz (see [AP] or Theorem X.3.1 in [L1]), or its reduced special group version (Theorem 7.31), respectively, to conclude that 2^k is hyperbolic. But then -1 is represented by 2^k, a contradiction.

If G is formally real and satisfies $[WMC]$, $2^k \in I^{k+1}$ implies $\langle 1 \rangle \in I$, a contradiction. □

2. The inductive limit of graded rings of exponent two

Recall that a **graded ring** is a sequence of abelian groups

$$\mathcal{H} = (H_1, H_2, \ldots, H_n, \ldots),$$

with an associative operation

$$H_n \times H_m \longrightarrow H_{n+m}, \quad (x, y) \mapsto x * y, \qquad n, m \geq 1$$

such that for all $x, x' \in H_n$ and $y \in H_m$,

$$(x + x') * y = (x * y) + (x' * y) \quad \text{and} \quad y * (x + x') = (y * x) + (y * x').$$

$\mathcal{H}$ is **commutative** if $*$ is commutative. The group H_n is called the group of degree n of $\mathcal{H}$.

DEFINITION 9.7. A sequence $\mathcal{H} = (H_1 \xrightarrow{h_1} H_2 \xrightarrow{h_2} \ldots H_n \xrightarrow{h_n} H_{n+1} \ldots)$ is an **inductive graded ring of exponent two** (IGR) if it satisfies, for all $n, m \geq 1$

1) H_n is a group of exponent 2, with a distinguished element $\top_n$.

2) h_n is a group homomorphism, such that $h_n(\top_n) = \top_{n+1}$.

3) $\mathcal{H}$ is a commutative graded ring.

4) For $1 \leq s \leq t$, define

$$h_s^t = \begin{cases} Id_{H_s} & \text{if } s = t \\ h_{t-1} \circ \ldots \circ h_{s+1} \circ h_s & \text{if } s < t. \end{cases}$$

Then, if $p \geq n$ and $q \geq m$, for all $x \in H_n$ and $y \in H_m$,

$$h_n^p(x) * h_m^q(y) = h_{n+m}^{p+q}(x * y). \qquad \diamond$$

If $\mathcal{H}$ is an IGR, let $\varinjlim \mathcal{H} = \langle H; \{\gamma_n : n \geq 1\}\rangle$ be its inductive limit. For ease of reference, we register the (well-known) basic properties of this construction in

REMARK 9.8. For each $n \geq 1$, we have a group homomorphism $\gamma_n : H_n \longrightarrow H$, satisfying :

1) For all $n \geq 1$, the following diagram is commutative :

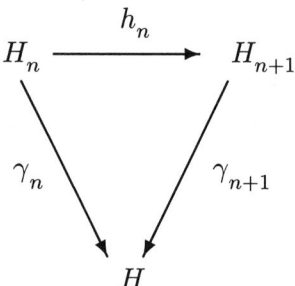

It follows readily that $\gamma_n = \gamma_m \circ h_n^m$ for all $1 \leq n \leq m$.

2) $H = \bigcup_{n \geq 1} \gamma_n(H_n)$.

If $w = \gamma_n(x)$, we say that $x \in H_n$ is a **representative** of $w \in H$.

3) For all $n, m \geq 1$ and all $x \in H_n$, $y \in H_m$, the following conditions are equivalent :

(i) $\gamma_n(x) = \gamma_m(y)$; \qquad (ii) There is $k \geq n, m$ such that $h_n^k(x) = h_m^k(y)$.

Note that the h_n's are injective iff the γ_n's are injective.

Write $\perp$ for the zero of H; it has $0 \in H_n$ as representative. $\qquad \diamond$

By Remark 9.8, addition in H can be described by representatives as follows :

For $w, z \in H$, let $x \in H_n$ and $y \in H_m$ be representatives of w and z, respectively, and let k be an integer greater than n and m. Then,

[sum] $\qquad w + z = \gamma_k(h_n^k(x) + h_m^k(y)) = \gamma_k(h_n^k(x)) + \gamma_k(h_m^k(y)).$

In particular, k may taken to be $\max\{n, m\}$. As an inductive limit of groups of exponent 2, H is an group of exponent 2. We now prove

THEOREM 9.9. *Suppose that $\mathcal{H}$ is an IGR such that*

[br 0] *For all $n \geq 1$, $\top_n \notin \ker h_n$.*

[br 1] *For $n \geq 1$ and $x \in H_n$, there is $k \geq n+1$ such that $h_{n+1}^k(\top_1 * x) = h_n^k(x)$.*

[br 2] *For $n \geq 1$ and $x \in H_n$, there is $k \geq 2n$ such that $h_n^k(x) = h_{2n}^k(x * x)$.*

Then, $H = \varinjlim \mathcal{H}$ is a non-trivial Boolean ring, that is, $\bot \neq \top =_{def} \gamma_1(\top_1)$ (its multiplicative identity) and for all $w \in H$,

$$w + w = \bot \quad \text{and} \quad w \cdot w = w.$$

PROOF. We start by verifying that H is a ring with identity $\top = \gamma_1(\top_1)$.

The graded ring structure in $\mathcal{H}$ induces a product in H, as follows :
For $w, z \in H$, let $x \in H_n$, $y \in H_m$ be representatives of w, z, respectively. Now set

[prd] $$w \cdot z = \gamma_{n+m}(x * y).$$

To show that this is well defined, let $x' \in H_p$ and $y' \in H_q$ be representatives of w, z, respectively. By Remark 9.8.(3), there is $k \geq n, m, p, q$ such that
$$h_n^k(x) = h_p^k(x') \quad \text{and} \quad h_m^k(y) = h_q^k(y').$$
By condition 4 in Definition 9.7, we have
$$h_{n+m}^{2k}(x * y) = h_n^k(x) * h_m^k(y) = h_p^k(x') * h_q^k(y') = h_{p+q}^{2k}(x' * y'),$$
and so Remark 9.8.(3) yields $\gamma_{n+m}(x * y) = \gamma_{p+q}(x' * y')$, proving that formula [prd] is independent of representatives. Since $\mathcal{H}$ is commutative, the same is true of the product in H.

A similar reasoning will show that H is a ring, that is, its product is associative and distributes over sum, with $\bot$ being its zero. [br 1] and Remark 9.8.(3) show that $\top = \gamma_1(\top_1)$ is the multiplicative identity in H, while [br 0] guarantees $\bot \neq \top$. Note that by condition 2 in Definition 9.7, the $\top_n$'s are representatives of $\top$.

Since the additive group of H is of exponent two, we have $w + w = \bot$, for all $w \in H$. That the product in H is idempotent follows from Remark 9.8.(3) and [br 2]. □

We register the following consequence of the proof of Theorem 9.9 :

COROLLARY 9.10. *For integers $n, m \geq 1$ and $x \in H_n$, $y \in H_m$, in $H = \varinjlim \mathcal{H}$ we have,*
$$\gamma_n(x) \cdot \gamma_m(y) = \gamma_{n+m}(x * y).$$
◇

We shall apply Theorem 9.9 to two situations, namely the mod 2 K-theory of a field of characteristic $\neq 2$, and the graded Witt ring of a special group.

We shall assume that the reader is familiar with the contents of [Mi], but the basic results we need are contained in section 1 of Chapter 8.

If F is a field of characteristic $\neq 2$, let

$$k_*F = (k_1 F, K_2 F, \ldots, k_n F, \ldots),$$

be the graded algebra over $\mathbb{F}_2$, corresponding to the mod 2 K-theory of F as in section 8.1. It follows directly from the definition of the $k_n F$ that they are groups of exponent 2. Notation for $k_* F$ will be as set down in section 8.1.

Note that Lemma 8.3.(a) assures that $l(a)$ depends only on the square class of a in F. Thus, we may write $l(c)$, for $c \in G(F)$, or $l(\bar{a})$, for $a \in \dot{F}$.

It follows from Lemma 8.3.(b) that $k_* F$ is a commutative graded ring. For each $n \geq 1$, there is a group homomorphism

$$\omega_n : k_n F \longrightarrow k_{n+1} F,$$

defined on generators as multiplication by $l(-1)$. For each $n \geq 1$, we set $\top_n = l(-1)^n$ as the distinguished element of $k_n F$. Thus, it is clear that the system

$$\mathcal{K}(F) = (k_1 F \xrightarrow{\omega_1} k_2 F \xrightarrow{\omega_2} \ldots k_n F \xrightarrow{\omega_n} k_{n+1} F \ldots),$$

satisfies conditions (1) – (3) in Definition 9.7. Note that for $1 \leq n \leq m$, ω_n^m is multiplication by $l(-1)^{m-n}$; since graded multiplication is commutative, condition (4) in Definition 9.7 is also satisfied and so we have

LEMMA 9.11. *If F is a field of characteristic $\neq 2$, then $\mathcal{K}(F)$ is an IGR.* ◇

If G is a special group, consider the sequence

$$\mathcal{W}_g(G) = (I/I^2, I^2/I^3, \ldots I^n/I^{n+1}, \ldots),$$

where I^n is the nth power of the fundamental ideal I of $W(G)$, defined in 1.26. Note that if $\varphi \in I^n$, then $\varphi \oplus \varphi = 2 \otimes \varphi \in I^{n+1}$, showing that I^n/I^{n+1} is a group of exponent 2 under addition. To ease exposition, we set $\overline{I^n} = I^n/I^{n+1}$ and write x/n, with $x \in I^n$, for an element of $\overline{I^n}$.

$\mathcal{W}_g(G)$ inherits a graded multiplication from the tensor product of forms. For $x/n \in \overline{I^n}$ and $y/m \in \overline{I^m}$, define

$$x/n * y/m = (x \otimes y)/n + m.$$

To show that this is well defined, suppose that $x - x' \in I^{n+1}$ and $y - y' \in I^{m+1}$. Then,

2. THE INDUCTIVE LIMIT OF GRADED RINGS OF EXPONENT TWO

$$(x \otimes y) - (x' \otimes y') \approx_G [(x \otimes y) - (x \otimes y')] \oplus [(x \otimes y') - (x' \otimes y')]$$
$$= [x \otimes (y - y')] \oplus [y' \otimes (x - x')] \in I^{n+m+1},$$

as needed. It follows directly from the analogous properties of sum and product of forms that $\mathcal{W}_g(G)$ is a commutative graded ring. For $n \geq 1$, we have group homomorphisms

$$t_n : \overline{I^n} \longrightarrow \overline{I^{n+1}},$$

given by $t_n(x/n) = (2x)/n+1 = (\langle 1, 1 \rangle \otimes x)/n+1$. Note that for $1 \leq n \leq m$, t_n^m is multiplication by 2^{m-n}. Thus, setting $\mathsf{T}_n = 2^n/n \in \overline{I^n}$, it is straightforward to see that we have:

LEMMA 9.12. *If G is a special group, then $\mathcal{W}_g(G)$ is an IGR.* ◇

Write
$$\begin{cases} \langle \mathcal{W}(G); \{\alpha_n : n \geq 1\} \rangle = \varinjlim \mathcal{W}_g(G) \\ \text{and} \\ \langle k(F); \{\beta_n : n \geq 1\} \rangle = \varinjlim \mathcal{K}(F). \end{cases}$$

We now state

THEOREM 9.13. *If F is a formally real field and G is an $\mathcal{AP}$ group, then*

a) $k(F)$ and $\mathcal{W}(G)$ are non-trivial Boolean rings.

b) The maps $\rho : G \longrightarrow \mathcal{W}(G)$ and $\kappa : G(F) \longrightarrow k(F)$, defined for $a \in G$ and $b \in G(F)$ by

$$\rho(a) = \alpha_1(\langle 1, -a \rangle / 1) \quad \text{and} \quad \kappa(b) = \beta_1(l(b)),$$

are morphisms of special groups, where $k(F)$ and $\mathcal{W}(G)$ have the natural special group structure of a Boolean ring.

c) κ and ρ are injective iff F is Pythagorean and G is reduced, respectively.

In preparation to the proof of 9.13 we show

PROPOSITION 9.14. *Let F be a field and G a special group. For an integer $n \geq 1$, let $\eta \in k_n F$ and $x \in I^n$. Then,*

a) $2^n x - (x \cdot x) \in I^{2n+1}$.

b) In $k_{2n} F$, $l(-1)^n \eta = \eta \cdot \eta$.

PROOF. a) Fix $x \in I^n$; x is the Witt-equivalence class of a linear combination of Pfister forms of degree n, say $\psi = \bigoplus_{i=1}^k a_i \varphi_i$, with $a_i \in G$, for $1 \leq i \leq k$. Thus, (a) amounts to proving that

$$2^n \psi - (\psi \otimes \psi) \text{ is in } I^{2n+1}. \tag{I}$$

Now define $\tau = \bigoplus_{i=1}^k \varphi_i$ and $\rho = \psi - \tau$.

Fact. If assertion (I) holds for τ, then it holds for ψ.

Proof. Note that $\rho = \bigoplus_{i=1}^{k} \langle a_i, -1 \rangle \varphi_i \in I^{n+1}$ and that $\psi \approx \tau \oplus \rho$. Thus,

$$2^n \psi - (\psi \otimes \psi) \approx (2^n \tau \oplus 2^n \rho) - ((\tau \oplus \rho) \otimes (\tau \oplus \rho))$$
$$= (2^n \tau - (\tau \otimes \tau)) \oplus 2^n \rho - 2(\tau \otimes \rho) - (\rho \otimes \rho).$$

Since $2^n \rho \in I^{2n+1}$, while $(\rho \otimes \rho), 2(\tau \otimes \rho) \in I^{2n+2}$, we conclude that ψ also satisfies (I). ◇

The Fact tells us that it is enough to verify (I) for sums $\psi = \bigoplus_{i=1}^{k} \varphi_i$ of Pfister forms of degree n. The proof will proceed by induction on the integer $k \geq 1$.

$\underline{k = 1}$. Then, $\psi = \bigotimes_{j=1}^{n} \langle 1, c_i \rangle$, with $c_i \in G$.

Now, we proceed by induction on $n \geq 1$. For $n = 1$, we have

$$2\langle 1, c \rangle - (\langle 1, c \rangle \otimes \langle 1, c \rangle) = \langle 1, c \rangle \oplus \langle 1, c \rangle - \langle 1, c \rangle - \langle 1, c \rangle, \quad \text{(II)}$$

showing that in this case $2\psi - (\psi \otimes \psi)$ is in fact hyperbolic. Now suppose the conclusion holds for n and let $\psi = \varphi \otimes \langle 1, c \rangle$, with φ a Pfister form of degree n. Then, using (II), we get

$$2^{n+1} \psi - (\psi \otimes \psi) = 2^{n+1} (\varphi \otimes \langle 1, c \rangle) - (\varphi \otimes \langle 1, c \rangle \otimes \varphi \otimes \langle 1, c \rangle)$$
$$= (2^n \varphi \otimes 2 \langle 1, c \rangle) - (\varphi \otimes \varphi \otimes \langle 1, c \rangle \otimes \langle 1, c \rangle)$$
$$= (2^n \varphi \otimes 2 \langle 1, c \rangle) - (\varphi \otimes \varphi \otimes 2 \langle 1, c \rangle)$$
$$= 2 \langle 1, c \rangle (2^n \varphi - (\varphi \otimes \varphi)) \in I^{2n+3}.$$

<u>Induction step.</u> Let $\psi = \varphi \oplus \tau$, where φ is a Pfister form of degree n and τ is the sum of less than k Pfister forms of degree n. Then,

$$2^n \psi - (\psi \otimes \psi) = (2^n \varphi \oplus 2^n \tau) - ((\varphi \oplus \tau) \otimes (\varphi \oplus \tau))$$
$$= (2^n \varphi - (\varphi \otimes \varphi)) \oplus (2^n \tau - (\tau \otimes \tau)) - 2(\varphi \otimes \tau).$$

Since $2(\varphi \otimes \tau) \in I^{2n+1}$, the induction hypothesis and the case $k = 1$ yield (I) for ψ, proving (a).

b) We first prove that (b) holds on generators $\eta = l(a_1) \ldots l(a_n)$ of $k_n F$. Since $k_n F$ is commutative, Proposition 8.2.(c) yields

$$l(-1)^n l(a_1) \ldots l(a_n) = [l(-1)l(a_1)][l(-1)l(a_2)] \ldots [l(-1)l(a_n)]$$
$$= l(a_1)^2 l(a_2)^2 \ldots l(a_n)^2 = \eta \cdot \eta.$$

If we assume that (b) holds for sums of $k \geq 1$ generators, let $\eta = \zeta + \xi$ where ζ is a generator of $k_n F$ and ξ is a sum of at most k generators. Then, the induction hypothesis and the fact that $k_n F$ is an additive group of exponent 2 yields

$$l(-1)^n \eta = l(-1)^n [\zeta + \xi] = l(-1)^n \zeta + l(-1)^n \xi$$
$$= (\zeta \cdot \zeta) + (\xi \cdot \xi) = (\zeta + \xi) \cdot (\zeta + \xi). \qquad \square$$

2. THE INDUCTIVE LIMIT OF GRADED RINGS OF EXPONENT TWO

Proof of (a) in Theorem 9.13 : We shall verify conditions $[br\ i]$ in Theorem 9.9, $0 \leq i \leq 2$.

For each $n \geq 1$, $\top_n = \begin{cases} l(-1)^n & \text{in } \mathcal{K}(F) \\ 2^n & \text{in } \mathcal{W}_g(G). \end{cases}$

Thus, $[br\ 0]$ is equivalent to $\begin{cases} l(-1)^{n+1} \neq 0 & \text{for } \mathcal{K}(F) \\ 2^{n+1} \notin I^{n+2} & \text{for } \mathcal{W}_g(G). \end{cases}$

If $l(-1)^{n+1} = 0$ in $k_{n+1}F$, then, by Theorem 8.7, we would have

$$\varepsilon_{n+1}(l(-1)^{n+1}) = \bigwedge_{i=1}^{n+1} \varepsilon_F(\pi(-1)) = \varepsilon_F(\pi(-1)) = \bot \text{ in } B(F).$$

Since ε_F is injective, we conclude that $\pi(-1) = 1$; but then $-1 \in \Sigma \dot{F}^2$ and F is not formally real. Since G is an $\mathcal{AP}$ group, it is clear that $2^{n+1} \notin I^{n+2}$. This proves $[br\ 0]$ for $\mathcal{K}(F)$ and $\mathcal{W}_g(G)$.

We register that Theorem 3.2 in [EL1] gives another proof of $[br\ 0]$ for the ring $\mathcal{K}(F)$.

For $[br\ 1]$, just notice that if $x/n \in I^n$ and $\eta \in k_n F$, then

$$t_{n+1}^{n+2}(\langle 1,1 \rangle/1 * x/n) = t_n^{n+2}(x/n) \quad \text{and} \quad \omega_{n+1}^{n+2}(l(-1) * \eta) = \omega_n^{n+2}(\eta).$$

The last condition to be verified, $[br\ 2]$, is a direct consequence of 9.14. This proves that $k(F)$ and $\mathcal{W}(G)$ are non-trivial Boolean rings. $\square$

Proof of (b) in Theorem 9.13.

I. The case of $\mathcal{W}_g(G)$. The proof will be done in two steps :

1. ρ is a group homomorphism, with $\rho(-1) = \top$. For $a, b \in G$, we have

$$(\langle 1, -a \rangle \oplus \langle 1, -b \rangle) - \langle 1, -ab \rangle = \langle 1, -a, -b, ab \rangle \oplus \langle 1, -1 \rangle =$$
$$= (\langle 1, -a \rangle \otimes \langle 1, -b \rangle) \oplus \langle 1, -1 \rangle,$$

proving that $\overline{\langle 1, -a \rangle}/1 + \overline{\langle 1, -b \rangle}/1 = \overline{\langle 1, -ab \rangle}/1$ in $\overline{I}^1$. Since α_1 is a group homomorphism, this shows that $\rho(ab) = \rho(a) + \rho(b)$. Finally,

$$\rho(1) = \alpha_1(\overline{\langle 1, -1 \rangle}/1) = \bot \quad \text{and} \quad \rho(-1) = \alpha_1(\overline{\langle 1, 1 \rangle}/1) = \top.$$

2. ρ is a special group morphism. For $a, b, c, d \in G$, let $u = \rho(a)$, $v = \rho(b)$, $w = \rho(c)$ and $z = \rho(d)$. It must be shown that

$$\langle a, b \rangle \equiv_G \langle c, d \rangle \quad \text{implies} \quad \langle u, v \rangle \equiv_{\mathcal{W}(G)} \langle w, z \rangle. \tag{I}$$

According to the definition of isometry in Boolean algebras (or rings), (I) is equivalent to

$$\langle a, b \rangle \equiv_G \langle c, d \rangle \quad \text{implies} \quad u + v = w + z \quad \text{and} \quad u \cdot v = w \cdot z. \tag{II}$$

It follows from [SG3] (the discriminant axiom) in 1.2, that $ab = cd$. Since ρ is a group homomorphism, this equation implies $u + v = w + z$. By the

definition of product in $\mathcal{W}(G)$, to prove $u \cdot v = w \cdot z$ it is enough to verify that
$$(\langle 1, -a \rangle \otimes \langle 1, -b \rangle) - (\langle 1, -c \rangle \otimes \langle 1, -d \rangle) \in I^3. \tag{III}$$
Indeed, if (III) holds, then $(\overline{\langle 1, -a \rangle} \cdot \overline{\langle 1, -b \rangle})/2 = (\overline{\langle 1, -c \rangle} \cdot \overline{\langle 1, -d \rangle})/2$, and so the value of α_2 at this point will give $u \cdot v = w \cdot z$.

To prove (III), compute as follows, recalling that $\langle a, b \rangle \equiv_G \langle c, d \rangle$, with $ab = cd = t$:
$$(\langle 1, -a \rangle \otimes \langle 1, -b \rangle) - (\langle 1, -c \rangle \otimes \langle 1, -d \rangle) =$$
$$= \langle 1, -a, -b, ab \rangle - \langle 1, -c, -d, cd \rangle =$$
$$= \langle 1, -1 \rangle \oplus (\langle c, d \rangle - \langle a, b \rangle) \oplus \langle t, -t \rangle,$$
which is hyperbolic.

II. The case of $\mathcal{K}(F)$. Since l takes product in $G(F)$ to sum in $k_1 F$ and β_1 is a homomorphism, it is clear that κ is a group homomorphism from $G(F)$ to $k(F)$. For the preservation of -1, just notice that $l(-1)$ in $k_1 F$ is a representative of $\top \in k(F)$. It remains to prove that κ is a morphism of special groups. Just as above, this reduces to check that for a, b, c, d in $\dot{F}$,
$$\langle \overline{a}, \overline{b} \rangle \equiv_{G(F)} \langle \overline{c}, \overline{d} \rangle \;\Rightarrow\; \begin{cases} \kappa(a) + \kappa(b) = \kappa(c) + \kappa(d) \\ \text{and} \\ \kappa(a) \cdot \kappa(b) = \kappa(c) \cdot \kappa(d). \end{cases} \tag{I}$$

Since $G(F)$ is a special group, Lemma 1.5.(a) yields
$$\langle \overline{a}, \overline{b} \rangle \equiv_{G(F)} \langle \overline{c}, \overline{d} \rangle \quad \text{iff} \quad \overline{ab} = \overline{cd} \;\text{and}\; \overline{ac} \in D_{G(F)}(1, \overline{cd}).$$

By the definition of isometry in $G(F)$ (1.28), we may restate this equivalence as
$$\langle \overline{a}, \overline{b} \rangle \equiv_{G(F)} \langle \overline{c}, \overline{d} \rangle \quad \text{iff} \quad \begin{array}{l} \exists\, x \in \dot{F} \text{ and } y, z \in F \text{ such that} \\ ab = cdx^2 \text{ and } ac = y^2 + cdz^2. \end{array} \tag{II}$$

It follows from the first equation in (II) and Proposition 8.3.(b) that
$$l(a) + l(b) = l(ab) = l(cdx^2) = l(cd) = l(c) + l(d). \tag{III}$$
Since β_1 is a homomorphism, the first equation in (I) is verified.

To complete the proof of (b), note that either y or z in the last equation in (II) must be in $\dot{F}$. The case of $z \neq 0$ is left to the reader, and we treat the case $y \neq 0$. It follows from (II) that $ac\,(1/y)^2 + -cd\,(z/y)^2 = 1$. Thus, 8.3.(a) and 8.2.(d) yield $l(ac)l(-cd) = 0$ in $k_2 F$. Now, (III) and 8.2.(c) yield,
$$0 = l(ac)l(-cd) = [l(a) + l(c)][l(-1) + l(c) + l(d)]$$
$$= l(a)l(-1) + l(a)[l(c) + l(d)] + l(c)l(-1) + l(c)[l(c) + l(d)]$$
$$= l(a)l(-1) + l(a)[l(a) + l(b)] + l(c)l(-1) + l(c)^2 + l(c)l(d)$$

2. THE INDUCTIVE LIMIT OF GRADED RINGS OF EXPONENT TWO

$$= l(a)l(-1) + l(a)^2 + l(a)l(b) + l(c)l(d) = l(a)l(b) + l(c)l(d).$$

Since $\kappa(a) \cdot \kappa(b) = \beta_2(l(a)l(b))$, it is clear that $\kappa(a)\kappa(b) = \kappa(c)\kappa(d)$, as desired. □

Proof of (c) in Theorem 9.13. Since any subgroup of a reduced special group must be reduced, it is enough to show the "only if" part of the equivalence in (c). So assume that F is Pythagorean and G is a reduced special group.

Note that, because F is Pythagorean, the morphism $G(F) \xrightarrow{\pi} G_{red}(F)$ is the identity.

Suppose that $a, b \in \dot{F}$ are such that $\kappa(a) = \kappa(b)$ in $k(F)$. By Remark 9.8.(3), there is $n \geq 1$ such that $l(-1)^n l(a) = l(-1)^n l(b)$ in $k_{n+1}F$. By Theorem 8.7, $\varepsilon_{n+1}(l(-1)^n l(a)) = \varepsilon_F(\overline{a})$; thus, we have $\varepsilon_F(\overline{a}) = \varepsilon_F(\overline{b})$. Since ε_F is injective (5.4), we conclude that κ must be injective. It should be mentioned that Theorem 3.2 in [EL1] can also be used to prove that κ is injective.

Now assume that for $a, b \in G$, $\rho(a) = \rho(b)$. Just as above, there is an integer $k \geq 1$ such that $2^{k-1}\langle 1, -a \rangle - 2^{k-1}\langle 1, -b \rangle \in I^{k+1}$. Note that

$$2^{k-1}\langle 1, -a \rangle - 2^{k-1}\langle 1, -b \rangle = 2^{k-1}\langle 1, -1 \rangle \oplus 2^{k-1}\langle -a, b \rangle,$$

and so $2^{k-1}\langle -a, b \rangle \in I^{k+1}$. It follows from the reduced special group version of the Arason-Pfister Hauptsatz (Theorem 7.31) that $2^{k-1}\langle -a, b \rangle$ must be hyperbolic. By Proposition 1.6.(e), $\langle -a, b \rangle$ must be hyperbolic, that is, $\langle -a, b \rangle \equiv_G \langle 1, -1 \rangle$, and the discriminant axiom [SG 3] yields $a = b$, completing the proof the proof of Theorem 9.13. □

As a direct consequence of Theorem 4.17.(4) and Theorem 9.13 we get the important

COROLLARY 9.15. *Let G be a reduced special group and let F be a Pythagorean field. Then, the SG-embeddings $\rho : G \longrightarrow W(G)$ and $\kappa : G(F) \longrightarrow k(F)$ of 9.13 have unique extensions to Boolean algebra homomorphisms $B(\rho) : B_G \longrightarrow W(G)$ and $B(\kappa) : B(F) \longrightarrow k(F)$, such that the following diagrams are commutative :*

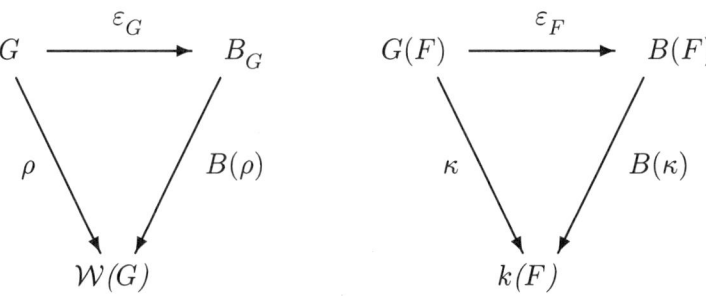

Moreover,

1) If $\{a_{ij} : 1 \leq i \leq k,\ 1 \leq j \leq n\} \subseteq G$ and $a = \triangle_{i=1}^{k} \bigwedge_{j=1}^{n} \varepsilon_G(a_{ij})$, then

$$B(\rho)(a) = \alpha_n(\overline{\varphi}/n), \qquad [SG]$$

where $\varphi = \bigoplus_{i=1}^{k} \bigotimes_{j=1}^{n} \langle 1, -a_{ij} \rangle \in I^n$.

2) If $\{a_{ij} : 1 \leq i \leq k,\ 1 \leq j \leq n\} \subseteq G(F)$ and $a = \triangle_{i=1}^{k} \bigwedge_{j=1}^{n} \varepsilon_F(a_{ij})$, then

$$B(\kappa)(a) = \beta_n(\eta), \qquad [KT]$$

where $\eta = \sum_{i=1}^{k} l(a_{i1})l(a_{i2}) \ldots l(a_{in}) \in k_n F$.

PROOF. Only formulas [SG] and [KT] remain to be checked. $B(\rho)$ and $B(\kappa)$ take symmetric difference to sum and meet to product in $\mathcal{W}(G)$ and $k(F)$, respectively.

To verify [SG], set $\varphi_i = \bigotimes_{j=1}^{n} \langle 1, -a_{ij} \rangle$; then Corollary 9.10 yields

$$B(\rho)(a) = \triangle_{i=1}^{k} \prod_{i=1}^{n} B(\rho)(\varepsilon_G(a_{ij})) = \sum_{i=1}^{k} \prod_{i=1}^{n} \rho(a_{ij})$$
$$= \sum_{i=1}^{k} \prod_{i=1}^{n} \alpha_1(\overline{\langle 1, -a_{ij}\rangle}/1) = \sum_{i=1}^{k} \alpha_n(\overline{\varphi_i}/n) = \alpha_n(\overline{\varphi}/n).$$

For [KT], write $\eta_i = l(a_{i1})l(a_{i2}) \ldots l(a_{in})$; by Corollary 9.10,

$$B(\kappa)(a) = \triangle_{i=1}^{k} \prod_{i=1}^{n} B(\kappa)(\varepsilon_F(a_{ij})) = \sum_{i=1}^{k} \prod_{i=1}^{n} \kappa(a_{ij})$$
$$= \sum_{i=1}^{k} \prod_{i=1}^{n} \beta_1(l(a_{ij})) = \sum_{i=1}^{k} \beta_n(\eta_i) = \beta_n(\eta). \qquad \square$$

3. The isomorphisms $k(F) \approx B(F)$ and $\mathcal{W}(G) \approx B_G$

In this section we prove the isomorphisms in title for Pythagorean fields and reduced special groups. F stands for a Pythagorean field and G for a reduced special group.

We first treat the case of a Pythagorean field F. One should keep in mind that $G(F) = G_{red}(F)$, that is, the canonical projection π is the identity.

Since $k(F) = \underrightarrow{lim}\, \mathcal{K}(F)$, Theorem 8.7.(b) implies that the sequence of homomorphisms $(\varepsilon_n)_{n \geq 1}$ induces a homomorphism $\varepsilon : k(F) \longrightarrow B(F)$, that can be described as follows :

For $z \in k(F)$, let $\eta = \sum_{i=1}^{k} l(a_{i1}) l(a_{i2}) \ldots l(a_{in}) \in k_n F$ be a representative of z; then,

$$\varepsilon(z) = \varepsilon_n(\eta) = \triangle_{i=1}^{k} \bigwedge_{j=1}^{n} \varepsilon_F(a_{ij}).$$

The map ε also makes the following diagram commutative, for all $n \geq 1$:

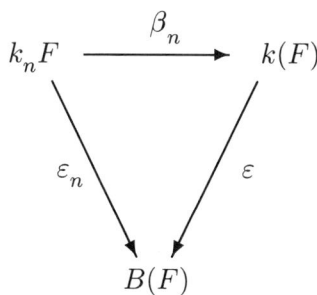

THEOREM 9.16. *Let F be a Pythagorean field. Then, the map $\varepsilon : k(F) \longrightarrow B(F)$ is a Boolean algebra isomorphism, whose inverse is the map $B(\kappa)$ of Corollary 9.15.*

PROOF. We shall show : $\begin{cases} \text{(I) } \varepsilon \circ B(\kappa) = Id_{B(F)} \\ \quad \text{and} \\ \text{(II) } B(\kappa) \circ \varepsilon = Id_{k(F)}. \end{cases}$

Proof of (I). Since

— $B(\kappa)$ and ε are Boolean algebra homomorphisms;

— $B(\kappa)$ extends κ (Corollary 9.15); and

— $G(F)$ generates $B(F)$ as a Boolean algebra (Proposition 4.10),

it is enough to show that for all $a \in G(F)$, $\varepsilon(\kappa(a)) = \varepsilon_F(a)$. But we have, by the commutative diagram above and Theorem 8.7,

$$\varepsilon(\kappa(a)) = \varepsilon(\beta_1(l(a))) = \varepsilon_1(l(a)) = \varepsilon_F(a).$$

Proof of (II). For $z \in k(F)$, let $\eta \in k_n F$ be a representative of z, that is, $z = \beta_n(\eta)$. Write $\eta = \sum_{i=1}^{k} l(a_{i1})l(a_{i2})\ldots l(a_{in})$. To prove (II), it is enough to show that $B(\kappa)(\varepsilon(z)) = \beta_n(\eta)$. But this follows directly from formula [KT] in Corollary 9.15. □

We now turn to the case of a reduced special group G. Recalling item (iii) of Fact 9.2, we define, for $x \in I^n$,

$$\mu_n(x) = \{\sigma \in X_G : sgn_\sigma(x) \text{ is not congruent to } 0 \bmod 2^n\}.$$

LEMMA 9.17. *With notation as above,*

a) *For all $n \geq 1$ and $x \in I^n$, $\mu_n(x) \in B_G$.*

b) *For all integers $1 \leq n, m \leq k$ and all $x \in I^n$, $y \in I^m$*

$$2^{k-n}x - 2^{k-m}y \in I^{k+1} \quad \text{implies} \quad \mu_n(x) = \mu_m(y).$$

c) *Let φ be a Pfister form of degree $n \geq 1$. For $a \in G$ and $x = \overline{a\varphi} \in I^n$,*

$$\mu_n(x) = -\mathcal{HT}_1(\varphi).$$

PROOF. a) Let $x = \overline{\psi}$, with $\psi = \bigoplus_{i=1}^{p} a_i \varphi_i$, where $a_i \in G$ and $\varphi_i = \bigotimes_{j=1}^{n} \langle 1, b_{ij} \rangle$ are Pfister forms of degree n. Suppose $\sigma \in \mu_n(x)$, that is, $sgn_\sigma(\psi) \not\equiv 0 \bmod 2^{n+1}$. Consider the clopen

$$U = \bigcap \{[b_{ij} = sgn_\sigma(b_{ij})] : 1 \leq i \leq p, 1 \leq j \leq n\} \cap$$
$$\cap \bigcap \{[a_i = sgn_\sigma(a_i)] : 1 \leq i \leq p\}.$$

It is clear that $\sigma \in U$, and that for all $\sigma \in U$ we have $sgn_\tau(\psi) = sgn_\sigma(\psi)$. But this means that U is a clopen neighbourhood of σ contained in $\mu_n(x)$. The same reasoning will show that the complement of $\mu_n(x)$ is open.

b) It must be shown that for all $\sigma \in X_G$,

$$sgn_\sigma(x) \equiv 0 \bmod 2^{n+1} \text{ iff } sgn_\sigma(y) \equiv 0 \bmod 2^{m+1}.$$

Our assumption and Fact 9.2.(iii) imply that, for some integer $p \geq 0$,

$$2^{k-n} sgn_\sigma(x) - 2^{k-m} sgn_\sigma(y) = p \, 2^{k+1}. \tag{I}$$

Now suppose that $sgn_\sigma(x)$ is a multiple of 2^{n+1}, say $sgn_\sigma(x) = q \, 2^{n+1}$. Then, (I) gives

$$q \, 2^{k+1} - 2^{k-m} sgn_\sigma(y) = p \, 2^{k+1},$$

which clearly implies $sgn_\sigma(y) \equiv 0 \bmod 2^{m+1}$. A similar reasoning shows that $sgn_\sigma(y) \equiv 0 \bmod 2^{m+1}$ implies $sgn_\sigma(x) \equiv 0 \bmod 2^{n+1}$.

c) Since $\varphi - a\varphi \in I^{n+1}$, (b) yields $\mu_n(\overline{\varphi}) = \mu_n(x)$. From 9.1.(e) we get

$$\mu_n(\overline{\varphi}) = \{\sigma \in X_G : sgn_\sigma(\varphi) \not\equiv 0 \bmod 2^{n+1}\}$$
$$= \{\sigma \in X_G : sgn_\sigma(\varphi) = 2^n\} = -\mathcal{HT}_1(\varphi). \qquad \square$$

It follows from Lemma 9.17 that for each $n \geq 1$ we have a map

$$\mu_n : \overline{I^n} \longrightarrow B_G, \text{ defined by } \mu_n(x/n) = \mu_n(x).$$

We now define a map $M : \mathcal{W}(G) \longrightarrow B_G$ as follows :

For $w \in \mathcal{W}(G)$, let $x/n \in \overline{I^n}$ be a representative of w. Then

$$M(w) = \mu_n(x).$$

It follows from Remark 9.8.(3) and Lemma 9.17.(b) that the definition of M is independent of representatives. The main result of this section reads

THEOREM 9.18. *The map $M : \mathcal{W}(G) \longrightarrow B_G$ is an isomorphism of Boolean algebras, whose inverse is the homomorphism $B(\rho)$ of Corollary 9.15.*

As part of the proof of Theorem 9.18, we first establish :

PROPOSITION 9.19. *With notation as above, let $n \geq 1$ be an integer. Then,*

a) M is a Boolean algebra homomorphism.

b) For $x/n \in \overline{I^n}$, let $w = \alpha_n(x/n)$ and $x = \overline{\psi}$, where $\psi = \bigoplus_{i=1}^{p} a_i \varphi_i$ is a linear combination of Pfister forms of degree n. Then,
$$M(w) = \triangle_{i=1}^{p} - \mathcal{HT}_1(\varphi_i) = (-1)^p \triangle_{i=1}^{p} \mathcal{HT}_1(\varphi_i).$$

PROOF. a) By Lemma 9.17.(c), we have
$$\begin{cases} M(\top) = \mu_1(\langle 1,1 \rangle) = - \mathcal{HT}_1(\langle 1,1 \rangle) = - \bot = \top \\ \quad\text{and} \\ M(\bot) = \mu_1(\langle 1,-1 \rangle) = - \top = \bot. \end{cases}$$

Next, we show that for $w, z \in W(G)$,
$$M(w+z) = M(w) \triangle M(z) \quad \text{and} \quad M(w \cdot z) = M(w) \wedge M(z). \quad (I)$$

Let x/n, y/m be representatives of w and z, respectively. We may suppose that $m \geq n$. Then, $M(w+z) = \mu_m(2^{m-n}x + y)$.

In any Boolean algebra $-(a \triangle b) = (a \wedge b) \vee (-a \wedge -b)$. Thus, if $t = 2^{m-n}x + y$, the verification the first equation in (I) reduces to showing that, for all $\sigma \in X_G$,

$sgn_\sigma(t) \equiv 0 \bmod 2^{m+1} \Leftrightarrow \sigma \in (\mu_n(x) \cap \mu_m(y)) \cup (-\mu_n(x) \cap -\mu_m(y))$. (II)

For $\sigma \in X_G$, write $sgn_\sigma(x) = \alpha \, 2^n$ and $sgn_\sigma(y) = \beta \, 2^m$, for integers $\alpha, \beta \geq 0$ (9.2.(iii)). Then,

— If $sgn_\sigma(t) = p \, 2^{m+1}$, then
$$p \, 2^{m+1} = sgn_\sigma(t) = 2^{m-n} sgn_\sigma(x) + sgn_\sigma(y) = 2^{m-n}\alpha 2^n + \beta 2^m$$
$$= (\alpha + \beta) 2^m,$$
and so $\alpha + \beta = 2p$. Thus, α, β are both even or both odd. If α, β are even, then $\sigma \in -\mu_n(x) \cap -\mu_m(y)$. If they are both odd, then $\sigma \in \mu_n(x) \cap \mu_m(y)$, proving part ($\Rightarrow$) of the equivalence.

— It is clear that if $\sigma \in -\mu_n(x) \cap -\mu_m(y)$, then $sgn_\sigma(t) \equiv 0 \bmod 2^{m+1}$. If $\sigma \in \mu_n(x) \cap \mu_m(y)$, then the integers α, β are both odd, say $\alpha = 2s + 1$, $\beta = 2r + 1$. Then,
$$sgn_\sigma(t) = 2^{m-n}(2s+1)2^n + (2r+1)2^m = 2^{m+1}s + 2^m + 2^{m+1}r + 2^m$$
$$\equiv 0 \bmod 2^{m+1},$$
ending the proof of (II).

For the second equation in (I), we have $M(w \cdot z) = \mu_{m+n}(x \cdot y)$. Set $t = x \cdot y$; by the same argument as above, the conclusion is equivalent to verifying that for all $\sigma \in X_G$,
$$sgn_\sigma(t) \equiv 0 \bmod 2^{m+n+1} \quad \Leftrightarrow \quad \sigma \in -\mu_n(x) \cup -\mu_m(y). \quad (III)$$

Fix $\sigma \in X_G$; as above, $sgn_\sigma(x) = \alpha \, 2^n$ and $sgn_\sigma(y) = \beta \, 2^m$. Then,

— If $sgn_\sigma(t) = sgn_\sigma(x)sgn_\sigma(y) = p2^{m+n+1}$, let r, s be the largest integers ≥ 0 such that 2^r divides $sgn_\sigma(x)$ and 2^s divides $sgn_\sigma(y)$. We have $r + s \geq m + n + 1$; thus, either $r \geq n+1$ or $s \geq m+1$, showing that $\sigma \in -\mu_n(x) \cup -\mu_m(y)$.

— If $\sigma \in -\mu_n(x)$, then α must be even, say $\alpha = 2p$. Consequently,
$$sgn_\sigma(t) = sgn_\sigma(x)sgn_\sigma(y) = p\beta 2^{n+m+1}.$$
Similarly, if $\sigma \in -\mu_m(y)$, then $sgn_\sigma(t)$ is a multiple of 2^{n+m+1}.

b) Lemma 9.17.(c) gives $\mu_n(a_i\varphi_i) = -\mathcal{HT}_1(\varphi_i)$. For $1 \leq i \leq p$, set $x_i = \overline{a_i\varphi_i}$ and $w_i = \alpha_n(x_i/n)$. Then, $x = \sum_{i=1}^p x_i$ and $w = \sum_{i=1}^p w_i$. Since M is a homomorphism, we get
$$M(w) = \sum_{i=1}^p M(w_i) = \sum_{i=1}^p \mu_n(x_i) = \bigtriangleup_{i=1}^p -\mathcal{HT}_1(\varphi_i). \qquad \square$$

The method of proof of Proposition 9.19 will also give

PROPOSITION 9.20. *If G is a reduced special group, then, for all $n \geq 1$,*

a) μ_n is a group homomorphism from $\overline{I^n}$ to B_G, such that $\mu_n(2^{n-1}\langle 1,1 \rangle/n) = \top$.

b) The following diagram is commutative :

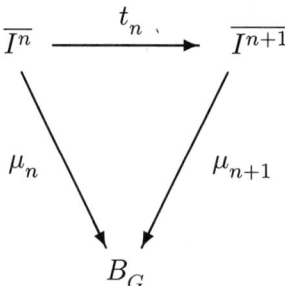

c) For all $x/n \in \overline{I^n}$ and $y/m \in \overline{I^m}$, $\mu_{n+m}(xy/(n+m)) = \mu_n(x) \wedge \mu_m(y)$.

d) If $x = \overline{\psi}$, where $\psi = \bigoplus_{i=1}^p a_i\varphi_i$ is a linear combination of Pfister forms of degree n, then
$$\mu_n(x) = \bigtriangleup_{i=1}^p -\mathcal{HT}_1(\varphi_i) = (-1)^p \bigtriangleup_{i=1}^p \mathcal{HT}_1(\varphi_i).$$

e) G satisfies $[WMC]$ iff for all $n \geq 1$, t_n is injective. $\diamond$

Proposition 9.20 can be used to give an alternative proof of Proposition 9.19.

Proof of Theorem 9.18. We prove that
$$\begin{cases} \text{(I) } M \circ B(\rho) = Id_{B_G} \\ \quad \text{and} \\ \text{(II) } B(\rho) \circ M = Id_{\mathcal{W}(G)}. \end{cases}$$

Proof of (I). Exactly as in the proof of (I) in Theorem 9.16, it suffices to show that for all $a \in G$, $M(\rho(a)) = \varepsilon_G(a)$. But Lemma 9.17.(c) yields

$$M(\rho(a)) = M(\overline{\langle 1, -a \rangle/1}) = \mu_1(\langle 1, -a \rangle) = -\mathcal{HT}_1(\langle 1, -a \rangle)$$
$$= -\varepsilon_G(-a) = \varepsilon_G(a),$$

as needed.

Proof of (II). For $w \in \mathcal{W}(G)$, let x/n be a representative of w, that is, $w = \alpha_n(x/n)$.

Let $\psi = \bigoplus_{i=1}^p a_i \varphi_i$ be a linear combination of Pfister forms of degree n, such that $x = \overline{\psi}$. Let $y = \overline{\tau}$, with $\tau = \bigoplus_{i=1}^p \varphi_i$. Since $\psi - \tau \in I^{n+1}$, $y/n = x/n$. Thus, Lemma 9.17.(c) gives

$$M(w) = \mu_n(x) = \mu_n(y).$$

To prove (II), it is therefore enough to show that $B(\rho)(M(w)) = \alpha_n(y/n)$. For $1 \leq i \leq p$, write $\varphi_i = \bigotimes_{j=1}^n \langle 1, b_{ij} \rangle$, with $b_{ij} \in G$. By Theorem 7.18,

$$-\mathcal{HT}_1(\varphi_i) = \bigwedge_{i=1}^n -b_{ij}.$$

By Proposition 9.19.(b), we have

$$M(w) = \triangle_{i=1}^p -\mathcal{HT}_1(\varphi_i) = \triangle_{i=1}^p \bigwedge_{i=1}^n -b_{ij}.$$

Now, formula [SG] in Corollary 9.15 yields

$$B(\rho)(M(w)) = B(\rho)(\triangle_{i=1}^p \bigwedge_{i=1}^n -b_{ij}) = \alpha_n(\overline{\tau}/n) = \alpha_n(y/n) = w,$$

which proves Theorem 9.18. $\square$

4. The equivalence of Marshall's conjecture to the weak Marshall conjecture for $\mathcal{AP}$ groups.

We start by proving the equivalence in the title for reduced special groups.

THEOREM 9.21. *Let G be a reduced special group. The following conditions are equivalent:*

(1) *G satisfies [MC].*

(2) *For all $n \geq 1$ and for all $\varphi \in I^n$,*

  If for all $\sigma \in X_G$, $sgn_\sigma(\varphi) \equiv 0 \bmod 2^{n+1}$, then $\varphi \in I^{n+1}$.

(3) *For all $n \geq 1$, $\mu_n : \overline{I^n} \longrightarrow B_G$ is injective.*

(4) *G satisfies [WMC].*

PROOF. It is clear that (1) $\Rightarrow$ (2).

(2) ⇒ (1). Let φ be a form over G such that $sgn_\sigma(\varphi) \equiv 0 \bmod 2^n$, for all $\sigma \in X_G$. If $n = 1$, then the Fact in the proof of Lemma 9.3 tells us that $\varphi \in I$, verifying [MC]. So assume that $n \geq 2$. We shall verify by induction that $\varphi \in I^{k+1}$, for all $1 \leq k \leq n-1$. For $k = 1$, since $\varphi \in I$ and $n \geq 2$, we have

$$\forall \sigma \in X_G \ (sgn_\sigma(\varphi) \equiv 0 \bmod 2^n) \ \Rightarrow \ \forall \sigma \in X_G \ (sgn_\sigma(\varphi) \equiv 0 \bmod 4),$$

and so (2) implies $\varphi \in I^2$. Now assume that $\varphi \in I^k$, for $k \leq n-1$. Then, just as above,

$$\forall \sigma \in X_G \ (sgn_\sigma(\varphi) \equiv 0 \bmod 2^n) \ \Rightarrow \ \forall \sigma \in X_G \ (sgn_\sigma(\varphi) \equiv 0 \bmod 2^{k+1}),$$

and so another application of (2) yields $\varphi \in I^{k+1}$.

(2) ⇒ (3). If G verifies (2) and $x/n \in \overline{I^n}$, we have $\mu_n(x/n) = 0$. This means that for all $\sigma \in X_G$, $sgn_\sigma(x) \equiv 0 \bmod 2^{n+1}$, and so (2) guarantees that $x/n \in I^{n+1}$, that is, $x/n = 0$ in $\overline{I^n}$. Since μ_n is a homomorphism (Proposition 9.20.(a)), μ_n must be injective.

(3) ⇒ (4). By the commutativity of the diagram in Proposition 9.20.(b), if μ_n is injective for all n, the same is true of all the t_n's. But this is equivalent to [WMC], according to Proposition 9.20.(e).

(4) ⇒ (2). For $n \geq 1$, assume that $\varphi \in I^n$ satisfies $sgn_\sigma(\varphi) \equiv 0 \bmod 2^{n+1}$, for all $\sigma \in X_G$. Let $w \in \mathcal{W}(G)$ be given by $\alpha_n(x/n)$, where $x = \overline{\varphi}$. Then, $M(w) = \bot$ in B_G; since M is injective (Theorem 9.18), we conclude that $w = \bot$ in $\mathcal{W}(G)$. Thus, $\alpha_n(x/n) = \bot = \alpha_n(2^{n-1}\langle 1, -1\rangle)$. By Remark 9.8.(3), there is $k \geq n$ such that

$$2^{k-n}\varphi - 2^{k-n}2^{n-1}\langle 1, -1\rangle = 2^{k-n}\varphi - 2^{k-1}\langle 1, -1\rangle \in I^{k+1}.$$

Thus, $2^{k-n}\varphi \in I^{k+1}$. By 9.3.(2), [WMC] yields $\varphi \in I^{n+1}$, as needed. □

Remark. An alternative proof of (4) ⇒ (1) appears in [DM1] (5.2, pp. 270 - 271); we have omitted it here. ◇

THEOREM 9.22. *If G is an $\mathcal{AP}$-special group, then*

$$G \text{ verifies } [MC] \quad \text{iff} \quad G \text{ verifies } [WMC].$$

PROOF. It is readily verified that [MC] implies [WMC].

Suppose G satisfies [WMC]. Then G_{red} also satisfies [WMC] (Lemma 9.4) and, by Theorem 9.21, it verifies [MC] as well. Let φ be a form over G such that $sgn_\sigma(\varphi) \equiv 0 \bmod 2^n$, for all σ in X_G. With $\pi : G \longrightarrow G_{red}$ denoting the quotient map, Lemma 3.2.(c) entails

$$\text{For all } \tau \in X_{G_{red}}, \ sgn_\tau(\pi \star \varphi) \equiv 0 \bmod 2^n. \qquad (I)$$

Hence, $\pi \star \varphi \in I^n(G_{red})$. Lifting this condition to G via π, there is a linear combination of Pfister forms of degree n <u>over G</u>, θ, such that $\pi \star \varphi \approx_{G_{red}}$

$\pi \star \theta$. By Proposition 2.21, there is $k \geq 0$ such that $2^k \varphi \approx_G 2^k \theta$, whence, $2^k \varphi \in I^{n+k}(G)$. Since G verifies $[WMC]$, Lemma 9.3 yields $\varphi \in I^n(G)$, as required. □

COROLLARY 9.23. *A formally real field verifies* $[MC]$ *iff it verifies* $[WMC]$. ◇

Note that a formally real field satisfying $[WMC]$ must be Pythagorean. To see this, it is enough to verify that $G(F)$ is reduced (Theorem 1.32). Let ψ be a form of dimension $2p$ over $G(F)$, such that 2ψ is hyperbolic. Then, $2\psi \in I^{2p+2}$, and so $[WMC]$ yields $\psi \in I^{2p+1}$. Now the Arason-Pfister Hauptsatz (field version) implies that ψ must be hyperbolic. Hence, $G(F)$ is reduced (Proposition 1.6.(e)).

5. Marshall's conjecture for Pythagorean fields

In this section we show that all Pythagorean fields verify Marshall's conjecture (Theorem 9.29).

We start with a result on the K-theory of quadratic extensions. This result is a consequence of Corollary 5.3 in [BT], which in turn comes from Theorem 2.3 in [Mi]. However, since the part that we need has a simple proof, we include it.

LEMMA 9.24. *Let F be a field of characteristic $\neq 2$ and $E = F(\sqrt{d})$ be a quadratic extension of F. Let n be an integer ≥ 2. Then, for each $\eta \in k_n E$, there is a finite set of indices I together with $a_1^i, \ldots, a_{n-1}^i \in \dot{F}$ and $b^i \in \dot{E}$, $i \in I$, such that*

$$\eta = \sum_{i \in I} l(a_1^i) \ldots l(a_{n-1}^i) l(b^i)$$

PROOF. Clearly, it is enough to prove the statement for generators $l(z_1)l(z_2)\ldots l(z_n)$ in $k_n E$. We show that the result holds for $n = 2$ and a straightforward induction will give the desired conclusion for all $n \geq 2$. Let $w = a + b\sqrt{d}$ and $z = x + y\sqrt{d}$ be elements of $F(\sqrt{d})$, with $a, x \in F$ and $b, y \in \dot{F}$. Then,

Claim. There are $s, t \in \dot{F}$ such that $sw + tz$ is either 0 or 1.

Proof. We have $w/b - z/y = a/b - x/y = \alpha \in F$. If $\alpha = 0$, we are done, with $s = 1/b$ and $t = -1/y$; if $\alpha \in \dot{F}$, then set $s = 1/\alpha b$ and $t = -1/\alpha y$, to get $sw + tz = 1$. ◇

The Claim and 8.2.(d) yield

$$0 = l(sw)l(tz) = [l(s) + l(w)][l(t) + l(z)],$$

from which it follows that $l(w)l(z)$ can writen in the desired form. □

In section 6 of [Mi], Milnor presents the construction, due to H. Bass and J. Tate, of a graded ring homomorphism from the mod 2 K-theory of a field F to the mod 2 cohomology ring of F. We recall the main points succintly.

For a field F, let F_s be the separable closure of F and G be the Galois group of F_s over F. F_s is a G-module under the operation $x \mapsto \sigma(x)$, for $\sigma \in G$. The exact sequence of G-modules

$$1 \longrightarrow \{\pm 1\} \longrightarrow \dot{F}_s \xrightarrow{(\cdot)^2} \dot{F}_s \longrightarrow 1, \tag{I}$$

leads to a long exact sequence in Galois cohomology (Proposition II.4.4, p. 115 in [Ri]) whose first terms are

$$H^0(G, \dot{F}_s) \xrightarrow{(\cdot)^2} H^0(G, \dot{F}_s) \xrightarrow{\delta} H^1(G, \{\pm 1\}) \longrightarrow H^1(G, \dot{F}_s), \tag{II}$$

where δ is the connecting homomorphism associated to the exact sequence (I). By Hilbert's Theorem 90 (Proposition V.1.2, p. 246 in [Ri]), $H^1(G, \dot{F}_s) = 0$. Moreover, since H^0 is the subgroup of fixed points of the G operation on $\dot{F}_s$, the sequence in (II) yields, with $\mathbb{Z}/2\mathbb{Z}$ in place of $\{\pm 1\}$, the exactness of

$$\dot{F} \xrightarrow{(\cdot)^2} \dot{F} \xrightarrow{\delta} H^1(G, \mathbb{Z}/2\mathbb{Z}) \longrightarrow 0. \tag{III}$$

Since $\dot{F}/\dot{F}^2$ may be identified with $k_1 F$ via l, the map

$$l(a) \in k_1 F \mapsto \delta(a) \in H^1(G, \mathbb{Z}/2\mathbb{Z})$$

is an isomorphism. We shall conform to standard practice and write (a) for $\delta(a)$. Note that for all $x \in \dot{F}$, $(ax^2) = (a)$ (compare 8.3.(b)).

By Lemma 6.1 in [Mi], the isomorphism $l(a) \mapsto (a)$ extends to a graded ring homomorphism

$$h^F = (h_n^F)_{n \geq 1} : k_* F \longrightarrow H^*(G, \mathbb{Z}/2\mathbb{Z}),$$

that sends a generator $l(a_1) l(a_2) \ldots l(a_n)$ to $(a_1) \cup (a_2) \cup \ldots \cup (a_n) =_{def} (a_1, \ldots, a_n)$, the cup product of the a_i's in $H^n(G, \mathbb{Z}/2\mathbb{Z})$.

For notational simplicity, write $H^n(F)$ for $H^n(G, \mathbb{Z}/2\mathbb{Z})$ and $H^*(F)$ for $H^*(G, \mathbb{Z}/2\mathbb{Z})$.

At the end of 1995, V. Voevodsky announced the proof of

THEOREM 9.25. (Theorem 1.1 in [V]) *If F is a field of characteristic $\neq 2$, then h_n^F is an* **isomorphism**, *for all $n \geq 1$.* $\diamond$

Among the many consequences of this beautiful result, we register

COROLLARY 9.26. *Let F be a field of characteristic $\neq 2$. Then*

a) $H^*(F)$ is generated, as a graded ring, by $H^1(F)$, that is, for all $n \geq 1$, if $\eta \in H^n(F)$, there is a finite set of indices J and elements $a_1^j, \ldots, a_n^j$ in $\dot{F}$ such that
$$\eta = \sum_{j \in J} (a_1^j, \ldots, a_n^j)$$

b) Let $E = F(\sqrt{d})$ be a quadratic extension of F. Then, for all $n \geq 1$ and all $\beta \in H^n(E)$, there is a finite set of indices I together with $a_1^i, \ldots, a_{n-1}^i$ in $\dot{F}$ and $b^i \in \dot{E}$, $i \in I$, such that
$$\beta = \sum_{i \in I} (a_1^i, \ldots, a_{n-1}^i, b^i).$$

PROOF. Item (a) follows directly from the fact that h_n^F is an isomorphism and that every element in $k_n F$ is a sum of generators of the type $l(a_1)l(a_2)\ldots l(a_n)$, taken by h_n^F to $(a_1, \ldots, a_n)$ in $H^n(F)$. Item (2) follows from Lemma 9.24 and the fact that h_n^E is an isomorphism. □

We now prove

PROPOSITION 9.27. *Let F be a Pythagorean field. Then, the map* $(-1) \cup (\cdot) : H^n(F) \longrightarrow H^{n+1}(F)$ *is an injection, for all $n \geq 1$.*

PROOF. Let $E = F(\sqrt{d})$ be a quadratic extension of F. By Corollary 4.6 in [A], there is a long exact sequence of mod 2 cohomology, which for each $n \geq 0$ consists of

$$\cdots H^n(F) \xrightarrow{Res} H^n(E) \xrightarrow{Cor} H^n(F) \xrightarrow{\mu} H^{n+1}(F) \xrightarrow{Res} \cdots, \qquad (Q)$$

where μ stands for cup product with (d), while Res and Cor are the restriction and corestriction maps in Galois cohomology, respectively. We have (see paragraph before Theorem 4.1 and paragraphs before Theorem 4.4 in [A]) :

(1) With the identification originating from the sequence (III) above, $Res : H^1(F) \longrightarrow H^1(E)$ is given, for each $a \in \dot{F}$, by $(a) \mapsto (a)$;

(2) The corestriction $Cor : H^1(E) \longrightarrow H^1(F)$ is given, for each $b \in \dot{E}$, by $(b) \mapsto (N_{E/F}(b))$, where $N_{E/F}$ is the field norm, that is, if $b = x + y\sqrt{d}$, then $N_{E/F}(b) = x^2 - y^2 d \in \dot{F}$.

Moreover, the maps Res and Cor have the following properties :

(3) For $\eta \in H^p(F)$ and $\zeta \in H^q(F)$, $Res(\eta \cup \zeta) = Res(\eta) \cup Res(\zeta)$.

(Proposition III.7.3, p. 191 in [Ri] or § 2.6.(a), p. I-14 in [S]).

(4) For $\eta \in H^p(F)$ and $\zeta \in H^q(E)$, $Cor(Res(\eta) \cup \zeta) = \eta \cup Cor(\zeta)$.

(See paragraph before Theorem 4.4 in [A], Proposition III.7.5, p. 192 in [Ri], or § 2.6.(a), p. I-14 in [S]).

For $\beta \in H^n(E)$, we may write, by Corollary 9.26.(b),

$$\beta = \sum_{i \in I} (a_1^i, \ldots, a_{n-1}^i, b^i),$$

with $a_k^i \in \dot{F}$, for $1 \leq k \leq n-1$ and all $i \in I$. Write $\eta_i = (a_1^i, \ldots, a_{n-1}^i)$. Then, recalling items (1) – (4) above

$$Cor(\beta) = Cor(\sum_{i \in I} \operatorname{Res}(\eta_i) \cup b^i) = \sum_{i \in I} Cor(\operatorname{Res}(\eta_i) \cup b^i)$$
$$= \sum_{i \in I} \eta_i \cup Cor(b^i) = \sum_{i \in I} \eta_i \cup (N_{E/F}(b^i)).$$

Now notice that if $d = -1$, then, for each $i \in I$, $N_{E/F}(b^i)$ is a non-zero sum of squares in F. Since F is Pythagorean, there is $x_i \in \dot{F}$ such that $N_{E/F}(b^i) = x_i^2$, for all $i \in I$. But, for each $i \in I$, we have $(x_i^2) = (1)$, the zero of $H^1(F)$. It follows that Cor is the zero map from $H^n(E)$ to $H^n(F)$. Now, the exactness of the sequence (Q) yields that cup product with (-1) is an injection from $H^n(F)$ to $H^{n+1}(F)$, as claimed. □

In section 2 we defined the homomorphisms $\omega_n : k_n F \longrightarrow k_{n+1} F$, consisting of multiplying by $l(-1)$. It is clear from the above considerations that the following diagram is commutative, for all $n \geq 1$:

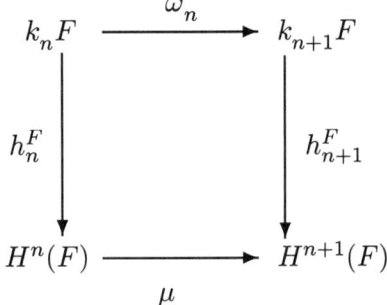

where μ is cup product by (-1). Thus, Theorem 9.25 and Proposition 9.27 yield

COROLLARY 9.28. $k_n F \xrightarrow{\omega_n} k_{n+1} F$ is injective, for all Pythagorean fields F and all $n \geq 1$. ◇

We now prove

THEOREM 9.29. *Every Pythagorean field verifies [MC].*

PROOF. In section 4 of [Mi], Milnor constructs, for each $n \geq 1$, a group homomorphism $s_n : k_n F \longrightarrow I^n/I^{n+1} = \overline{I^n}$, that sends a generator $l(a_1) l(a_2) \ldots l(a_n)$ to the class of the form $\bigotimes_{i=1}^n \langle 1, -a_i \rangle$ modulo I^{n+1}.

Since the ω_n are injective for all n, it follows from Remark 4.2 in [Mi] that s_n is an isomorphism, for all $n \geq 1$. Now notice that, for each $n \geq 1$, the following diagram is commutative :

5. MARSHALL'S CONJECTURE FOR PYTHAGOREAN FIELDS

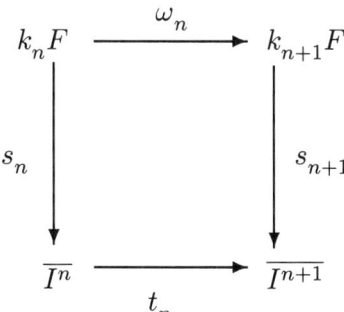

where t_n is multiplication by $2 = \langle 1, 1 \rangle$. Corollary 9.28 and the fact that the s_n's are bijective imply that t_n is injective for all n, i.e., F satisfies $[WMC]$. But then Theorem 9.21 (or Corollary 9.23) guarantees that F verifies $[MC]$. □

For a Pythagorean field F, let $B(F)$ be the Boolean hull of $G(F)$ as in Chapter 4. For each $n \geq 1$, write $B_F(n)$ for the subgroup of $B(F)$ (under $\triangle$), generated by the meets of at most n elements of $G(F)$. We can then state

COROLLARY 9.30. *Let F be a Pythagorean field. Then, for all $n \geq 1$,*

a) The homomorphism ε_n of 8.7 is an isomorphism from $k_n F$ onto $B_F(n)$.

b) The homomorphism μ_n of 9.20 is an isomorphism from $\overline{I^n}$ onto $B_F(n)$.

c) $H^n(F)$ is isomorphic to $B_F(n)$.

PROOF. a) It follows from the construction of ε_n that its image is $B_F(n)$. With notation as in sections 2 and 3, we have a commutative diagram

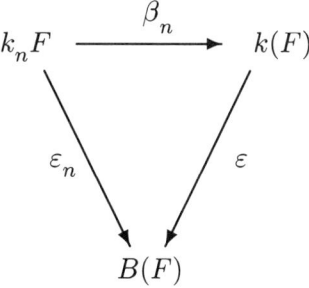

Now, each β_n is injective, because the same is true of each ω_n (Corollary 9.28). Since ε is an isomorphism (Theorem 9.16), we conclude that ε_n is also injective, as asserted.

b) By items (d) and (e) of 9.1 and Proposition 9.20.(d), the image of μ_n is precisely $B_F(n)$. But Theorems 9.29 and 9.21.(3) tell us that μ_n is injective for all $n \geq 1$.

c) Clear, by Theorem 9.25 and item (b). □

Another consequence of Theorem 9.29 is

COROLLARY 9.31. *For a Pythagorean field F and all integers $n \geq 1$,*
$$I^n(F) = I^n(B(F)) \cap W(F).$$
◇

In [AEJ] the question was raised of characterizing, for a Pythagorean field F and a given integer $n \geq 1$, the kernel of the "total signature" map :
$$\tau : I^n(F) \longrightarrow \prod_{\sigma \in \chi(F)} \overline{I^n}(F_\sigma),$$
given by :
$$\tau(\varphi) = \langle \varphi \otimes F_\sigma / I^{n+1}(F_\sigma) \ : \ \sigma \in \chi(F) \rangle,$$
where $\chi(F)$ is the space of orders of F, F_σ denotes the real closure of $\langle F, \sigma \rangle$, and $\varphi \otimes F_\sigma$ is the image of φ in F_σ. Marshall's conjecture for F obviously entails that $\ker(\tau) = I^{n+1}(F)$, proving this "local-global" principle as well. Thus, Theorem 9.25 implies that it also holds for the graded cohomology ring of F.

CHAPTER 10

The category of special groups

In this Chapter we study several classes of special groups and morphisms, from a categorical point of view. We shall consider two categories :

∗ The category **SG** of (not necessarily reduced) special groups, with SG-morphisms.

∗ The category **RSG** of <u>reduced</u> special groups, with SG-morphisms.

The definitions of the notions considered here (monics, epics, injective and projective objects, etc.) can be found in [McL].

1. Monics and Epics

We start with the following simple result.

PROPOSITION 10.1. *The monics in both* **SG** *and* **RSG** *are the injective morphisms.*

PROOF. Let $f : G \longrightarrow H$ be a non-injective SG-morphism; let x, y be distinct elements of G such that $f(x) = f(y)$. Let $F_2 = \{1, -1, a, -a\}$ be the fan of rank 1. Define maps g_1, g_2 from F_2 to G by sending -1 to -1 and a to x and y, respectively. By Remark 1.13, g_1, g_2 are SG-morphisms. Clearly, $f \circ g_1 = f \circ g_2$, although $g_1 \neq g_2$. Thus, f cannot be monic. □

Our next result is a characterization of epics (or right-cancelable morphisms) in the category **RSG**. We shall use the notion of Boolean hull (Chapter 4, section 2), particularly the commutativity of diagram [BH] in Theorem 4.17.

THEOREM 10.2. *Let G, H be rsg's and $G \xrightarrow{f} H$ be a special group morphism. The following are equivalent :*

(1) *f is epic (in **RSG**).*

(2) *$B(f)$ is epic (in the category **BA** of Boolean algebras).*

(3) *$B(f)$ is surjective.*

(4) *For every $u \in B_H$ there are finitely many $g_{ij} \in G$ so that*

$$u = \bigcup_i \bigcap_j \varepsilon_H(f(g_{ij})) \qquad (in\ B_H),$$

that is, B_H is generated, as a BA, by $Im(\varepsilon_H \circ f)$.

(5) For all $\sigma, \mu \in X_H$, $\sigma_{|Im(f)} = \mu_{|Im(f)}$ implies $\sigma = \mu$.

PROOF. (1) $\Rightarrow$ (2). We are given a situation $B_G \xrightarrow{B(f)} B_H \overset{F_2}{\underset{F_1}{\rightrightarrows}} B$,

where B is a BA, F_1, F_2 are BA-morphisms and $F_1 \circ B(f) = F_2 \circ B(f)$. Consequently, $F_1 \circ B(f) \circ \varepsilon_G = F_2 \circ B(f) \circ \varepsilon_G$. Therefore, using the commutativity of diagram [BH], $F_1 \circ \varepsilon_H \circ f = F_2 \circ \varepsilon_H \circ f$. Since f is epic, we get $F_1 \circ \varepsilon_H = F_2 \circ \varepsilon_H$. Since $Im(\varepsilon_H)$ is a set of Boolean generators for B_H (Proposition 4.10.(b)), this last identity entails $F_1 = F_2$.

(2) $\Leftrightarrow$ (3). This is Proposition 4.9.(b).

(3) $\Rightarrow$ (4). This is clear from the fact that B_G is generated, as a BA, by $Im(\varepsilon_G)$ (Proposition 4.10.(b)) and the commutativity of diagram [BH].

(4) $\Rightarrow$ (5). Let σ, μ be characters of H. We know from Corollary 5.4 that σ and μ extend uniquely to BA-morphisms from B_H to $\mathbb{Z}_2$, still denoted by σ and μ, respectively. Note that the hypothesis implies $\sigma \circ f = \mu \circ f$. Thus, $\sigma = \mu$ follows directly from (4).

(5) $\Rightarrow$ (1). Suppose we have a situation such as

$$G \xrightarrow{f} H \overset{h}{\underset{g}{\rightrightarrows}} K \xrightarrow{\sigma} \mathbb{Z}_2,$$

where g, h are morphisms of rsg's such that $g \circ f = h \circ f$. Let σ be a character in X_K; then $\lambda = \sigma \circ g$ and $\mu = \sigma \circ h$ are both in X_H (Lemma 1.12). Note that $\lambda \circ f = \mu \circ f$, and so by (5), $\lambda = \mu$. We have shown that for all $\sigma \in X_K$, $\sigma \circ g = \sigma \circ h$. Thus, for all $a \in H$ and all $\sigma \in X_H$ we have $\sigma(g(a)h(a)) = 1$. By Fact 3.4, $g = h$, and f is indeed epic. $\square$

COROLLARY 10.3. *If G is a reduced special group, then $\varepsilon_G : G \longrightarrow B_G$ is epic.*

PROOF. Follows directly from condition (4) in Theorem 10.2 and Proposition 4.10.(b). $\square$

Remark. The map ε_G is not surjective, unless $G = B_G$ (equivalently, G is a BA). $\diamond$

We now turn our attention to the category **SG** (-1 may or may not be equal to 1).

PROPOSITION 10.4. *Let $K \xrightarrow{f} H$ be a morphism of special groups. Then, f is epic in **SG** iff it is surjective (even with the additional condition $-1 \neq 1$).*

PROOF. Suppose that $G = Im(f)$ is not H and let Δ be a subgroup of H such that every element of H can be written uniquely as a product of elements of G and Δ. To obtain such a Δ, choose a basis $\mathcal{B}_0$ for G (containing -1), and extend it to a basis $\mathcal{B}$ of H. Then, take Δ as the subgroup generated by $\mathcal{B} \setminus \mathcal{B}_0$.

Let G_t denote G with its trivial special group structure (Example 1.9) and let $G_t[\Delta]$ be the extension of G_t by Δ, as in Example 1.10. We have :

i) $-1 = (-1, 1)$;

ii) $\langle g_1 \cdot \delta_1, g_2 \cdot \delta_2 \rangle \equiv_{G_t[\Delta]} \langle g_3 \cdot \delta_3, g_4 \cdot \delta_4 \rangle$ iff $\begin{cases} g_1 g_2 = g_3 g_4 \\ \text{and} \\ \delta_1 \delta_2 = \delta_3 \delta_4. \end{cases}$

As a group, $G_t[\Delta] = H$: for each $a \in H$ there is a unique pair $(g_a, \delta_a) \in G_t[\Delta]$ such that $a = g_a \cdot \delta_a$. Define $g, h : H \longrightarrow G_t[\Delta]$, by $g(a) = (g_a, \delta_a)$ and $h(a) = (g_a, 1)$. It is clear that g and h are group homomorphisms that take $-1 \in G$ to $(-1, 1) = -1$ in $G_t[\Delta]$. Using (ii), it is easily checked that g and h are SG-morphisms; moreover, since $G \neq H$, we have $g \neq h$. Now note that for $x \in K$, $g(f(x)) = (f(x), 1) = h(f(x))$, because $f(x)$ is in $G = Im(f)$. But this contradicts the assumption that f is epic. □

2. Free, injective and projective objects

It follows from Proposition 5.19 and Corollary 5.20 that the **free objects** in the categories **SG** and **RSG** are the fans. However, remark that abstract order spaces which are fans may not be free objects in the dual category **AOS** (in fact, only the trivial one- and two-element fans are free **AOS**'s). Moreover, Proposition 5.14 shows that **RSG** has no injectives. However, given a class of embeddings, we may consider the notion of injectivity with respect to that class. Thus, besides the usual notion of injectivity — relative to the family of all SG-embeddings, treated in section 1 of Chapter 5 — one may consider the class of **complete injective** reduced special groups, that is, those K such that if we are given a SG-morphism $f : G \longrightarrow K$ and G is **complete** subgroup of H, then f has an extension to H. Contrary to the abovementioned result, we show next that there are many complete injective rsg's.

PROPOSITION 10.5. *The complete-injective reduced special groups are exactly the complete BA's.*

PROOF. Recall that the injective BA's (in fact, the injective complete Heyting algebras are the complete ones ([HBA], Theorem 5.13, p. 71). Since the embedding $\varepsilon_{F_1} : F_1 \longrightarrow B_{F_1}$ in diagram (*) of the proof of Proposition 5.14 is complete (Corollary 5.4), the first half of that proof establishes that a complete-injective rsg must be a complete BA.

Conversely, let us show that any complete BA, B, is a complete-injective in **RSG**. Consider a diagram such as (*) below, where $G \overset{\iota}{\hookrightarrow} H$ is a complete embedding, and $f : G \longrightarrow B$ a SG-morphism.

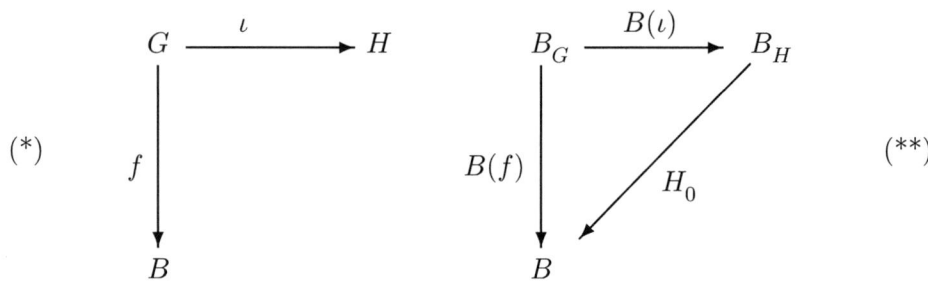

Applying the Boolean hull functor yields diagram (**), in the category of BA's. Since B is complete, there is a BA-morphism H_0, completing the diagram, as shown. The SG-morphism $H_0 \circ \varepsilon_H$ completes (*) to a commutative diagram, as needed. □

We remark the following interesting corollary :

COROLLARY 10.6. *A complete BA is both completely embedded and a quotient of any reduced special group containing it.*

PROOF. Let $B \overset{f}{\longrightarrow} G$ be an embedding of the complete BA B into the rsg G. We know that f is a complete embedding (Corollary 5.4). Since B is complete-injective, the identity id_B extends to a SG-morphism $G \overset{g}{\longrightarrow} B$ that satisfies $g \circ f = id_B$. Lemma 5.17.(iv) shows that $\widehat{g} : G/\ker(g) \longrightarrow B$, the map induced by g, is an isomorphism. □

Next, we characterize the projective objects in **SG** and **RSG**.

PROPOSITION 10.7. $\mathbb{Z}_2$ *is the only projective object in the categories* **SG** *and* **RSG**.

PROOF. (i) $\mathbb{Z}_2$ is a projective object of **SG** (and of **RSG**).

Given SG-morphisms $G \overset{f}{\longrightarrow} H$ and $\mathbb{Z}_2 \overset{h}{\longrightarrow} H$, the only possible SG-morphism $g : \mathbb{Z}_2 \longrightarrow G$ makes diagram (*) below commute, proving (i).

(ii) Every projective object in **SG** and **RSG** is a fan.

Let P be a projective special group; let κ be a cardinal such that there is a surjective SG-morphism $F_\kappa \overset{f}{\longrightarrow} P$ (Corollary 5.20.(c)). Let $P \overset{h}{\longrightarrow} F_\kappa$ be a SG-morphism making commutative diagram (**) below. This forces h to be injective, and hence P is embeddable in a fan. But this implies that P is itself a fan (5.21).

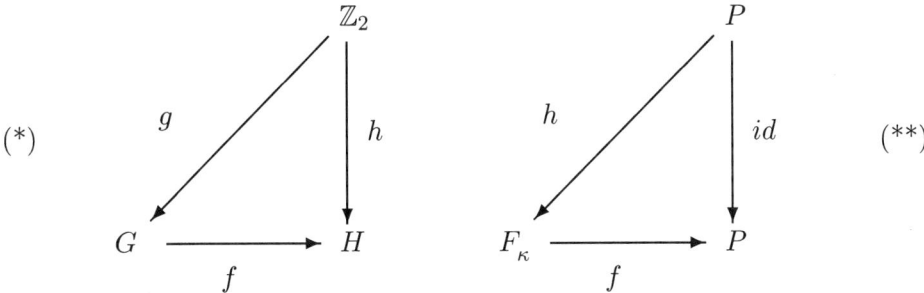

(iii) No fan of rank ≥ 2 is projective.

Let $\kappa \geq 2$. Consider the canonical embedding $\varepsilon_\kappa : F_\kappa \longrightarrow B_\kappa$, where B_κ is the Boolean hull of the fan F_κ of rank κ; ε_κ is a non-surjective epimorphism. Let $\mathcal{B}$ be a $\mathbb{F}_2$-basis of F_κ, containing -1. Since $\kappa \geq 2$, there is $x \in \mathcal{B} \setminus \{-1\}$. Let $a \in B_\kappa \setminus Im(\varepsilon_\kappa)$. The map $f_0 : \mathcal{B} \longrightarrow B_\kappa$ given by

$$f_0(y) = \begin{cases} -1 & \text{if } y = -1 \\ a & \text{if } y = x \\ \text{arbitrary} & \text{if } y \in \mathcal{B} \setminus \{-1, x\} \end{cases}$$

can be extended to a SG-morphism f from F_κ to B_κ, by Proposition 5.19.(b). But note that there cannot be a SG-morphism $g : F_\kappa \longrightarrow F_\kappa$ such that $\varepsilon_\kappa \circ g = f$, because $f(x) = a \notin Im(\varepsilon_\kappa)$. □

Note. The same result (with the same proof) holds for the category **SG*** of those special groups for which $-1 \neq 1$. ◇

By duality we obtain :

COROLLARY 10.8. *The one-point space is the only injective object in the category* **AOS**. ◇

3. Products and Coproducts

As already remarked in section 5.2 (see paragraph before 5.22), the categories **SG** and **RSG** have products, given by defining all pertinent concepts coordinatewise. We shall see that the situation is quite different concerning the dual notion : the coproduct of two reduced special groups does not exist, in general.

LEMMA 10.9. *Let G and H be rsg's. Assume that the coproduct of G and H exists in the category* **RSG**, *denoted by (T, α_G, α_H), with SG-morphisms α_G, α_H from G and H, respectively, into T. Then, $(B(T), B(\alpha_G), B(\alpha_H))$ is the coproduct (or free product) of B_G and B_H in the category* **BA** *of Boolean algebras.*

PROOF. We have to check the following universal property : given a BA, A, and a BA-morphisms g, h, from B_G and B_H, respectively, into A, there is a unique BA-morphism $t : B_T \longrightarrow A$ such that the following diagram is commutative :

(*)
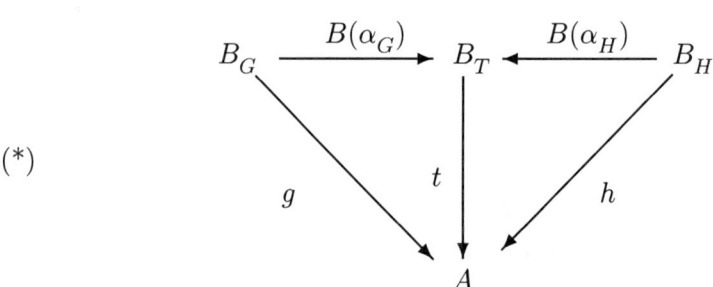

Since we have maps

$$G \xrightarrow{\varepsilon_G} B_G \xrightarrow{g} A \quad \text{and} \quad H \xrightarrow{\varepsilon_H} B_H \xrightarrow{h} A,$$

the universal property of (T, α_G, α_H) yields a <u>unique</u> SG-morphism $p : T \longrightarrow A$ such that diagram (**) is commutative :

(**)
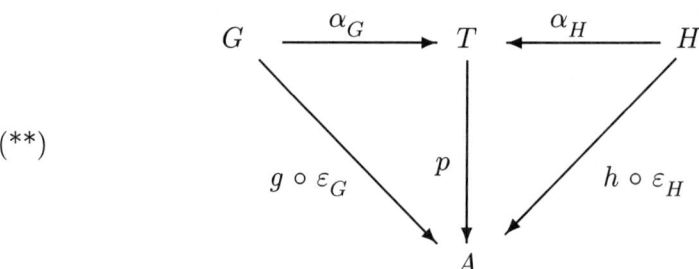

We shall show that $t = B(p)$ makes diagram (*) commute, i.e., :

i) $B(p) \circ B(\alpha_G) = g$;   (ii) $B(p) \circ B(\alpha_H) = h$.

We illustrate the argument by proving (i); the proof of (ii) is similar.

Let $b \in B_G$. Since $Im(\varepsilon_G)$ generates B_G as a BA, there are finitely many $g_{ij} \in G$ such that $b = \bigvee_i \bigwedge_j \varepsilon_G(g_{ij})$. Since, $B(\alpha_G)$ is a BA-morphism and $\varepsilon_T \circ \alpha_G = B(\alpha_G) \circ \varepsilon_G$, we have

$$B(\alpha_G)(b) = \bigvee_i \bigwedge_j [B(\alpha_G) \circ \varepsilon_G](g_{ij}) = \bigvee_i \bigwedge_j [\varepsilon_T \circ \alpha_G](g_{ij}).$$

Since $B(p)$ is also a BA-morphism and $p = B(p) \circ \varepsilon_T$, commutativity of the left-hand side triangle in (**) yields :

$$[B(p) \circ B(\alpha_G)](b) = \bigvee_i \bigwedge_j [B(p) \circ \varepsilon_T \circ \alpha_G](g_{ij}) = \bigvee_i \bigwedge_j [p \circ \alpha_G](g_{ij})$$

$$= \bigvee_i \bigwedge_j [g \circ \varepsilon_G](g_{ij}) = g(\bigvee_i \bigwedge_j \varepsilon_G(g_{ij})) = g(b).$$

Uniqueness of the map t in (*) is an immediate consequence of the uniqueness of p in (**) and the fact that $Im(\varepsilon_T)$ generates B_T as a BA. □

LEMMA 10.10. *Assume that the coproduct (T, α_G, α_H) of the rsg's G and H exists in* **RSG**. *Let $G \xrightarrow{f} H$ be a SG-morphism. Then,*

a) B_T *is isomorphic to B_H, as BA's.*

b) *If f is a complete embedding, then $B_T \approx B_G \approx B_H$.*

PROOF. a) Consider the maps $\chi(T) \xrightarrow{\alpha_H^*} \chi(H)$ and $\chi(T) \xrightarrow{\alpha_G^*} \chi(G)$ given, respectively, by

$$\tau \mapsto \tau \circ \alpha_H \quad \text{and} \quad \tau \mapsto \tau \circ \alpha_G \quad (\tau \in \chi(T)).$$

Clearly, these are continuous group homomorphisms. Further, if $\tau \in X_T$, then $\alpha_H^*(\tau) \in X_H$ and $\alpha_G^*(\tau) \in X_G$. Thus, the restrictions of these maps to X_T take values, respectively, in X_H and X_G.

Now, given $\sigma \in X_H$, consider the situation

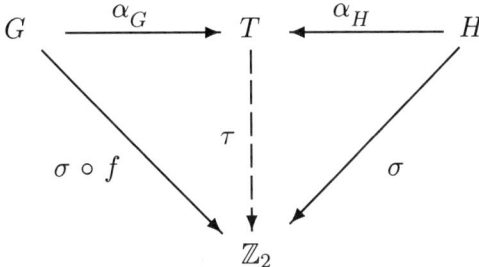

Since (T, α_G, α_H) is the coproduct of G and H, there is a unique $\tau \in X_T$, completing the above diagram to a commutative one; hence $\sigma = \alpha_H^*(\tau)$. Thus, α_H^* is surjective, with the uniqueness of τ guaranteeing that α_H^* is also injective. Consequently, α_H^* induces a homeomorphism between X_T and X_H, and its Stone dual gives the desired isomorphism between B_T and B_H, proving (a).

(b) If f is a complete embedding, every character in X_G extends to a character in X_H, i.e. for every $\gamma \in X_G$ there is $\sigma \in X_H$ such that $\gamma = \sigma \circ f$ (Proposition 5.3.(1)). The same argument used above will now show that α_G^* is a homeomorphism between X_T and X_G. Thus, B_T is isomorphic to both B_G and B_H. □

Note. $\alpha_{H|X_T}^*$ is not necessarily an isomorphism between X_T and X_H as AOS's, since its inverse is not a priori induced by the inverse of α_H^* (this may not even exist!).

PROPOSITION 10.11. *Coproducts do not always exist in* **RSG**.

PROOF. Follows from Lemmas 10.9 and 10.10, by considering any of the following situations :

(i) Any rsg, G, such that its Boolean hull, B, is not isomorphic to the coproduct (in the category of BA's) $B \oplus B$, and $H = Sg(B)$; for example, take $G = H = Sg(B)$, for any such B. Note that there are many BA's B such that $B \not\approx B \oplus B$, as, for instance, any finite BA of cardinality ≥ 4.

(ii) Any pair of rsg's G, H, such that G is a complete subgroup of H, but their Boolean hulls are not isomorphic. There are many such pairs; for instance :

a) Two fans of different ranks, say $G = F_\kappa$, $H = F_\lambda$, where κ, λ are cardinals with $\kappa < \lambda$.

b) $G = F_2$, and any H containing F_2, such that $\text{card}(B_H) \geq 8$. Note that the inclusion of F_2 into any H is complete, because F_2 is a BA (Corollary 5.6). □

Remark. Lemmas 10.9 and 10.10 prove, in fact, that the coproduct of two rsg's exists only in exceptional circumstances. One such case is that of a fan of infinite rank with itself, whose proof we propose as an exercise for the reader. ◇

CHAPTER 11

Some Model Theory of Special Groups

In this chapter we present some basic model-theoretic results concerning reduced special groups. These results were obtained at an early stage of our research; the matter is further pursued in Appendix A below. More advanced results appear in V. Astier's doctoral dissertation [As], which deals with (not necessarily reduced) special groups of finite chain length, and in the forthcoming paper [DM3].

Recall that L_{SG} is the language of special groups, consisting of a binary operation ($\cdot$), two individual constants (1 and -1) and a quaternary relation ($\equiv$). The quaternary relation can be replaced by a binary relation $R(\cdot,\cdot)$ interdefinable with $\equiv$ as follows :

$$R(x,y) \leftrightarrow \langle x, xy \rangle \equiv \langle 1, x \rangle,$$
$$\langle a,b \rangle \equiv \langle c,d \rangle \leftrightarrow ab = cd \wedge R(ac, cd).$$

$R(x,y)$ is usually denoted by $x \in D(1, y)$ in the theory of quadratic forms, a convention which we will continue to adhere to.

The symbols **SG** and **SG**$_{red}$ will denote the first-order theory of special groups and reduced special groups, respectively, in the language L_{SG}. Thus, they include the axioms for groups of exponent 2, the statements [SG0] - [SG6] and the reduction axiom in the case of **SG**$_{red}$, cf. Definition 1.2.

1. The first-order theory of fans

The aim of this section is to establish some model-theoretic properties of the first-order theory of infinite fans, including quantifier elimination in the language of special groups.

NOTATION 11.1. We call **FAN** the first-order theory in the language L_{SG} determined by the following axioms :

* The axioms for groups of exponent 2.

* The axioms [SG0] - [SG5] for special groups (Definition 1.1).

* [SGF1] $1 \neq -1$.

* [SGF2] $\forall\, a\, b\, (b \neq -1 \wedge a \in D(1, b) \rightarrow a = 1 \vee a = b)$.

FAN$_\infty$ is the extension of **FAN** which asserts that all models are infinite.

Remarks. The axiom [SG6] of special groups is a consequence of the axioms in **FAN** (see [LP3]). Likewise, **FAN** proves the reduction axiom [red] (Definition 1.2). ◇

Proposition 5.19 yields at once

FACT 11.2. *a) The theory **FAN** is categorical in all powers (finite or infinite).*

*b) The theory **FAN**$_\infty$ is categorical in all infinite powers.*

*c) **FAN**$_\infty$ is a complete theory.* ◇

Since the axioms of **FAN**$_\infty$ are universal or existential, Lindstrom's theorem (see [Mac], Thm 23, pp. 162-163) implies :

FACT 11.3. **FAN**$_\infty$ *is model-complete.* ◇

We even have

PROPOSITION 11.4. **FAN**$_\infty$ *admits elimination of quantifiers in the language* L_{SG}.

PROOF. By Lemma 12, items a and b, in [Mac] (p. 155), it suffices to show

Claim. **FAN** has the amalgamation property.

This is done by the same kind of argument used in the proof of items (b) and (c) of 5.19. Given injective SG-morphisms $f_i : F_0 \longrightarrow F_i$, $i = 1, 2$, where the F_i's are fans, we shall prove that if F_3 is any fan of rank $\geq rank(F_1) + rank(F_2) - rank(F_0)$, there are SG-embeddings $g_i : F_i \longrightarrow F_3$ such that $g_1 \circ f_1 = g_2 \circ f_2$. Recall that $rank(F) = dim_{\mathbb{F}_2}(|F|) - 1$, where $|F|$ denotes the group underlying F.

Let $\mathcal{B}_0$ be a basis of F_0 as a $\mathbb{F}_2$-vector space containing -1; then, $f_i[\mathcal{B}_0]$ is an $\mathbb{F}_2$-independent subset of F_i, which can be identified with $\mathcal{B}_0$. Let $\mathcal{B}_i$ be $\mathbb{F}_2$-basis of F_i extending $\mathcal{B}_0$, $i = 1, 2$, and $\mathcal{B}_3$ a basis of F_3 containing -1. The assumption on the rank of F_3 implies :

(i) There is a $\mathcal{B}'_0 \subseteq \mathcal{B}_3$ such that $-1 \in \mathcal{B}'_0$ and $\mathrm{card}(\mathcal{B}'_0) = \mathrm{card}(\mathcal{B}_0)$.

(ii) There is $\mathcal{B}'_1 \subseteq \mathcal{B}_3$ such that $\mathrm{card}(\mathcal{B}'_1) = \mathrm{card}(\mathcal{B}_1)$.

(iii) $\mathrm{card}(\mathcal{B}_3 \setminus \mathcal{B}'_1) \geq \mathrm{card}(\mathcal{B}_2 \setminus f_2[\mathcal{B}_2])$.

Hence there are

∗ A bijection $g'_1 : \mathcal{B}_1 \longrightarrow \mathcal{B}'_1$ such that $g'_1[f_1[\mathcal{B}_0]] = \mathcal{B}'_0$ and $g'_1(-1) = -1$.

∗ An injection $g'_2 : (\mathcal{B}_2 \setminus f_2[\mathcal{B}_0]) \longrightarrow (\mathcal{B}_3 \setminus \mathcal{B}'_1)$.

Extend g'_2 to $g''_2 : \mathcal{B}_2 \longrightarrow \mathcal{B}_3$ by setting, for $b_0 \in \mathcal{B}_0$:

$$g_2''(f_2(b_0)) = g_1'(f_1(b_0)).$$

Note that g_2'' is injective (since $g_1' \circ f_1$ is) and $g_2''(-1) = g_1'(-1) = -1$. The maps g_1' and g_2'' can be extended by linearity to group homomorphisms $g_i : F_i \longrightarrow F_3$ such that $g_i(-1) = -1$ ($i = 1, 2$). Since g_1' and g_2'' are injective, so are the g_i's. Proposition 6.10.(a) implies that these maps are morphisms of special groups. Clearly we also have

$$g_2(f_2(x)) = g_1(f_1(x)), \quad x \in F_0$$

since this holds, by construction, for elements of the $\mathbb{F}_2$-basis $\mathcal{B}_0$. □

2. Existentially closed special groups

Here we characterize the existentially closed models of the theory of reduced special groups; as a corollary we prove that this theory has a model-companion. The reader is referred to [Mac] or [Ch] for the definitions of the model-theoretic notions used here and in later sections.

PROPOSITION 11.5. *A reduced special group is existentially closed (in the language L_{SG}) if and only if there is an atomless Boolean algebra A such that $G = Sg(A)$.*

PROOF. ($\Rightarrow$) Let us first prove that $G = Sg(B)$ for some Boolean algebra B. Since G is embedded (as a special group) in a BA, namely, B_G, it suffices to prove that the meet (or the join) operation is **existentially definable** in the language L_{SG} for the theory of rsg's, i.e., that there is an existential L_{SG}-formula $\bigwedge(x,y,z)$ (resp., $\bigvee(x,y,z)$) such that, if H is a reduced special group, then

(*) $\quad H \models \bigwedge(x,y,z) \quad \text{iff} \quad \begin{cases} z \text{ is the g.l.b. of } x \text{ and } y \\ \text{in the partial order} \\ x \leq y \text{ iff } x \in D(1,y). \end{cases}$

(resp., $\bigvee(x,y,z)$ iff z is the l.u.b. of x and y). Note that the right-hand side of the equivalence (*) is a first-order formula of L_{SG}.

By Proposition 7.17 it suffices to set

$$\bigwedge(x,y,z): \quad \langle 1, x, y, -xy \rangle \equiv \langle 1, -1, z, z \rangle$$

We may also set $\bigvee(x,y,z)$ equal to $\bigwedge(x,y,xyz)$ or $\bigwedge(-x,-y,-z)$.

Next, let us check that if B is a BA such that $Sg(B)$ is an existentially closed special group, then B is atomless. B can be embedded in an atomless BA : take, for instance, $A = B \oplus C$ where C is any atomless BA. The existential L_{SG}-formula

$$\neg\, At(x): \quad \exists v(v \in D(1,x) \land v \neq 1 \land v \neq x),$$

expresses the property 'x is not an atom'. Clearly, $Sg(A) \models \forall x(\neg\, At(x))$. Since $Sg(B)$ is existentially closed, we must have $Sg(B) \models \neg\, At(b)$, for every $b \in B$. This proves that B is atomless.

($\Leftarrow$) For the converse, let A be an atomless BA and φ an existential L_{SG}-sentence with parameters in A; assume that $H \models \varphi$, for some reduced special group H such that $Sg(A) \subseteq H$. We have $H \subseteq B_H$, and we may embed B_H into an atomless BA, say C. Since $H \subseteq Sg(C)$, we have that $Sg(C) \models \varphi$.

We invoke now the well-known result that the first-order theory of atomless BA's is model complete in the standard language L_{BA} for Boolean algebras (or any reasonable variant of it (see [Mac]); this also follows easily from Vaught's theorem (see [HBA], pp. 72-74), which asserts the $\aleph_0$-categoricity of the theory of atomless BA's. Since the primitive notions of L_{SG} are quantifier free definable in L_{BA}, the class $\{Sg(A) : A \text{ is an atomless BA}\}$ is model-complete in L_{SG}. Hence, $Sg(A) \prec Sg(C)$ and so, $Sg(A) \models \varphi$. □

Notation. **ASG** denotes the first-order theory of the class

$$\{Sg(A) : A \text{ is an atomless BA}\}$$

in the language L_{SG}. ◇

The proof of Proposition 11.5 also yields :

COROLLARY 11.6. *a) The theory* **ASG** *is axiomatized by the axioms for reduced special groups plus the statements :*

$$\forall x\, y\, \exists z\, \bigwedge(x,y,z) \qquad \forall x\, [x \neq 1 \to \neg\, At(x)]$$

b) **ASG** *is the model-companion of the theory of reduced special groups.*

c) **ASG** *is $\aleph_0$-categorical.*   *d)* **ASG** *is complete.*

PROOF. a) See the proof of 11.5. Item (b) is a consequence of :

i) Every rsg is embedded in a BA, namely its Boolean hull, which in turn can be embedded in an atomless BA.

ii) **ASG** is model-complete (immediate from the model-completeness of atomless BA's; see proof Proposition 11.5).

c) Any two countable models of **ASG** are isomorphic since the underlying atomless BA's are (Vaught's theorem).

d) An immediate consequence of (c), since **ASG** has no finite models. ◇

3. The model-completion of the theory of reduced special groups

It is natural to ask whether Corollary 11.6 can be improved so as to read "**ASG** is the model-completion of the theory of reduced special groups". We recall the well-known fact that if a theory T^* is the model-companion of another theory T, then a necessary and sufficient condition for T^* to be the model-completion of T is that the amalgamation property holds for models of T (see Lemma 12.(b) in [Mac]). Corollary 5.15 shows that the amalgamation property fails for models of $\mathbf{SG}_{red}$ — construed as L_{SG}-structures — thus showing that the answer to our question is negative, so long as we confine ourselves to the language L_{SG}.

However, it was shown in Proposition 5.9 that the amalgamation property holds for **complete embeddings** of rsg's. On the other hand, the characterization of complete embeddings given by clause (6) of Theorem 5.2, shows how the language L_{SG} may be enriched so as to make the monomorphisms in the new language coincide with the complete embeddings.

We enlarge the language L_{SG} by adding countably many relation symbols $\{P_n : n \geq 2\}$, where P_n is $(n+1)$–ary. The interpretation of P_n in a special group G is, for $x, a_1, \ldots, a_n \in G$

$$G \models P_n[x, a_1, \ldots, a_n] \quad \text{iff} \quad x \in D_G(\bigotimes_{i=1}^n \langle 1, a_i \rangle).$$

Let $L_{SG}^{\mathcal{P}}$ be the resulting language. The preceding remarks show that the $L_{SG}^{\mathcal{P}}$- monomorphisms of rsg's are precisely the complete embeddings, and also prove

PROPOSITION 11.7. *Let* $\mathbf{ASG}^{\mathcal{P}}$ *and* $\mathbf{SG}_{red}^{\mathcal{P}}$ *denote, respectively, the first order theories* $\mathbf{ASG}$ *and* $\mathbf{SG}_{red}$, *in the language* $L_{SG}^{\mathcal{P}}$ *enlarged by the axioms*

(*) $\quad \forall\, x\, y_1, \ldots, y_n\, [P_n(x, y_1, \ldots, y_n) \quad \longleftrightarrow \quad x \in D(\bigotimes_{i=1}^n \langle 1, y_i \rangle)]$,

for $n = 2, 3, \ldots$ *Then* $\mathbf{ASG}^{\mathcal{P}}$ *is the model-completion of* $\mathbf{SG}_{red}^{\mathcal{P}}$. ◊

Note that the right-hand side of (*) is a (long) existential L_{SG}-formula.

4. The atomless hull of a Boolean algebra

We shall introduce a 'canonical' way of embedding a given Boolean algebra in an atomless one. Our construction has the properties of a 'closure', and hence, will be suitable for model-theoretic considerations (we apply it in sections 5 and 6). The principle of the construction is as follows

CONSTRUCTION 11.8. Let B be a BA. Let $At(B)$ denote the (possibly empty) set of atoms of B. We fix an atomless BA, A, such that $B \subseteq A$, and a function $c : At(B) \longrightarrow A$ such that $\bot \neq c(a) < a$, for $a \in At(B)$. Let $B^* = B^*(A, c) =$ subalgebra of A generated by $B \cup \{c(a) : a \in At(B)\}$. ◊

The **atomless hull** of B, $AH(B)$, will be obtained by iterating countably many times the construction above (choosing arbitrarily a function c at each step). Before proceeding to the formal definition of $AH(B)$, we prove

PROPOSITION 11.9. *Let B, C be BA's and A, D be atomless BA's containing B and C respectively. Let $c : At(B) \longrightarrow A$, $d : At(C) \longrightarrow D$ be functions as in 11.8 above. Let $B^* = B^*(A, c)$, $C^* = C^*(D, d)$ denote the BA's defined in 11.8. Then, every BA-monomorphism $f : B \longrightarrow C$ such that $f[At(B)] \subseteq At(C)$, extends to a monomorphism $f^* : B^* \longrightarrow C^*$ such that $f^* \circ c = d \circ f$. If f is an isomorphism, so is f^*.*

We will need two lemmas for the proof of Proposition 11.9.

LEMMA 11.10. *Let A, B be BA's, $B \subseteq A$, with A atomless, and let $c : At(B) \longrightarrow A$ be a function as in 11.8 above. Then :*

a) $-a_1 \vee a_2 \vee \ldots \vee a_n = -a_1$, *for distinct* $a_1, \ldots, a_n \in At(B)$.

b) $-c(a_1) \vee a_2 \vee \ldots \vee a_n \neq \top$, *for distinct* $a_1, \ldots, a_n \in At(B)$.

c) $x \in B$, $a \in At(B)$ *and* $a_1, \ldots, a_n \in At(B)$ *be distinct atoms. Then*

(i) $x \geq c(a)$ *iff* $x \geq a$.

(ii) $x \wedge c(a) = \bot$ *iff* $x \wedge a = \bot$.

(iii) $x \leq \bigvee c(a_i)$ *iff* $x = \bot$.

PROOF. a) It is enough to verify that $a_2 \vee \ldots \vee a_n \leq -a_1$; otherwise, $a_1 \leq (a_2 \vee \ldots \vee a_n)$, and so $a_1 = \bigvee_{i=2}^{n} (a_1 \wedge a_i) = \bot$, which is absurd.

b) If (b) was false then $c(a_1) \leq a_2 \vee \ldots \vee a_n$, and so
$$c(a_1) = \bigvee_{i=2}^{n} (c(a_1) \wedge a_i) \leq \bigvee_{i=2}^{n} (a_1 \wedge a_i) = \bot,$$
an impossibility.

c) (i) If $a \not\leq x$, then $x \wedge a < a$. Since a is an atom, $x \wedge a = \bot$; but $x \geq c(a)$ implies $\bot = x \wedge a \geq x \wedge c(a) = c(a)$, a contradiction. A similar reasoning proves (ii).

(iii) If $x \leq \bigvee c(a_i) \leq \bigvee a_i$. If $x \neq \bot$, then, since the a_i's are atoms, $x = a_k$, for some $k \leq n$. But then $a_k = \bigvee (a_k \wedge c(a_i)) = c(a_k)$, which is impossible. $\square$

Before the proof of of the next Lemma we remark certain properties of terms in $B^*(A, c)$ that will be useful in the sequel.

REMARK 11.11. (1) With notation as in 11.8, let $a_1, \ldots, a_n \in At(B)$ and $y \in B$. Consider a term t in A given by $\bigwedge_{i=1}^{n} \varepsilon_i c(a_i) \wedge y$, where ε_i is a sequence of 1's and -1's (corresponding to the fact that $c(a_i)$ or $-c(a_i)$

appears in t) and the a_i's are all distinct. Let $\alpha = card(\{i : \varepsilon_i = 1\})$ be the cardinal of the set of coefficients 1. Then

(*) $$ t = \begin{cases} 1 & \alpha \geq 2 \\ c(a_i) \wedge y & \text{if } \alpha = \{i\} \\ \bigwedge_{i=1}^n -c(a_i) \wedge y & \text{if } \alpha = 0 \end{cases} $$

Formula (*) comes from the following observations:

(i) If a, b are distinct atoms of B, then $c(a) \wedge c(b) \leq a \wedge c(b) \leq a \wedge b = \bot$; from this it follows that if $\alpha \geq 2$, $t = \bot$.

(ii) We also have $c(a) \leq -c(b)$, and so $c(a)$ is less than or equal to any intersection of complements of atoms distinct from b.

Whenever convenient we shall denote a term such as t by $t^c(\varepsilon, \vec{a}, y) = \bigwedge_{i=1}^n \varepsilon_i c(a_i) \wedge y$, with the understanding that $\vec{a}$ always denotes a sequence of distinct atoms.

(2) With notation as in 11.9, corresponding to a term t_c, we have a term

$$ t^d(\varepsilon, f(\vec{a}), f(y)) = \bigwedge_{i=1}^n \varepsilon_i d(f(a_i)) \wedge f(y) \text{ in } C^* \subseteq D. $$

Note that formula (*) holds also for t^d, since f takes atoms in B to atoms in C. Moreover, the final or reduced forms of t^c and t^d, given by (*), correspond to each other. $\diamond$

LEMMA 11.12. *With notation as in Proposition 11.9 and Remark 11.11, let $t_1^c, \ldots, t_n^c, w_1^c, \ldots, w_m^c$ be terms in B^*. The following are equivalent:*

(1) $\bigvee_{i=1}^n t_i^c = \bigvee_{j=1}^m w_i^c$ *(in A)*

(2) $\bigvee_{i=1}^n t_i^d = \bigvee_{j=1}^m w_i^d$ *(in D)*

PROOF. By symmetry and the fact that f is monic, it is enough to show that (1) $\Rightarrow$ (2). We begin by proving the following

Fact. $t^c = \bot$ (in A) iff $t^d = \bot$ (in D).

Proof. By (*) and Remark 11.11.(2) we have only to check two cases:

(i) $t = c(a) \wedge y$ (ii) $t = \bigwedge_{i=1}^n -c(a_i) \wedge y$.

(i) If $c(a) \wedge y = \bot$, then $a \wedge y = \bot$ (11.10.(c)). Since f is a BA-homomorphism, we have $f(y) \leq f(a) = \bot$, and so $t^d = f(y) \wedge d(f(a)) = \bot$.

(ii) In this case, we have $y \leq \bigvee_{i=1}^n c(a_i)$, which implies $y = \bot$ (11.10.(c)). Thus, $f(y) = \bot$, and $t^d = \bot$.

Note that if $t^c = t^c(\varepsilon_1, \vec{a}, y)$ and $w^c = w^c(\varepsilon_2, \vec{b}, z)$ are terms as above, then $v = t^c \wedge w^c$ is a term of the same type, i.e.
$$v = v(\varepsilon_1, \varepsilon_2, \vec{a}, \vec{b}, y \wedge z).$$
Moreover, v^d is precisely $t^d \wedge w^d$. Thus, it is enough to prove that if $t^c = \bigvee_{i=1}^m w_m^c$, then $t^d = \bigvee_{i=1}^m w_m^d$, which we proceed to do by induction on m.

So suppose $w^c = \bigwedge_{i=1}^k \varepsilon_i c(b_i) \wedge z$, and that $t^c = w^c$. Then, we have

(I) $\quad -\varepsilon_j c(b_j) \wedge t^c = \bot \quad (\forall\, j \leq k) \quad \text{and} \quad -z \wedge t^c = \bot.$

It follows from the Fact above, the observation concerning the meet of the terms we are considering, as well as that f is a BA-homomorphism, that (I) implies

(II) $\quad -\varepsilon_j d(f(b_j)) \wedge t^d = \bot \quad (\forall\, j \leq k) \quad \text{and} \quad -f(z) \wedge t^d = \bot.$

Now, (II) yields $(\bigvee_{i=1}^k -\varepsilon_i d(f(b_i)) \vee -z) \wedge t^d = -w^d \wedge t^d = \bot$, and so $t^d \leq w^d$. Since the argument is symmetric, we conclude $t^d = w^d$, as required.

Now suppose

(III) $\quad t^c = w^c \vee \bigvee_{i=1}^m w_i^c, \quad \text{with } w^c = \bigwedge_{i=1}^k \varepsilon_i c(b_i) \wedge z.$

Assume also that our conclusion holds for joins of length $\leq m$.

Note that $t^c \wedge w^c = w^c$ and $t^c \wedge w_i^c = w_i^c$, for all $i \leq m$. Thus, induction gives
$$w^d \vee \bigvee_{i=1}^m w_i^d \leq t^d.$$
To prove the reverse inequality, observe that (III) implies

(IV) $\quad \begin{cases} -\varepsilon_i c(b_i) \wedge t^c = \bigvee_{i=1}^m -\varepsilon_i c(b_i) \wedge w_i^c \quad (\forall\, i \leq k) \\ \text{and} \\ -z \wedge t^c = \bigvee_{i=1}^m -z \wedge w_i^c. \end{cases}$

The equations in (IV) and the induction hypothesis give

(V) $\quad \begin{cases} -\varepsilon_i d(f(b_i)) \wedge t^d = \bigvee_{i=1}^m -\varepsilon_i df((b_i)) \wedge w_i^d, \quad \forall\, i \leq k \\ \text{and} \\ -f(z) \wedge t^d = \bigvee_{i=1}^m -f(z) \wedge w_i^d. \end{cases}$

Taking joins on both sides of the equations in (V) yields
$$-w^d \wedge t^d = -w^d \wedge \bigvee_{i=1}^m w_i^d,$$

wherefrom it readily follows that $t^d \leq w^d \vee \bigvee_{i=1}^{m} w_i^d$, ending the proof. $\square$

Proof of Proposition 11.9. Set $f^*|_B = f$, $f^*(c(a)) = d(f(a))$, for $a \in At(B)$, and extend to B^* by linearity : by the normal form theorems for BA's (see [HBA], pp. 50-54), the general form of an element in B^* is

$$x = \bigvee_i (\bigwedge_j \varepsilon_{ij} c(a_{ij}) \wedge y_i) = \bigvee_i t_i^c,$$

where the t_i^c are terms in A just as above. Now define,

$$f^*(x) = \bigvee_i ((\bigwedge_j \varepsilon_{ij} d(f(a_{ij})) \wedge f(y_i) = \bigvee_i t_i^d.$$

By Lemma 11.12, f^* is well defined. In checking that f^* is a BA-morphism, it is easily verified that $f^*(x \vee y) = f^*(x) \vee f^*(y)$, for all $x, y \in B^*$.

In order to verify that $f^*(-x) = -f^*(x)$, with x as above, we write

$$-x = -\bigvee_{i=1}^{n}(\bigwedge_{j=1}^{k_i} \varepsilon_{ij} c(a_{ij}) \wedge y_i) = \bigwedge_{i=1}^{n}(\bigvee_{j=1}^{k_i} -\varepsilon_{ij} c(a_{ij}) \vee -y_i)$$

in disjunctive normal form, by setting $I_i = \{1, \ldots, k_i\}$, $x_{ij} = -\varepsilon_{ij} c(a_{ij})$, for $1 \leq j \leq k_i$ and $x_{i,k_i+1} = -y_i$, to get

$$-x = \bigvee_{h \in \prod_{i=1}^{n} I_i} \bigwedge_{i=1}^{n} x_{i,h(i)}.$$

By the definition of f^*,

$$f^*(-x) = \bigvee_{h \in \prod_{i=1}^{n} I_i} \bigwedge_{i=1}^{n} f^*(x_{i,h(i)}).$$

Since

$$f^*(x_{ij}) = \begin{cases} -\varepsilon_{ij} d(f(a_{ij})) & \text{for } 1 \leq j \leq k_i \\ -f(y_i) & \text{for } j = k_i + 1, \end{cases}$$

we obtain

$$f^*(-x) = \bigvee_{h \in \prod_{i=1}^{n} I_i} \left((\bigwedge_{i=1}^{n} -\varepsilon_{i,h(i)} d(f(a_{i,h(i)}))\wedge -f(y_i)\right) = -f^*(x).$$

Now, Lemma 11.12 guarantees that the kernel of f^* is $\bot$. Thus, f^* is injective.

Finally, if f is a BA-isomorphism, then $f[At(B)] = At(C)$ and so f^* is surjective (because C^* is generated by $C \cup \{d(f(a)) : a \in At(B)\}$), which shows that f^* is an isomorphism as well, ending the proof. $\square$

Now we define the **atomless hull** of a BA, B, by iterating the construction above inside an arbitrary atomless BA, A, containing B.

$$B_0 = B; \quad B_{n+1} = B_n^*(A, c_n),$$

where $c_n : At(B_n) \longrightarrow A$ is a map such that $\bot \neq c_n(a) < a$, for $a \in At(B)$. Finally,

$$AH(B, A, (c_n)_{n\in \omega}) = \bigcup_{n\in\omega} B_n.$$

Now, iterating Proposition 11.9 through this inductive definition, we show that the latter is independent — up to B-isomorphism — of the choice of A and the functions $(c_n)_{n\in\omega}$, needed at each step. The argument also establishes that the BA thus constructed has the properties of a (model-theoretic) 'closure'.

THEOREM 11.13. *Let B, C be BA's, $B \xrightarrow{f} C$ a monomorphism such that $f[At(B)] \subseteq At(C)$. Let A, D be atomless BA's containing B, C respectively. For $n \in \omega$, let*

$$c_n : At(B_n) \longrightarrow A \quad \text{and} \quad d_n : At(C_n) \longrightarrow D$$

be the choice functions as in the inductive construction above. Then,

a) f extends to a BA-monomorphism

$$\overline{f} : AH(B, A, (c_n)_{n\in\omega}) \longrightarrow AH(C, D, (d_n)_{n\in\omega}),$$

such that for each $n \in \omega$, $\overline{f} \circ c_n = d_n \circ (f_{|At(B_n)})$. If f is an isomorphism, so is $\overline{f}$. In particular,

b) The atomless hull of B, $AH(B)$, does not depend on the atomless BA containing B nor on the choice functions $(c_n)_{n\in\omega}$. If A, D are atomless BA's containing B, then there is an isomorphism

$$f : AH(B, A, (c_n)) \longrightarrow AH(B, D, (d_n))$$

such that $f_{|B} = id$.

c) Let C be an atomless BA, and $g : B \longrightarrow C$ a BA-monomorphism. Then, g extends to a BA-monomorphism $\overline{g} : AH(B) \longrightarrow C$.

PROOF. a) Standard iteration of Proposition 11.9. Let $f_0 = f$. If $f_n : B_n \longrightarrow C_n$ has been constructed so that $f_{n|B_i} = f_i$, for $0 \leq i \leq n$, let $f_{n+1} = f_n^* : B_{n+1} \longrightarrow C_{n+1}$ be the map obtained from f_n by the construction of Proposition 11.9; thus, $f_n^* \circ c_{n+1} = d_{n+1} \circ f_{n|At(B_n)}$. Set $f = \bigcup_{n\in\omega} f_n$.

b) Apply (a) to $C = B$ and $f = id_B$.

c) Since $B \subseteq A$ is isomorphic to $g[B] \subseteq C$ (via g), applying (b) yields an isomorphism $AH(B) \xrightarrow{\gamma} AH(g[B])$ extending g. Now, just set $\overline{g} = \iota \circ \gamma$, where ι is the inclusion map from $AH(g[B])$ to C. □

REMARK 11.14. a) If B is atomless, then $AH(B) = B$. Indeed, $At(B) = \emptyset$, and then $B_n = B$ for all $n \in \omega$ (see the construction of B^* in 11.8).

(b) It can be shown, with notation as in 11.8, that

$$At(B^*) = \{c(a), a \wedge -c(a) : a \in At(B)\},$$

i. e., each atom of B splits into two atoms of B^*. Since this is immaterial for our present purposes, we leave its proof as an exercise for the reader. ◇

We now show that the atomless hull construction is monotone in the following sense:

PROPOSITION 11.15. *Let B, C be BA's. Any embedding $B \xrightarrow{f} C$ extends to an embedding $\overline{f} : AH(B) \longrightarrow AH(C)$.*

PROOF. Since, obviously, $B \approx f[B]$, Theorem 11.13(1) shows that $AH(B) \approx AH(f[B])$ by an isomorphism extending f. Thus, we need only prove that

$$B \subseteq B' \Rightarrow AH(B) \subseteq AH(B').$$

Let A be an atomless BA containing B' (and hence B). If, for $n \in \omega$, $B_n \subseteq C_n$ (for $n = 0$ this is our assumption), it suffices to produce choice functions $c_n : At(B_n) \longrightarrow A$, $c'_n : At(B'_n) \longrightarrow A$ as in 11.8, such that $B_n^*(A, c_n) \subseteq B'^*_n(A, c'_n)$. This guarantees that $B_k \subseteq B'_k$, for all $k \in \omega$, and so $AH(B) \subseteq AH(B')$, since the construction is independent of the A and the choice sequences used in the process (Theorem 11.13).

So let $C \subseteq D$ be subalgebras of A. We split the atoms of C into two sets, as follows:

$$X = \{a \in At(C) : a \text{ is not an atom in } D\}$$

and

$$Y = \{a \in At(C) : a \text{ is an atom in } D\}.$$

Given any choice function $d : At(D) \longrightarrow A$ so that $d(b) < b$, for all $b \in At(D)$, we define a choice function $c : At(C) \longrightarrow A$ by

$$c(a) = \begin{cases} \text{an element } b' \in D \text{ such that } \bot \neq b' \leq a & \text{if } a \in X \\ d(a) & \text{if } a \in Y \end{cases}$$

Thus, for $a \in At(C)$, $c(a) < a$; it is clear that $Im(c) \subseteq D \cup Im(d)$. Since $C \subseteq D$, we conclude that $C^* \subseteq D^*$ (see 11.8), as needed. □

COMMENTS AND OPEN QUESTIONS 11.16. Another construction is known in the theory of BA's which associates to each BA an atomless algebra containing it, namely $B \mapsto B \oplus C$, where C is the unique countable atomless BA; for more details about coproducts see [HBA], Ch. 4, §11, p. 157ff.

Theorem 11.13.(3) applied to the canonical embedding $B \hookrightarrow B \oplus C$ shows that $AH(B)$ is isomorphic to a subalgebra of $B \oplus C$. In general, $AH(B)$ is not B-isomorphic to $B \oplus C$; for example, if B is atomless, then $AH(B) = B$, but B **is not B-isomorphic** to $B \oplus C$. It would be interesting to find a significant class of BA's with atoms for which $AH(B)$ is not B-isomorphic to $B \oplus C$, as well as of non-trivial examples of BA's for which

$AH(B)$ is B isomorphic to $B \oplus C$. For example, do atomic BA's have this property ?

At any rate, not only the atomless hull construction is natural, but $AH(B)$ seems more 'tightly' linked to B than $B \oplus C$. ◇

5. The atomless hull of a reduced special group

Using the construction of the preceding section, we define the atomless hull of a reduced special group G as follows :

$$AH(G) =_{def} Sg(AH(B_G)).$$

The main properties of this construction are given by :

PROPOSITION 11.17. *Let G, K be a reduced special groups. Then,*

*a) There is a canonical **complete** embedding of G into $AH(G)$.*

b) Any complete embedding $G \xrightarrow{f} K$ extends to a (necessarily complete) embedding $\widehat{f} : AH(G) \longrightarrow AH(K)$.

c) Any complete embedding $G \xrightarrow{f} Sg(A)$, where A is an atomless BA, extends to a (necessarily complete) embedding $\widehat{f} : AH(G) \longrightarrow Sg(A)$.

PROOF. (a) The embedding is $\iota \circ \varepsilon_G$, where $\varepsilon_G : G \longrightarrow B_G$ denotes the canonical embedding of G into its Boolean hull (see Corollary 5.4), and $\iota : B_G \longrightarrow AH(B_G)$ is the inclusion map. It is complete since it is a composition of complete embeddings.

(b) Let $f : G \longrightarrow K$ be a complete embedding. By Theorem 5.2, the map $B(f) : B_G \longrightarrow B_K$ is an embedding of BA's. By Proposition 11.15, $B(f)$ extends to an embedding $\widehat{B(f)} : AH(B_G) \longrightarrow AH(B_K)$; whence, $\widehat{B(f)}$ is a complete embedding of special groups.

(c) Use (b) with $K = Sg(A)$. Note that

$$AH(Sg(A)) = Sg(AH(B_{Sg(A)})) = Sg(AH(A)) = Sg(A),$$

since A is atomless. □

6. Quantifier Elimination.

Our aim is to prove

THEOREM 11.18. *The theory $\mathbf{ASG}^{\mathcal{P}}$ admits quantifier elimination in the language $L_{SG}^{\mathcal{P}}$.*

We shall use the following

6. QUANTIFIER ELIMINATION.

LEMMA 11.19. *Let C, D be BA's. We construe $Sg(C)$, $Sg(D)$ as L_{SG}^P-structures, with the predicates P_n as in section 3 : for $x, a_1, \ldots, a_n \in C$*

(*) $\qquad C \models P_n[x, a_1, \ldots, a_n] \quad \text{iff} \quad x \in D_{Sg(C)}(\bigotimes_{i=1}^n \langle 1, a_i \rangle),$

with a corresponding statement holding for D. Let G be an L_{SG}^P-substructure of C and D. Let B_1, B_2 be the Boolean subalgebras of C and D, respectively, generated by G. Then B_1 and B_2 are G-isomorphic.

Remark. G is not necessarily a special group; hence the BA's B_i a priori have no relation with the BA's associated in section 4 to a reduced special group. The lemma just says that a substructure G of a BA, C, in the language L_{SG}^P is rich enough to determine a smallest BA containing it, which is **independent of C, up to G-isomorphism.** ◇

PROOF. Since G is closed under complements, the elements of B_i are of the form

(+) $\qquad \bigvee_i \bigwedge_j g_{ij} \quad \text{with } g_{ij} \in G,$

(finite joins and meets in B_i; non-unique representation). Let $B_1 \xrightarrow{f} B_2$ be defined as the correspondence associating to an element in B_1 of the form (+), the homonymous element in B_2. The only delicate point is showing that f is well-defined. This follows from

(**) $\qquad C \models \bigvee_i \bigwedge_j g_{ij} = \bot \Rightarrow D \models \bigvee_i \bigwedge_j g_{ij} = \bot.$

Proof of (**). It suffices to show (**) for elements of the form $\bigwedge_{i=1}^n g_i$, with $g_1, \ldots, g_n \in G$. If $C \models \bigwedge_{i=1}^n g_i = \bot$, then $C \models \bigvee_i -g_i = \top$, which means that the Pfister form $\bigotimes_{i=1}^n \langle 1, -g_i \rangle$ is isotropic (Corollary 7.20.(i)), and hence hyperbolic in $Sg(C)$. Thus, $-1 \in D_{Sg(C)}(\bigotimes_{i=1}^n \langle 1, -g_i \rangle)$. Using (*), we have $C \models P_n(-1, -g_1, \ldots, -g_n)$. Then, $G \models P_n(-1, -g_1, \ldots, -g_n)$, which implies that $D \models P_n(-1, -g_1, \ldots, -g_n)$. Using (*) again, $-1 \in D_{Sg(D)}(\bigotimes_{i=1}^n \langle 1, -g_i \rangle)$, and therefore

$$D \models \bigvee_i -g_i = \top, \quad \text{i.e.,} \quad D \models \bigwedge_{i=1}^n g_i = \bot,$$

ending the proof of (**).

To prove well-definedness of f assume $x = y$, with $x, y \in B_1$ represented as in (+); then $x \triangle y = \bot$. Applying (**) to the disjunctive normal form of $x \triangle y$ in terms of the elements of G representing x and y, and unraveling it in D, shows that $f(x) \triangle f(y) = \bot$, whence, $f(x) = f(y)$.

Straightforward calculations show that f is a BA-morphism. Clearly, $f_{|G}$ is the identity. Observing that (**) is symmetric shows that f is also bijective. □

Proof of Theorem 11.18. Using a well-known model-theoretic result (see [Ch] or [Mac]) we need to prove that given models A_1, A_2 of $\mathbf{ASG}^\mathcal{P}$, an $L_{SG}^\mathcal{P}$-substructure G of both A_1 and A_2, a primitive $L_{SG}^\mathcal{P}$ formula $\varphi(\bar{v})$ on the free variables $\bar{v} = \langle v_1, \ldots, v_n \rangle$, and $\bar{g} = \langle g_1, \ldots, g_n \rangle$ in G^n,

$$A_1 \models \varphi[\bar{g}] \quad \Rightarrow \quad A_2 \models \varphi[\bar{g}].$$

The formula $\varphi(\bar{v})$ is $\exists w \psi(\bar{v}, w)$, where $\psi(\bar{v}, w)$ is a conjunction of atomic and negated atomic formulas of the language $L_{SG}^\mathcal{P}$, on the displayed variables.

Let B_i be the Boolean subalgebra of A_i generated by G ($i = 1, 2$). By Lemma 11.19 (with $C = A_1$ and $D = A_2$), there is a G-isomorphism $f : B_1 \longrightarrow B_2$. By 11.13, there is $\widehat{f} : AH(B_1) \longrightarrow AH(B_2)$, a G-isomorphism extending f, where $AH(B_i)$ is the atomless hull of B_i in A_i.

Assume that $A_1 \models \exists w \psi(\bar{g}, w)$. Since the theory $\mathbf{ASG}^\mathcal{P}$ is model-complete (11.7), $AH(B_1) \models \exists w \; \psi(\bar{g}, w)$. Let $a \in AH(B_1)$ be such that $AH(B_1) \models \psi[\bar{g}, a]$; since $\widehat{f}$ is a G-isomorphism, $AH(B_2) \models \psi[\bar{g}, \widehat{f}(a)]$, which implies $A_2 \models \exists w \; \psi(\bar{g}, w)$, as required. $\square$

APPENDIX A

The Universal Theory of Reduced Special Groups

by

M. Dickmann and A. Petrovich

In this appendix we give an axiom system for the set of universal L_{SG}-sentences holding in all reduced special groups.

PRELIMINARIES A.1. In the sequel G stands for a reduced **pre**-special group (abbreviated: rpsg), see Definition 1.2. The language is L_{SG}. The notion of a saturated subgroup of G is defined as in 2.3. The intersection of an arbitrary family of saturated subgroups of G is saturated (if the family is empty, the intersection is G). Hence, the smallest saturated subgroup containing a subset X of G exists; it is denoted by $\overline{[X]}$; cf. Definition 2.7. It may be improper (equivalently, contain -1); also, $\overline{[\emptyset]} = \{1\}$ (G is reduced). Note that:

(*) $\qquad \overline{[X]} = \bigcup \{\overline{[b_1, \ldots, b_n]} \mid n \in \mathbb{N} \text{ and } b_1, \ldots, b_n \in X\}.$

The following is an inductive characterization of the saturation $\overline{[X]}$. Let $\Delta = [X]$ be the subgroup of G generated by X. Given a subgroup Γ, let Γ' denote the subgroup generated by $\{x \in G \mid \text{There is } y \in \Gamma \text{ such that } x \in D_G(\langle 1, y \rangle)\}$. Let

$$\Delta^{(0)} = \Delta \qquad \text{and} \qquad \Delta^{(k+1)} = (\Delta^{(k)})' \qquad \text{for } k \in \mathbb{N}.$$

Routine checking shows:

(**) $\qquad \overline{[X]} = \bigcup_{k \in \mathbb{N}} \Delta^{(k)}.$ $\diamond$

We shall prove:

THEOREM A.2. *a) A reduced pre-special group G is embeddable in a reduced special group if and only if for all $n \in \mathbb{N}$ it satisfies the law:*

$(\dagger)_n \quad$ For all $x, b, b_1, \ldots, b_n \in G$,

$$x \in \bigcap_{f \in \{\pm 1\}^n} \overline{[b, f(1)b_1, \ldots, f(n)b_n]} \Rightarrow x \in D_G(\langle 1, b \rangle).$$

b) For a fixed integer $n \in \mathbb{N}$, the law $(\dagger)_n$ is equivalent (modulo the axioms [SG0] – [SG5] and the reduction axiom [red]; cf. Definition 1.2) to countably many universal L_{SG}-sentences.

Before starting the proof we note :

FACT A.3. $\mathbf{SG}_{red} \models (\dagger)_n$ for all $n \in \mathbb{N}$. More generally, if $G \models \mathbf{SG}_{red}$, φ is a Pfister form over G, and $x, b_1, \ldots, b_n \in G$, then

$$(\dagger\dagger)_n \qquad x \in \bigcap_{f \in \{\pm 1\}^n} D_G(\varphi \otimes \bigotimes_{i=1}^n \langle 1, f(i)b_i \rangle) \Rightarrow x \in D_G(\varphi).$$

PROOF. Propositions 2.6 and 2.17 show, for $a_1, \ldots, a_k \in G$:

$$(1) \qquad \overline{[a_1, \ldots, a_n]} = D_G(\bigotimes_{i=1}^k \langle 1, a_i \rangle).$$

This shows that $(\dagger)_n$ is a particular case of $(\dagger\dagger)_n$, for $\varphi = \langle 1, b \rangle$. Let B_G be the Boolean hull of G. Corollary 7.20(i) and (1) give :

$$(2) \qquad x \in \overline{[a_1, \ldots, a_n]} \Leftrightarrow x \leq \bigvee_{i=1}^k a_i \quad (\text{in } B_G).$$

Now, we prove $(\dagger\dagger)_n$ by induction on n. For $n = 0$ there is nothing to prove (we take $\bigotimes_{i=0}^0 \langle 1, b_i \rangle = \langle 1 \rangle$).

$n \Rightarrow n+1$. Assume the left-hand side of $(\dagger\dagger)_n$ holds. Given $g \in \{\pm 1\}^{n-1}$, by considering the functions $g\hat{}\langle 1 \rangle$ and $g\hat{}\langle -1 \rangle$ of $\{\pm 1\}^n$, we get

$$x \in D_G(\psi \otimes \langle 1, b_n \rangle) \cap D_G(\psi \otimes \langle 1, -b_n \rangle),$$

where $\psi = \varphi \otimes \bigotimes_{i=1}^{n-1} \langle 1, g(i)b_i \rangle$. If $\varphi = \bigotimes_{j=1}^k \langle 1, a_j \rangle$, setting $c = \bigvee_{j=1}^k a_j \vee \bigvee_{i=1}^{n-1} g(i)b_i$ ($\in B_G$), then (1) and (2) yield

$$x \leq c \vee b_n \quad \text{and} \quad x \leq c \vee -b_n,$$

whence $x \leq c$. Using (1) and (2) again, we obtain $x \in D_G(\psi)$. Since $g \in \{\pm 1\}^{n-1}$ is arbitrary, the antecedent of $(\dagger\dagger)_n$ gives

$$x \in \bigcap_{g \in \{\pm 1\}^{n-1}} D_G(\varphi \otimes \bigotimes_{i=1}^{n-1} \langle 1, g(i)b_i \rangle);$$

by induction hypothesis, $x \in D_G(\varphi)$. □

REMARK A.4. a) The law $(\dagger\dagger)_n$ is not equivalent to a universal L_{SG}-sentence unless the Pfister form φ has degree 1.

b) The argument employed in the induction step of A.3 can easily be adapted to show $(\dagger)_{n+1} \Rightarrow (\dagger)_n$ for all $n \in \mathbb{N}$.

c) $(\dagger)_0$ is equivalent to each of the following :

 i) $D_G(\langle 1, b \rangle) = \overline{[b]}$ for every $b \in G$.

 ii) $D_G(\langle 1, b \rangle)$ is saturated for every $b \in G$.

 iii) The binary relation $x \in D_G(\langle 1, y \rangle)$ is transitive.

d) $(\dagger)_1$ implies the statement :

[SG4'] $\qquad x \in D_G(\langle 1, y \rangle) \Rightarrow -y \in D_G(\langle 1, -x \rangle).$

Indeed, the antecedent implies $x \in \overline{[y]} \subseteq \overline{[-x, y]}$; then, $-1 = (-x)x \in \overline{[-x, y]}$, which yields $-y \in \overline{[-x, y]}$. Since obviously $-y \in \overline{[-x, -y]}$, then $(\dagger)_1$ implies $-y \in D_G(\langle 1, -x \rangle)$. Using the axioms [SG i] for i = 0, 1, 3, 5, it is easily shown that [SG4'] $\Leftrightarrow$ [SG4]. ◇

Turning now to the proof of Theorem A.2, a moment's thought will convince the reader that in order to prove the implication ($\Rightarrow$) of item (a) it suffices to show that there is a Boolean algebra B and an L_{SG}-monomorphism embedding G into B. The implication ($\Leftarrow$) follows from (b) and the Fact A.3. Beginning the proof of item (a) we define :

DEFINITION A.5. A saturated subgroup Δ of a pre-special group G is called **binary** iff

(i) $-1 \notin \Delta$.

(ii) For each $x \in G$, either $x \in \Delta$ or $-x \in \Delta$.

The set of all binary subgroups of G will be denoted by $X(G)$. ◇

REMARK A.6. a) If Δ is a binary subgroup of a psg G, then G/Δ endowed with the relation $\equiv^*_{G/\Delta}$ defined in 2.13 is isomorphic to $\mathbb{Z}_2$, as pre-special groups. Hence, the binary subgroups of a psg G are exactly the kernels of (pre-)special group characters of G.

b) A binary subgroup of a psg G clearly is maximal saturated. If G is a rsg, the converse holds as well (Proposition 2.10).

c) It is not obvious, *a priori*, that every rpsg possesses a binary subgroup; this will follow from Theorem A.7 below. ◇

Assuming, momentarily, that $X(G)$ is non-empty, it can be endowed with the topology induced by $\{\pm 1\}^G$, exactly as in the case of rsg's, see Definition 3.1. $X(G)$ is a closed subset of $\{\pm 1\}^G$, hence a Boolean space. Let $\mathcal{B}(G)$ denote the Boolean algebra of clopens of $X(G)$. As in the case of rsg's, treated in Chapter 4, we define a map $\varepsilon : G \longrightarrow \mathcal{B}(G)$ as follows : for $g \in G$,

$$\varepsilon(g) = \{\Delta \in \mathcal{B}(G) \mid -g \in \Delta\}.$$

The argument of Proposition 4.10(a) shows that ε is a group homomorphism into $Sg(\mathcal{B}(G))$ sending -1 to $-1 (= X(G))$ and preserving isometry. That $X(G)$ is non-empty, ε is injective, and

$$\varepsilon(g) \subseteq \varepsilon(g') \Rightarrow g \in D_G(\langle 1, g' \rangle),$$

all follow easily from :

THEOREM A.7. (The separation theorem revisited). *Let G be a reduced pre-special group satisfying the law* $(\dagger)_n$ *for all $n \in \mathbb{N}$. Then, for all $a, b \in G$ such that $a \notin D_G(\langle 1, b \rangle)$ there is $\Delta \in X(G)$ such that $b \in \Delta$ and $a \notin \Delta$.*

PROOF. Given a, b as in the statement, let $G_b = G \setminus \{b\}$. For each finite, non-empty $F \subseteq G_b$, say $F = \{b_1, \ldots, b_r\}$, we set
$$T_F = \{h \in \{\pm 1\}^{G_b} \mid a \notin \overline{[b, h(b_1)b_1, \ldots, h(b_r)b_r]}\}.$$
We claim that T_F is a non-empty closed subset of $\{\pm 1\}^{G_b}$. Since G satisfies $(\dagger)_r$ and $a \notin D_G(\langle 1, b \rangle)$, there is $f \in \{\pm 1\}^r$ so that $a \notin \overline{[b, f(1)b_1, \ldots, f(r)b_r]}$; setting $h(b_i) = f(i)$ for $1 \leq i \leq r$, and h arbitrary elsewhere, shows that $h \in T_F$. To show that T_F is closed, let $h \in \{\pm 1\}^{G_b} \setminus T_F$; then, $U_h = \{h' \in \{\pm 1\}^{G_b} : h_{|F} = h'_{|F}\}$ is an open neighborhood of h contained in the complement of T_F. Further, it is clear that $\bigcap_{i=1}^k T_{F_i} = T_{\bigcup_{i=1}^k F_i} \neq \emptyset$, i.e., the family $\{T_F : \emptyset \neq F \subseteq G_b, F \text{ finite}\}$ has the finite intersection property. By compactness there is an h in all the T_F's. Let Δ be the saturated subgroup of G generated by $\{b\} \cup \{h(g)g \mid g \in G_b\}$. Clearly Δ verifies condition A.5(ii) and contains b. If $a \in \Delta$, the equality (*) in A.1 shows that $a \in \overline{[b, h(g_1)g_1, \ldots, h(g_r)g_r]}$ for some finite subset $F = \{g_1, \ldots, g_r\}$ of G_b (non-empty, since $a \notin D_G(\langle 1, b \rangle)$), contradicting that $h \in T_F$. In particular, $-1 \notin \Delta$, and $\Delta \in X(G)$. □

The proof of item (b) in Theorem A.2 relies on the construction, for every integer $n \in \mathbb{N}$ and every finite tree T with a single root, of an *existential* L_{SG}-formula on $n+1$ free variables, $\sigma_T^n(v, w_1, \ldots, w_n)$, such that,

(††) For every reduced pre-special group G and every $a, b_1, \ldots, b_n \in G$,
$$a \in \overline{[b_1, \ldots, b_n]} \Leftrightarrow \text{ there is a finite, single}-\text{rooted tree } T \text{ so that}$$
$$G \models \sigma_T^n[a, b_1, \ldots, b_n].$$
Intuitively speaking, $\sigma_T^n(\cdot, b_1, \ldots, b_n)$ defines the set of all elements of G having T as "family tree" in the inductive construction of $\overline{[b_1, \ldots, b_n]}$ explained in A.1.

Note. In the sequel "tree" means finite, single-rooted tree; the root of T is denoted by r_T. We omit the index n in σ_T^n whenever unnecessary.

A.8. Inductive definition of the formulas σ_T.

If $n = 0$, we set $\sigma_T^0(v) : \quad v = 1, \quad$ for any tree T.

Fix $n \geq 1$. We define the formulas σ_T by induction on the height of T, $h(T)$. $h(T) = 0$. Then, $T = \{r_T\}$, and we set

$$\sigma_T(v, w_1, \ldots, w_n) : \quad \bigvee_{k=1}^{n} \bigvee_{i_1, \ldots, i_k \in \{1, \ldots n\}} (v = \textstyle\prod_{j=1}^k w_{i_j}).$$

Thus, σ_T asserts that v belongs to the subgroup generated by $w_1, \ldots, w_n$.

A. THE UNIVERSAL THEORY OF REDUCED SPECIAL GROUPS   235

Induction step. Let $h(T) = \ell \geq 1$. Let $t_1, \ldots, t_k$ be the immediate successors of r_T in T and, for $1 \leq i \leq k$, let $T_i = \{t \in T \mid t_i \leq t\}$; T_i is the subtree of T formed by the successors of t_i; it is a finite tree with t_i as its root, and $h(T_i) = \ell - 1$. By induction hypothesis the existential formulas $\sigma_{T_i}(v, w_1, \ldots, w_n)$ are defined. We define $\sigma_T(v, w_1, \ldots, w_n)$ to be the formula

$$\exists x_1 \ldots x_k y_1 \ldots y_k [v = \prod_{i=1}^k x_i \wedge \bigwedge_{i=1}^k \sigma_{T_i}(y_i, w_1, \ldots, w_n) \wedge x_i \in D(\langle 1, y_i \rangle)].$$

Obviously, σ_T is equivalent to the existential L_{SG}-formula obtained by renaming the (existentially) quantified variables of $\sigma_{T_1}, \ldots, \sigma_{T_k}$ into disjoint blocks not containing $x_1, \ldots, x_k, y_1, \ldots, y_k$, and putting it in prenex form. $\diamond$

Next we show (††) :

PROPOSITION A.9. *Let G be a reduced pre-special group, and $b_1, \ldots, b_n \in G$. Let $\Delta = \bigcup_{T \text{ tree}} \sigma_T^n(\cdot, b_1, \ldots, b_n)^G$. Then, Δ is the smallest saturated subgroup of G containing $b_1, \ldots, b_n$, i.e., $\Delta = \overline{[b_1, \ldots, b_n]}$.*

PROOF. (a) $[b_1, \ldots, b_n] \subseteq \Delta$.

This is obvious: take T to be the one-element tree (note this also holds for $n = 0$).

(b) Δ is a subgroup of G.

Δ contains 1 by (a). Let $a_1, a_2 \in \Delta$. Let T_1, T_2 be trees so that $G \models \sigma_{T_i}[a_i, b_1, \ldots, b_n]$, $i = 1, 2$. Let T be the tree :

T :
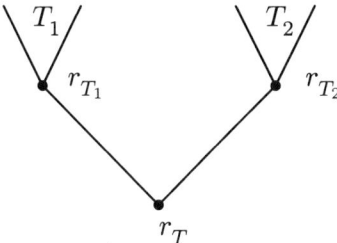

Since $a_i \in D_G(\langle 1, a_i \rangle)$, the inductive definition of σ_T gives (for $x_i, y_i \longrightarrow a_i$, $i = 1, 2$), $G \models \sigma_T[a_1 a_2, b_1, \ldots, b_n]$. Hence, $a_1 a_2 \in \Delta$.

(c) Δ is saturated.

Let $a_1 \in D_G(\langle 1, a_2 \rangle)$, $a_1 \in G$, $a_2 \in \Delta$. Let T be a tree such that $G \models \sigma_T[a_2, b_1, \ldots, b_n]$. Let T' be the tree :

T' :

Then, $G \models \sigma_{T'}[a_1, b_1, \ldots, b_n]$ (by the assignment $x_1 \longrightarrow a_1, y_1 \longrightarrow a_2$), which shows that $a_1 \in \Delta$.

(d) If Γ is a saturated subgroup of G containing $b_1, \ldots, b_n$, then $\Delta \subseteq \Gamma$.

We may assume $n \geq 1$. We have to show, for $a \in G$ and any tree T :

$$G \models \sigma_T[a, b_1, \ldots, b_n] \quad \Rightarrow \quad a \in \Gamma.$$

Induction on $h(T) = \ell$. If $\ell = 0$, then $T = \{r_T\}$; hence $G \models \sigma_T[a, b_1, \ldots, b_n]$ iff $a \in [b_1, \ldots, b_n]$, which implies $a \in \Gamma$, by assumption.

$\ell - 1 \Rightarrow \ell$. Let $t_1, \ldots, t_k$ be the immediate successors of r_T, and let T_i be the tree of successors of t_i ($i = 1, \ldots, k$), as in A.8. By assumption there are elements $c_1, \ldots, c_k, d_1, \ldots, d_k$ in G so that $a = \prod_{i=1}^{k} c_i$, $c_i \in D_G(\langle 1, d_i \rangle)$ and $G \models \sigma_{T_i}[d_i, b_1, \ldots, b_n]$, for $i = 1, \ldots, k$. Since $h(T_i) = \ell - 1$, the induction hypothesis implies that $d_1, \ldots, d_k \in \Gamma$. Since Γ is saturated, $c_1, \ldots, c_k \in \Gamma$, which clearly yields $a \in \Gamma$. □

Now we complete the proof of Theorem A.2 :

FACT A.10. *For a fixed integer* $n \in \mathbb{N}$, *the law* $(\dagger)_n$ *is equivalent (modulo the axioms* [SG 0]-[SG 5] *and* [red]*) to the conjuntion of infinitely many* universal *L_{SG}-formulas; namely, for any set of 2^n finite trees,* $\mathcal{T} = \{T_f : f \in \{\pm 1\}^n\}$,

$(\dagger)_{\mathcal{T}} \quad [\bigwedge_{f \in \{\pm 1\}^n} \sigma_{T_f}^{n+1}(v, y, f(1)w_1, \ldots, f(n)w_n)] \longrightarrow v \in D(\langle 1, y \rangle).$

PROOF. Firstly, by trivial propositional manipulations, $(\dagger)_{\mathcal{T}}$ is equivalent to a universal formula, since the σ_T's are existential.

The equivalence of $(\dagger)_n$ and $\{(\dagger)_{\mathcal{T}} \mid \mathcal{T} \text{ a set of } 2^n \text{ trees}\}$ follows at once from Proposition A.9; indeed, for any rpsg G and $x, b, b_1, \ldots, b_n \in G$ this result implies the equivalence of the following statements:

— $x \in \bigcap_{f \in \{\pm 1\}^n} \overline{[b, f(1)b_1, \ldots, f(n)b_n]}$;

— For every $f \in \{\pm 1\}^n$ there is a tree T_f such that

$$G \models \sigma_{T_f}^{n+1}[x, b, f(1)b_1, \ldots, f(n)b_n].$$

□

APPENDIX B

Table of References for [DM1] and [DM2]

The version of the present monograph on which [DM1] and [DM2] were based is distinct from the present one. For the benefit of the readers of the aforementioned papers, we include here the main distinctions.

1. The **multiplicative Horn-Tarski invariants** that appear in [DM1] and [DM2] were rebaptized as **Boolean Stiefel-Whitney invariants** and their symbols have been changed from $\mathcal{HT}_k^*$ to $\mathcal{SW}_k$. Hence, if G is a reduced special group and $\varphi = \langle a_1, \ldots, a_n \rangle$ is a form over G, for all $1 \leq k \leq n$, we have

$$\mathcal{HT}_k^*(\varphi) = \mathcal{SW}_k(\varphi).$$

2. The following table gives the correspondence between the results mentioned in [DM1] and [DM2] — using a previous version of this monograph — and their numbering in the present version.

In [DM1] or [DM2]	Present version
Section 1	Chapter 1
Section 2	Chapter 2
Section 3	Chapter 3
Section 4	Chapter 4, Section 1
Section 5	Chapter 4, Section 2
Section 6	Chapter 5, Section 1
Section 7	Chapter 7
Definition 1.1	Definition 1.2
Lemma 1.5	Lemma 1.5
Proposition 1.6	Proposition 1.6
Example 1.7	Section 1.3
Theorem 1.17	Theorem 1.23
Definition 2.3	Definition 2.3
Lemma 2.4	Lemma 2.4
Corollary 2.7	Corollary 2.8
Theorem 2.10	Theorem 2.11
Proposition 2.19	Proposition 2.21
Proposition 2.22	Proposition 2.28
Proposition 2.24	Proposition 2.30
Theorem 2.25	Theorem 2.31

(cont.)

In [DM1] or [DM2]	Present version
Proposition 3.3	Proposition 3.7
Proposition 3.3*	Proposition 3.7*
Definition 4.1	Definition 4.1
Corollary 4.4	Corollary 4.4
Proposition 5.1	Proposition 4.10
Theorem 5.7	Theorem 4.17
Corollary 6.4	Corollary 5.4
Theorem 7.1	Theorem 7.1
Theorem 7.18	Theorem 7.18
Theorem 7.30	Theorem 7.31

Bibliography

[A]   J. Kr. Arason, *Cohomologische Invarianten quadratischer Formen*, J. Algebra, **36** (1975), 448–491.

[ABR] C. Andradas, L. Bröcker, J. M. Ruiz, **Constructible Sets in Real Geometry**, Ergebnisse der Math. und ihrer Grenzgebiete **33**, Springer-Verlag, Berlin, Heidelberg, New York, 1996.

[AEJ] J. Kr. Arason, R. Elman, B. Jacob, *On Quadratic Forms and Galois Cohomology*, Rocky Mountain Journal Math. **19** (1989), 575 – 588.

[AP] J. Kr. Arason, A. Pfister, *Beweis des Krullschen Durchschnittsatzes für den Wittring*, Inventiones Math. **12** (1971), 173–176.

[As] V. Astier, *Théorie des modèles des groupes spéciaux de longueur de chaîne finie*, Ph. D. thesis, University of Paris VII, 1999.

[BD] R. Balbes, P. Dwinger, **Distributive Lattices**, Indiana University Press, 1988

[BT] H. Bass, J. Tate, *The Milnor ring of a global field*, in **Algebraic K-Theory II**, Lecture Notes in Math., **342**, 349 – 446, Springer Verlag, 1973.

[Be] E. Becker, **Hereditarily Pythagorean Fields and Orderings of Higher Level**, Monografias de Matematica **29**, IMPA, Rio de Janeiro, 1978.

[CK] C. C. Chang, H. J. Kiesler, **Model Theory**, North-Holland Publ. Co., Amsterdam, 1983.

[Ch] G. Cherlin, **Model Theoretic Algebra : Selected Topics**, Lect. Notes Math. **521**, Spriger, Berlin, 1976.

[Cr] T. Craven, *The Boolean space of orderings of a field*, Trans. Amer. Math. Soc. **209** (1975), 225–235.

[D]   M. Dickmann, *Anneux de Witt abstraits et groupes speciaux*, Séminaire de Structures Algébriques Ordonnées 1991-1992, (F.Delon, M.Dickmann, D.Gondard, eds.) Paris VII - CNRS, Logique, Prépublications **42**, Paris, 1993.

[DM1] M. Dickmann, F. Miraglia *On quadratic forms whose total signature is zero mod 2^n. Solution to a problem of M. Marshall*, Inventiones Math., **133** (1998), 243 – 278 .

[DM2] _____ *Algebraic K-Theory of Fields and Special Groups*, to appear in Contemporary Math, 1999.

[DM3] _____, *Marshall's Conjecture in a General Setting*, 1996, to appear.

[DM4] _____, *Algebraic K-Theory and Special Groups*, 1997, to appear.

[DM5] _____, *Orders and Relative Pythagorean Closures*, 1998, to appear.

[Di] L. E. Dickson, **History of the Theory of Numbers**, vol. I (Divisibility and Primality), Chelsea Publ. Co., New York, 1952.

[E]   R. Engelking, **Topology**, Sigma Series in Pure Mathematics, vol. 6, Heldermann Verlag, Berlin, 1989.

[EL1] R. Elman, T. Y. Lam, *Pfister forms and K-theory of fields*, J. Algebra **23** (1972), 181–213.

[EL2] R. Elman, T. Y. Lam, *Quadratic forms over formally real and pythagorean fields*, Amer. J. Math. **94** (1972), 1155–1194.

[FS] M. Fourman, D. S. Scott, *Sheaves and Logic*, in **Applications of Sheaves**, (M. Fourman, C. Mulvey and D. S. Scott, ed.) Lecture Notes in Math. **753**, Springer-Verlag, 302 – 401 (1979).

[HBA] **Handbook of Boolean Algebras**, vol.1 by S. Koppelberg, (J. Donald Monk, R. Bonnet, eds.) North-Holland Publ. Co., Amsterdam, 1989.

[HJ] D. Haran, M. Jarden, *The absolute Galois group of a pseudo real closed field*, Ann. Sc. Norm. Sup. Pisa, serie IV, **12** (1985), 449–489.

[HT] A. Horn, A. Tarski, *Measures in Boolean Algebras*, Trans. Amer. Math. Soc., **64**, 1948, 467 – 497

[Ho] W. Hodges, **Model Theory**, Encyclopedia of Mathematics and its Applications, vol. 42, Cambridge Univ. Press, 1993.

[KS] M. Knebush, W. Scharlau, **Algebraic theory of quadratic forms**, (Generic Methods and Pfister forms), DMV Seminar 1, Birkhauser, Boston, 1980.

[KMS] M. Kula, M. Marshall, A. Sladek, *Direct limits of finite spaces of orderings*, Pacific J. Math. **112** (1984), 391–406.

[KSS] M. Kula, L. Szczepanik, K. Szymiczek, *Quadratic form scemes and quaternionic schemes*, Fund. Math. **130** (1988), 181–190.

[L1] T. Y. Lam, **The algebraic theory of quadratic forms**, W. A. Benjamin, Mass., 1973.

[L2] ———, **Orderings, valuations and quadratic forms**, Regional Conf. Maths. **52**, A.M.S., 1983.

[L3] ———, **Ten Lectures on Quadratic Forms over Fields**, in Conf. on Quadratic Forms (G. Orzech, ed.), Queen's Papers on Pure and Applied Math. **46** (1977), 1 – 102, Queen's University, Ontario, Canada.

[LP1] A. Lira de Lima, *Les Groupes Spéciaux*, Séminaire de Structures Algébriques Ordonnées 1991-1992, (F.Delon, M.Dickmann, D.Gondard, eds.) Paris VII - CNRS, Logique, Prépublications **42**, Paris, 1993.

[LP2] ———, *Espaces d'Ordres Abstraits*, Séminaire de Structures Algébriques Ordonnées 1991-1992, (F.Delon, M.Dickmann, D.Gondard, eds.) Paris VII - CNRS, Logique, Prépublications **42**, Paris, 1993.

[LP3] ———, *Les Groupes Spéciaux*, Ph.D. thesis, University of Paris VII, 1996.

[McL] S. MacLane, **Categories for the Working Mathematician**, Graduate Texts in Mathematics **5**, Springer-Verlag, New York, 1971.

[Mac] A. Macintyre, *Model completeness*, Chapter A4 in **Handbook of Mathematical Logic** (J. Barwise, ed.), North-Holland Publ. Co., Amsterdam 1977, 139–180.

[Ma1] M. Marshall, **Abstract Witt Rings**, Queens Papers in Pure and Applied Math. **57** (1980), Queen's University, Ontario, Canada.

[Ma2] ———, *Classification of finite spaces of orderings*, Can. J. Math.**31** (1979), 320–330.

[Ma3] ———, *Quotients and inverse limits of spaces of orderings*, Can. J. Math. **31** (1979), 604–616.

[Ma4] ———, *Spaces of Orderings IV*, Can. J. Math. **32** (1980), 603–627.

[Ma5] ———, *The Witt ring of a space of orderings*, Trans. Amer. Math. Soc. **258** (1980), 505–521.

[Ma6] ———, *A reduced theory of quadratic forms*, in Conf. on Quadratic Forms (G. Orzech, ed.), Queen's Papers on Pure and Applied Math. **46** (1977), Queen's University, Ontario, Canada, 569 – 579.

[Ma7] ———, **Spaces of Orderings and Abstract Real Spectra**, Lect. Notes Math. **1636**, Springer, Berlin, 1996.

[MP] M. Marshall, V. Powers, *Higher level form schemes*, Comm. Algebra, **21** (1993), 4083–4102.

[Mi] J. Milnor, *Algebraic K-Theory and Quadratic Forms*, Inventiones Math., **9** (1970), 318–344.

[Pf] A. Pfister, *Quadratische Formen in beliebigen Körpern*, Inventiones Math. **1** (1966), 116–132.

[Pon] L. Pontrjagin, **Topological Groups**, 2nd ed., Gordon and Breach Sci. Publ., 1966.

[P1] A. Prestel, **Lectures on Formally Real Fields**, Monografias de Matematica **22**, IMPA, Rio de Janeiro, 1975 (also published in Lect. Notes Math. **1093**, Springer, Berlin, 1984).

[P2] ———, *Pseudo real closed fields*, in **Set Theory and Model Theory**; Proceedings, Bonn 1979 (R.B. Jensen, A. Prestel, eds.), Lecture Notes Math. **872**, Springer, Berlin, 1981, 127–156.

[Ri] L. Ribes, **Introduction to Profinite Groups and Galois Cohomology**, Queen's Papers in Pure and Applied Math. **24** (1970), Queen's University, Ontario, Canada.

[S] J. P. Serre, **Cohomologie Galoisienne**, Lecture Notes in Math. **5**, Springer-Verlag, 3rd edition, 1965.

[Sch] W. Scharlau, **Quadratic and Hermitian Forms**, Springer-Verlag, Berlin, 1985.

[V] V. Voevodsky, *The Milnor Conjecture*, 51 pp.; preprint of December 1996 distributed by e-mail (http://www.math.uiuc.edu).

Appendix co-author :   A.Petrovich
                              Departamento de Matemáticas
                              Facultad de Ciencias Exactas y Naturales
                              Universidad de Buenos Aires
                              1428 **BUENOS AIRES - ARGENTINA**
                              e-mail : apetrov@dm.uba.ar

Index

*-sum, *see also* special group
K-theoretic
   group mod 2, 175
3-transitivity axiom, 2

abstract order space, ix, 52
   category of, 55, 57
   injective, 213
   morphism of, 55
   no projective, 86
abstract Witt ring, vii, 47
algebraic
   K-theory, xi, 174–178
      basic properties, 175–176
   field extension
      maximal, 80
amalgamation, *see also* complete embedding, first-order theory, special group morphism
anisotropic, 3, 20, *see also* form
anitropy-preserving morphism, *see also* morphism
Arason-Pfister, *see also* special group
   Hauptsatz, xi, 171, 183
atomless Boolean algebra, *see also* Boolean algebra, reduced special group

Baer-Krull Theorem, 26–32
basic
   element, 12
      special subgroup of, 11–14, 88
Boolean
   algebra, 59–60
      as reduced special group, 60–63
      atomless, 219–220
      atomless hull of, 221–227
      category of, 59–60, 66
      complete, 91, 211
      free, 89
      isometry in, 60, 81
      non-trivial, 60
      of sections, 133–134
      space of orders of, 65

      special group characterization of, 153–154, 167–168
   hull, ix, 59, 72, 140
      of quotient, 86
      of reduced special group, 68–73
      of SG-filtered power, 125–134
   ideal
      and saturated subgroup, 64–65
   morphism
      and special group morphism, 63–64
   polynomial, 117
   power
      filtered, 102
   projective filtration, 127
   ring, 188–189
   space, ix, 66, 100
      category of, 66
   -theoretic
      Stiefel-Whitney invariants, *see also* Stiefel-Whitney invariants

categoricity, *see also* first-order theory
   aleph-zero, 220
chain-equivalent, 15
character, *see also* special group
   continuous, 56
cohomology, *see also* Galois, group
compactification
   one-point, 110
compatible, *see also* pre-order, valuation
complete
   -injective, *see also* reduced special group
   Boolean algebra, 91, 211
   embedding, x, 74, 75, 78
      amalgamation, 82, 221
      Boolean characterization, 75
   lattice, *see also* lattice
   subgroup, 75
      Boolean characterization, 77–78
      of Boolean hull, 78
completeness, *see also* first-order theory
conjecture, *see also* Marshall

continuous
  filtration, 104
  function, x, 91, 100–101
coproduct, *see also* reduced special group
cup-product, 204
  injectivity of, 205

dimension, 3, *see also* form
direct sum, 3, *see also* form
directed
  down, 1
  up, 1
  upward, 40
discriminant, 3
Duality Theorem, 57

element
  basic, 12
embedding
  complete, 74, 75
  diagonal, 90, 91
  isotropy-reflecting, 74, *see also* morphism
  pure, 74, *see also* pure
  with retract, 74, *see also* morphism
epic, *see also* special group morphism
equivalent
  chain, 15
  simply, 15
  Witt, 19
evaluation, 56
extension, 10, 88

fan, 9, 83, 89–90
  as free special group, 89, 211
  Boolean hull of, 71, 89
  first-order theory of, *see also* first-order theory
  infinite
    first-order theory of, *see also* first-order theory
  rank of, 89
field
  and abstract order space, 78
  extension, 79
    maximal algebraic, 80
    odd-degree, 96
    pseudo real closed, 80
    SAP, 80
    totally quadratic, 96–97
  formally real, 21, 26, 44, 78–80, 92, 180–181, 185
  K-theory, 174–175
  K-theory mod 2, 175

Pythagorean, 78–80, 180–181, 185, 205
rational function, 92
reduced special group of, 20
SG-subgroup of, 21
special group of, 20, 78–80
filtered
  Boolean power, x, 102
  power, 100–103
    SG-, 103–110, *see also* SG-filtered power
filtration, 102
  continuous, 104
  projective, 126
    Boolean, 127
  SG-, 104
  trivial, 102
first-order
  language
    for special groups, 74, 217
first-order theory
  of atomless Boolean algebra
    as reduced special group, 220
  of fans, 217–219
    amalgamation, 218
    axioms, 217
    categoricity, 218
  of infinite fans, 217–218
    categoricity, 218
    completeness, 218
    model-completeness, 218
    quantifier elimination, 218
  of reduced special groups, xii
    model-companion of, 220
    model-completion of, 221
    universal, *see also* reduced special group
form, 3
  anisotropic, 3, 20
  dimension of, 3
  direct sum, 3
  discriminant of, 3
  hyperbolic, 3
  indefinite, 151, 155
    Boolean characterization, 152
  isotropic, 3
    characterization of, 4
  n-form, 3
  Pfister, *see also* Pfister
  scheme, *see also* higher level form scheme
  signature of, 51, 182
  tensor product, 3
  universal, 3
  Witt-equivalent, 19

formally real
    field, *see also* field
    special group, *see also* special group
formula
    addition, *see also* Horn-Tarski, Stiefel-Whitney
    positive-existential, 87, 91–92
    positive-primitive, 91–92
    product, *see also* Stiefel-Whitney
free
    Boolean algebra, *see also* Boolean algebra
    special group, *see also* fan
fully compatible, *see also* pre-order, valuation
function
    characteristic, 101
    continuous, *see also* continuous, special group
fundamental ideal, 20
    powers of, 182–208

Galois cohomology, xii, 180
    group of field, 183–184
    long exact sequence, 183–184, 204
germ, 111, 112
graded ring, 187
    inductive of exponent two, 187–188
    and Boolean ring, 188–189
    Witt, *see also* Witt
group
    K-theoretic, *see also* K-theoretic, algebraic K-theory
        mod 2, *see also* K-theoretic, algebraic K-theory
    character
        continuous, 56
    exponent two, 1, 10
    Galois cohomology, 183–184
    of characters, 1
    special, *see also* special group

Haupsatz, *see also* Arason-Pfister
higher level form scheme, viii
homomorphism
    Milnor, 183–184, 204
    of pre-special group, *see also* pre-special group
    of special group, *see also* special group
Horn-Tarski
    conditions for isometry, 136–140
    invariants, x–xi, 140
        addition formulas, 144

    additive, x–xi, *see also* Horn-Tarski invariants
    basic properties, 141–149
    multiplicative, x–xi, *see also* Stiefel-Whitney invariants
    of linear combination of Pfister forms, 163–164
    of multiple of Pfister form, 154–155
    of Pfister form, 154–155, *see also* Pfister
hull, 72
    atomless
        of Boolean algebra, *see also* Boolean algebra
        of reduced special group, *see also* reduced special group
    Boolean, *see also* Boolean, 72
hyperbolic, 3, *see also* form

ideal
    fundamental, 20
indefinite, *see also* form
injective, *see also* abstract order space, reduced special group
isometry, 2
    and K-theoretic Stiefel-Whitney invariants, 178–181
    in $C(X, G)$, 102
    in Boolean algebra, 60
    preservation by direct sum, 4
    preservation by tensor product, 4
    quotient, 39, 44
    strong, 52
    weak, 52
isomorphism
    of pre-special group, *see also* pre-special group
    of special group, *see also* special group
isotropic, 3, *see also* form
isotropy
    theorem, *see also* Marshall
isotropy-reflecting, *see also* morphism

K-theoretic
    group, 174–175
    group mod 2
        inductive limit of, 183, 190–191
    Stiefel-Whitney invariants, *see also* isometry
K-theory
    algebraic, *see also* algebraic

language
    first-order

for special groups, 74, 217
lattice, 103
  complete, 104, 107, 126
logarithm, 174

Marshall
  conjecture, 182–208
    equivalence to weak, 201–203
    for formally real special groups, 182
    for Pythagorean fields, xi–xii, 182, 203–207
    weak, xii, 182, 185
  Isotropy Theorem, 97–98
Milnor, see also algebraic K-theory, homomorphism
model-companion, see also first-order theory
model-completeness, see also first-order theory
model-completion, see also first-order theory
monic, see also special group morphism
morphism
  anisotropy-preserving, see also isotropy-reflecting
  isotropy-reflecting, 74, 95–99
    and complete embedding, 96
    and pure embedding, 96
    examples, 96–99
    of Boolean algebra, 98–99
  of pre-special group, see also pre-special group
  of special group, see also special group
  regular, 43
  with retract, 74, 87–91

odd-degree field extension, 96
one-point compactification
  of discrete topology, 110
order space, see also space of orders

partition, 101
  finer, 101
Pfister, see also Arason
  form, 33
    basic properties of, 33–34
    degree of, 33
    Horn-Tarski invariants, 154–155
    linear combination of, 163, see also Horn-Tarski, Stiefel-Whitney
    pure subform of, 33, 156, 163
    Stiefel-Whitney invariants, 159
  local-global principle, ix, 46–48, 51–52
  neighbour, 156

over, 33
quotient, 39–44
subgroup, 40, 49
Pontrjagin Duality Theorem, ix
positive-existential, see also formula
positive-primitive, see also formula
power
  filtered, see also filtered
  quasi-Boolean, see also quasi-Boolean
  of compact support, see also quasi-Boolean
  SG-filtered, see also filtered
pre-order, 21
  compatible with valuation, 27
  fully compatible with valuation, 27
pre-special group, 2–4
  axioms, 2
  homomorphism, 10
  isomorphism, 10
  morphism, 10
  reduced, 2, 231–236
    embeddable in reduced special group, 231–232
pre-special relation, 2
product
  of special groups, 90, 213
projective
  filtration, 126
    Boolean, 127
  special group, see also special group
pseudo real closed field, 80
pure
  embedding, 74, 91–95
    examples, 92–95
  subform, see also Pfister
Pythagorean closure
  Ω-, 97
  relative, 97
Pythagorean field, see also field, Marshall

quantifier elimination
  for theory $\mathbf{ASG}^{\mathcal{P}}$, 228–230
  for theory of infinite fans, 218
quasi-Boolean
  power, 109
    of compact support, 109
  space, 109
quaternionic
  scheme, vii
  structure, vii, viii
quotient
  criterion, 43

isometry, 39, 44
Pfister, *see also* Pfister
space of orders, *see also* space of orders
special group, 39, 44
    Boolean hull of, 86

reduced special group, viii, 3, *see also* special group
    and Boolean algebra, *see also* Boolean algebra
    atomless hull of, 228
    Boolean hull of, 68–73
    category of, 11, 57, 85–86, 209–216
    characterizations of, 4
    complete-injective, 211
    coproduct, 213–216
        non-existence of, 215
    existentially closed, 219–220
        and atomless Boolean algebra, 219–220
    no injective, 85
    of atomless Boolean algebra, 219–220
    of field, 20
    universal first-order theory of, 231–236
        axioms, 231–232, 236
regular, 43
representative, 188
represented, 3
    in Boolean algebra, 149
    in field, 21
    weakly, 52
retract, *see also* morphism
ring
    Boolean, 188–189
    graded, *see also* graded ring
    semi-local, 25
    Witt, *see also* Witt

SAP space, 59, *see also* strong approximation property
saturated subgroup, *see also* subgroup
saturation, 36, 231
sentence, 91
Separation Theorem, ix, 38, 233–234
SG-
    filtered power, 103–110, *see also* filtered
        Boolean hull, 125–134
        space of orders, 119–125
    homomorphism, *see also* special group
    isomorphism, *see also* special group
    morphism, *see also* special group
    subgroup of field, 21
signature, *see also* form

    total, 208
simply equivalent, 15
space
    Boolean, *see also* Boolean
    quasi-Boolean, *see also* quasi-Boolean
    SAP, *see also* SAP, strong
space of orders
    abstract, 52, *see also* abstract
    of Boolean algebra, 65
    of SG-filtered power, 119–125
    of special group, 50
    quotient, 78
special group, vii, 2
    *-sum, 93–95, 110
    $\mathcal{AP}$ (Arason-Pfister), 182–183, 187, 191, 202–203
    axioms, 2
    categories of, 11, 209–216
    character, 46
        unique extension to Boolean hull, 78
    characterizations of, 14–19
    coproduct, *see also* reduced special group
    extension of, 10
        has retract, 88
    first-order language, 74, 217
    formally real, 50, 182
    free, *see also* fan
    homomorphism, 10
    inductive limit of, 114–115
    isomorphism, 10
    model theory of, 217–230
    morphism, 10
        epic, 209–210
        monic, 209
        no amalgamation for, 85
        with retract, 87–91
    of atomless Boolean algebra, 220
    of continuous functions, 91, 100–134
    of field, 20
    product, 90, 93–95, 213
    projective, 212
    quotient, 39, 44
    reduced, 3
    reduced quotient of, 44
    space of orders, 50
        basic properties, 50–51
    trivial, 10, 39
special relation, 2
    trivial, 9
special subgroup
    complete, 75, 129
    example of non-complete, 83
    of basic elements, 11

extension, 13
stalk, 112
Stiefel-Whitney
  Boolean-theoretic invariants, x–xi, *see also* Stiefel-Whitney invariants
  invariants, x–xi, 141
    K-theoretic, 176, *see also* isometry
    addition formulas, 144
    basic properties, 141–149
    of linear combination of Pfister forms, 168
    of multiple of Pfister form, 162
    of Pfister form, 159, *see also* Pfister
    product formula, 146
    total, 176
Stone Duality Theorem, ix
strong
  approximation property (SAP), 59
  isometry, 52
subgroup
  Pfister, *see also* Pfister
  saturated, viii, 35
    binary, 233
    maximal, 38
    properties of, 35
  special, *see also* special subgroup
symmetric difference, 60

tensor product, 3, *see also* form
torsion-free, 180
total signature, 208
totally quadratic field extension, 96–97

universal, 3, *see also* form
  first-order theory, *see also* reduced special group

valuation, 26
  compatible with pre-order, 27
  fully compatible with pre-order, 27
Voevodsky's Theorem, xii, 183–184, 204

weak
  isometry, 52
  Marshall conjecture, *see also* Marshall
weakly represented, 52
Witt
  -equivalent, 19
  Cancellation Theorem, 4
  graded ring, 190–191
    inductive limit of, 183, 191
  ring
    abstract, vii, 47
    of special group, 19

Editorial Information

To be published in the *Memoirs*, a paper must be correct, new, nontrivial, and significant. Further, it must be well written and of interest to a substantial number of mathematicians. Piecemeal results, such as an inconclusive step toward an unproved major theorem or a minor variation on a known result, are in general not acceptable for publication. *Transactions* Editors shall solicit and encourage publication of worthy papers. Papers appearing in *Memoirs* are generally longer than those appearing in *Transactions* with which it shares an editorial committee.

As of February 29, 2000, the backlog for this journal was approximately 6 volumes. This estimate is the result of dividing the number of manuscripts for this journal in the Providence office that have not yet gone to the printer on the above date by the average number of monographs per volume over the previous twelve months, reduced by the number of issues published in four months (the time necessary for preparing an issue for the printer). (There are 6 volumes per year, each containing at least 4 numbers.)

A Copyright Transfer Agreement is required before a paper will be published in this journal. By submitting a paper to this journal, authors certify that the manuscript has not been submitted to nor is it under consideration for publication by another journal, conference proceedings, or similar publication.

Information for Authors and Editors

Memoirs are printed by photo-offset from camera copy fully prepared by the author. This means that the finished book will look exactly like the copy submitted.

The paper must contain a *descriptive title* and an *abstract* that summarizes the article in language suitable for workers in the general field (algebra, analysis, etc.). The *descriptive title* should be short, but informative; useless or vague phrases such as "some remarks about" or "concerning" should be avoided. The *abstract* should be at least one complete sentence, and at most 300 words. Included with the footnotes to the paper, there should be the 2000 *Mathematics Subject Classification* representing the primary and secondary subjects of the article. This may be followed by a list of *key words and phrases* describing the subject matter of the article and taken from it. A list of the numbers may be found in the annual index of *Mathematical Reviews*, published with the December issue starting in 1990, as well as from the electronic service e-MATH [**telnet e-MATH.ams.org** (or **telnet 130.44.1.100**). Login and password are **e-math**]. For journal abbreviations used in bibliographies, see the list of serials on the web at `http://www.ams.org/msnhtml/serials-list/annser_frames.html`. When the manuscript is submitted, authors should supply the editor with electronic addresses if available. These will be printed after the postal address at the end of each article.

Electronically prepared papers. The AMS encourages submission of electronically prepared papers in $\mathcal{A}_{\mathcal{M}}\mathcal{S}$-TEX or $\mathcal{A}_{\mathcal{M}}\mathcal{S}$-LATEX. The Society has prepared author packages for each AMS publication. Author packages include instructions for preparing electronic papers, the *AMS Author Handbook*, samples, and a style file that generates the particular design specifications of that publication series for both $\mathcal{A}_{\mathcal{M}}\mathcal{S}$-TEX and $\mathcal{A}_{\mathcal{M}}\mathcal{S}$-LATEX.

Authors with FTP access may retrieve an author package from the Society's Internet node `e-MATH.ams.org` (130.44.1.100). For those without FTP access, the author package can be obtained free of charge by sending e-mail to pub@ams.org (Internet) or from the Publication Division, American Mathematical Society, P.O. Box 6248, Providence, RI 02940-6248. When requesting an author package, please specify $\mathcal{A}_{\mathcal{M}}\mathcal{S}$-TeX or $\mathcal{A}_{\mathcal{M}}\mathcal{S}$-LaTeX, Macintosh or IBM (3.5) format, and the publication in which your paper will appear. Please be sure to include your complete mailing address.

Submission of electronic files. At the time of submission, the source file(s) should be sent to the Providence office (this includes any TeX source file, any graphics files, and the DVI or PostScript file).

Before sending the source file, be sure you have proofread your paper carefully. The files you send must be the EXACT files used to generate the proof copy that was accepted for publication. For all publications, authors are required to send a printed copy of their paper, which exactly matches the copy approved for publication, along with any graphics that will appear in the paper.

TeX files may be submitted by email, FTP, or on diskette. The DVI file(s) and PostScript files should be submitted only by FTP or on diskette unless they are encoded properly to submit through e-mail. (DVI files are binary and PostScript files tend to be very large.)

Files sent by electronic mail should be addressed to the Internet address pub-submit@ams.org. The subject line of the message should include the publication code to identify it as a Memoir. TeX source files, DVI files, and PostScript files can be transferred over the Internet by FTP to the Internet node e-math.ams.org (130.44.1.100).

Electronic graphics. Figures may be submitted to the AMS in an electronic format. The AMS recommends that graphics created electronically be saved in Encapsulated PostScript (EPS) format. This includes graphics originated via a graphics application as well as scanned photographs or other computer-generated images.

If the graphics package used does not support EPS output, the graphics file should be saved in one of the standard graphics formats—such as TIFF, PICT, GIF, etc.—rather than in an application-dependent format. Graphics files submitted in an application-dependent format are not likely to be used. No matter what method was used to produce the graphic, it is necessary to provide a paper copy to the AMS.

Authors using graphics packages for the creation of electronic art should also avoid the use of any lines thinner than 0.5 points in width. Many graphics packages allow the user to specify a "hairline" for a very thin line. Hairlines often look acceptable when proofed on a typical laser printer. However, when produced on a high-resolution laser imagesetter, hairlines become nearly invisible and will be lost entirely in the final printing process.

Screens should be set to values between 15% and 85%. Screens which fall outside of this range are too light or too dark to print correctly.

Any inquiries concerning a paper that has been accepted for publication should be sent directly to the Editorial Department, American Mathematical Society, P. O. Box 6248, Providence, RI 02940-6248.

Editors

This journal is designed particularly for long research papers (and groups of cognate papers) in pure and applied mathematics. Papers intended for publication in the *Memoirs* should be addressed to one of the following editors. In principle the Memoirs welcomes electronic submissions, and some of the editors, those whose names appear below with an asterisk (*), have indicated that they prefer them. However, editors reserve the right to request hard copies after papers have been submitted electronically. Authors are advised to make preliminary e-mail inquiries to editors about whether they are likely to be able to handle submissions in a particular electronic form.

Ordinary differential equations, partial differential equations, and applied mathematics to PETER W. BATES, Department of Mathematics, Brigham Young University, 292 TMCB, Provo, UT 84602-1001; e-mail: peter@math.byu.edu.

Harmonic analysis, representation theory, and Lie theory to ROBERT J. STANTON, Department of Mathematics, The Ohio State University, 231 West 18th Avenue, Columbus, OH 43210-1174; e-mail: stanton@math.ohio-state.edu.

Ergodic theory and dynamical systems to ROBERT F. WILLIAMS, Department of Mathematics, University of Texas at Austin, Austin, TX 78712-1082; e-mail: bob@math.utexas.edu

Real and harmonic analysis and geometric partial differential equations to *WILLIAM BECKNER, Department of Mathematics, University of Texas at Austin, Austin, TX 78712-1082; e-mail: beckner@math.utexas.edu.

Algebra to CHARLES CURTIS, Department of Mathematics, University of Oregon, Eugene, OR 97403-1222 e-mail: cwc@darkwing.uoregon.edu

Algebraic topology and cohomology of groups to STEWART PRIDDY, Department of Mathematics, Northwestern University, 2033 Sheridan Road, Evanston, IL 60208-2730; e-mail: s_priddy@math.nwu.edu.

Differential geometry and global analysis to *CHUU-LIAN TERNG, Department of Mathematics, Northeastern University, Huntington Avenue, Boston, MA 02115-5096; e-mail: terng@neu.edu.

Probability and statistics to *KRZYSZTOF BURDZY, Department of Mathematics, University of Washington, Box 354350, Seattle, WA 98195-4350; e-mail: burdzy@math.washington.edu.

Combinatorics and Lie theory to PHILIP J. HANLON, Department of Mathematics, University of Michigan, Ann Arbor, MI 48109-1003; e-mail: hanlon@math.lsa.umich.edu.

Logic to *THEODORE SLAMAN, Department of Mathematics, University of California at Berkeley, Berkeley, CA 94720-3840; e-mail: slaman@math.berkeley.edu.

Number theory to MICHAEL J. LARSEN, Department of Mathematics, Indiana University, Bloomington, IN 47405; e-mail: larsen@math.indiana.edu.

Complex analysis and complex geometry to DANIEL M. BURNS, Department of Mathematics, University of Michigan, Ann Arbor, MI 48109-1003; e-mail: dburns@math.lsa.umich.edu.

Algebraic geometry and commutative algebra to LAWRENCE EIN, Department of Mathematics, University of Illinois, 851 S. Morgan (M/C 249), Chicago, IL 60607-7045; e-mail: ein@uic.edu.

Geometric topology, knot theory, hyperbolic geometry, and general topoogy to JOHN LUECKE, Department of Mathematics, University of Texas at Austin, Austin, TX 78712-1082; e-mail: luecke@math.utexas.edu.

Partial differential equations and applied mathematics to BARBARA LEE KEYFITZ, Department of Mathematics, University of Houston, 4800 Calhoun, Houston, TX 77204-3476; e-mail: keyfitz@uh.edu

Operator algebras and functional analysis to BRUCE E. BLACKADAR, Department of Mathematics, University of Nevada, Reno, NV 89557; e-mail: `bruceb@math.unr.edu`

All other communications to the editors should be addressed to the Managing Editor, WILLIAM BECKNER, Department of Mathematics, University of Texas at Austin, Austin, TX 78712-1082; e-mail: `beckner@math.utexas.edu`.

Selected Titles in This Series

(*Continued from the front of this publication*)

657 **Richard F. Bass and Krzysztof Burdzy,** Cutting Brownian paths, 1999

656 **W. G. Bade, H. G. Dales, and Z. A. Lykova,** Algebraic and strong splittings of extensions of Banach algebras, 1999

655 **Yuval Z. Flicker,** Matching of orbital integrals on $GL(4)$ and $GSp(2)$, 1999

654 **Wancheng Sheng and Tong Zhang,** The Riemann problem for the transportation equations in gas dynamics, 1999

653 **L. C. Evans and W. Gangbo,** Differential equations methods for the Monge-Kantorovich mass transfer problem, 1999

652 **Arne Meurman and Mirko Primc,** Annihilating fields of standard modules of $\mathfrak{sl}(2,\mathbb{C})^\sim$ and combinatorial identities, 1999

651 **Lindsay N. Childs, Cornelius Greither, David J. Moss, Jim Sauerberg, and Karl Zimmermann,** Hopf algebras, polynomial formal groups, and Raynaud orders, 1998

650 **Ian M. Musson and Michel Van den Bergh,** Invariants under Tori of rings of differential operators and related topics, 1998

649 **Bernd Stellmacher and Franz Georg Timmesfeld,** Rank 3 amalgams, 1998

648 **Raúl E. Curto and Lawrence A. Fialkow,** Flat extensions of positive moment matrices: Recursively generated relations, 1998

647 **Wenxian Shen and Yingfei Yi,** Almost automorphic and almost periodic dynamics in skew-product semiflows, 1998

646 **Russell Johnson and Mahesh Nerurkar,** Controllability, stabilization, and the regulator problem for random differential systems, 1998

645 **Peter W. Bates, Kening Lu, and Chongchun Zeng,** Existence and persistence of invariant manifolds for semiflows in Banach space, 1998

644 **Michael David Weiner,** Bosonic construction of vertex operator para-algebras from symplectic affine Kac-Moody algebras, 1998

643 **Józef Dodziuk and Jay Jorgenson,** Spectral asymptotics on degenerating hyperbolic 3-manifolds, 1998

642 **Chu Wenchang,** Basic almost-poised hypergeometric series, 1998

641 **W. Bulla, F. Gesztesy, H. Holden, and G. Teschl,** Algebro-geometric quasi-periodic finite-gap solutions of the Toda and Kac-van Moerbeke hierarchies, 1998

640 **Xingde Dai and David R. Larson,** Wandering vectors for unitary systems and orthogonal wavelets, 1998

639 **Joan C. Artés, Robert E. Kooij, and Jaume Llibre,** Structurally stable quadratic vector fields, 1998

638 **Gunnar Fløystad,** Higher initial ideals of homogeneous ideals, 1998

637 **Thomáš Gedeon,** Cyclic feedback systems, 1998

636 **Ching-Chau Yu,** Nonlinear eigenvalues and analytic-hypoellipticity, 1998

635 **Magdy Assem,** On stability and endoscopic transfer of unipotent orbital integrals on p-adic symplectic groups, 1998

634 **Darrin D. Frey,** Conjugacy of Alt_5 and $SL(2,5)$ subgroups of $E_8(\mathbb{C})$, 1998

633 **Dikran Dikranjan and Dmitri Shakhmatov,** Algebraic structure of pseudocompact groups, 1998

632 **Shouchuan Hu and Nikolaos S. Papageorgiou,** Time-dependent subdifferential evolution inclusions and optimal control, 1998

For a complete list of titles in this series, visit the
AMS Bookstore at **www.ams.org/bookstore/**.